# Interdisciplinary Applied Mathematics

Volume 61

Problems in engineering, computational science, and the physical and biological sciences are using increasingly sophisticated mathematical techniques. Thus, the bridge between the mathematical sciences and other disciplines is heavily traveled. The correspondingly increased dialog between the disciplines has led to the establishment of the series: Interdisciplinary Applied Mathematics.

The purpose of this series is to meet the current and future needs for the interaction between various science and technology areas on the one hand and mathematics on the other. This is done, firstly, by encouraging the ways that mathematics may be applied in traditional areas, as well as point towards new and innovative areas of applications; and secondly, by encouraging other scientific disciplines to engage in a dialog with mathematicians outlining their problems to both access new methods as well as to suggest innovative developments within mathematics itself.

The series will consist of monographs and high-level texts from researchers working on the interplay between mathematics and other fields of science and technology.

Kiyohiro Ikeda • Yuki Takayama

# Bifurcation Theory of Pattern Formation in Economic Geography

Springer

Kiyohiro Ikeda
Civil and Environmental Engineering
Tohoku University
Sendai, Japan

Yuki Takayama
Civil and Environmental Engineering
Institute of Science Tokyo
Tokyo, Japan

ISSN 0939-6047 ISSN 2196-9973 (electronic)
Interdisciplinary Applied Mathematics
ISBN 978-3-031-71685-0 ISBN 978-3-031-71686-7 (eBook)
https://doi.org/10.1007/978-3-031-71686-7

Mathematics Subject Classification: 20C35, 91A22, 91D25, 37G10, 91-10, 91B72



This Springer imprint is published by the registered company Springer Nature Switzerland AG
The registered company address is: Gewerbestrasse 11, 6330 Cham, Switzerland

# Preface

This book proposes an interdisciplinary study of pattern formation in economic agglomerations. The methodology proposed in this book targets readers in economic geography and spatial economics and advances the theoretical foundation for mathematicians interested in studying economic geography and spatial economics.

Economic geography studies economic agglomerations in some places by gathering workers through migrations from other places. In the framework of economic geography, Krugman found that two places with the same economic and trade environments, as well as the same population size, can undergo bifurcation that breaks uniformity to engender the so-called core-periphery pattern comprising a place with a small population and the other place with a large population. This finding is a fundamental mechanism of the self-organization of economic agglomeration patterns out of uniformity, underpinned by micro-economic mechanisms. Subsequently, various economic models have been developed using the spatial platform of two identical places. However, this platform of two places is too restrictive and must be upgraded to a large number of places to address realistic cases undergoing pattern formation.

This book proposes a bifurcation theory of pattern formation in economic geography. This book project started as follows: Prof. Kono at Tohoku University told us fairy tales of central place theory and the importance of bifurcation in economic agglomeration. It motivated us to conduct a theoretical study of economic geography models centered on bifurcation. We studied the pattern formation of spatial platforms with various symmetries:

- An economy of two identical places has a bilateral symmetry.
- A long-narrow economy also has a bilateral symmetry.
- A racetrack economy with a chain of places equally spread on the circumference of a circle has a dihedral group symmetry.
- Two-dimensional spaces modeled by square and hexagonal lattices have a local dihedral symmetry and a global translational symmetry.
- An equidistant economy with a series of places located equidistantly has a symmetry expressed by a symmetric group.

We were not complete newcomers to this study, as these spatial platforms possess symmetries and undergo symmetry-breaking bifurcations. Our experience of group-theoretic bifurcation analysis of various symmetric systems is presented in a series book:

*Imperfect Bifurcation in Structures and Materials*, 3rd ed. Springer (2019) by K. Ikeda, K. Murota.

In physics and nonlinear mathematics, bifurcation is well-acknowledged as a source of pattern formation, i.e., the self-organization of patterns from a uniform state. Given the symmetry of the spatial platform in question, we can theoretically predict how bifurcation engenders a pattern with reduced symmetry. In particular, a uniform object with high symmetry can exhibit diverse patterns through bifurcation. Our first study in economic geography examined the racetrack economy. The authors' experience with hierarchical bifurcation of a system with dihedral group symmetry was helpful in successfully describing the bifurcation of the racetrack economy in the Krugman model.

Part I of this book studies the pattern formation of economic agglomerations on various spatial platforms. We noted that hexagonal patterns predicted by central place theory exist on the hexagonal lattice with periodic boundary conditions. The theory of such existence blossomed into a monograph:

*Theory for Hexagonal Agglomeration in Economic Geography*. Springer (2014) by K. Ikeda, K. Murota.

Based on the theory in this monograph, we proposed a spectrum analysis procedure to capture characteristic agglomeration patterns in the real population data of Germany and the Eastern USA. This analysis is influenced by central place theory, which envisages the emergence of hexagonal hierarchical economic agglomerations, such as megalopolises, cities, towns, and so on.

Subsequently, we have found various unique properties of economic agglomerations on symmetric spatial platforms. In particular, we introduce an invariant pattern that is a stationary point of the dynamics for any value of the bifurcation (transportation cost) parameter. There are a large number of invariant patterns in the so-called replicator dynamics on symmetric spatial platforms. In Part II of this book, invariant patterns, together with the hierarchical bifurcations of symmetric platforms, elucidate the global view of economic agglomerations. In particular, a global–local spatial platform is advanced as a systematic means to construct a hierarchical structure of megalopolises, cities, towns, and so on.

We dedicate many chapters to the study of square and hexagonal lattices, encouraged by the recent success of Quantitative Spatial Economics, which targets economic agglomeration analysis in two-dimensional space (Redding and Rossi-Hansberg, 2017, *Annual Review of Economics*). We hope such a study will become mainstream in spatial economics in the years to come.

The advice of Dr. Jose M. Gaspar, Prof. Tatsuhito Kono, Prof. Kazuo Murota, and Prof. Dao-Zhi Zeng is valuable in writing this book. We greatly appreciate the advice on Chap. 7 provided by Prof. Daisuke Oyama.

The authors would like to thank their many coauthors: Dr. Hiroki Aizawa, Prof. Takashi Akamatsu, Dr. Jose M. Gaspar, Dr. Yosuke Kogure, Prof. Tatsuhito Kono,

Dr. Daisuke Murakami, Prof. Kazuo Murota, Mikihisa Onda, Dr. Minoru Osawa, and Prof. Jacques-François Thisse.

The authors would like to thank the students at Tohoku University who ardently contributed to the studies of economic agglomerations: Kazuki Akiyoshi, Motohiro Kamei, Daichi Komatsu, Kenji Sakamoto, Shunji Yamaki, and Akito Yanagimoto.

Our final thanks go to Tomoko Ikeda and Miho Takayama for warm and encouraging support.

Sendai, Japan<br>Tokyo, Japan<br>July 8, 2024

Kiyohiro Ikeda<br>Yuki Takayama

# Contents

# Chapter 1
# Introduction to Bifurcation Analysis in Economic Geography

**Abstract** This chapter provides an overview of this book. We introduce various spatial platforms, such as two places, line segment, equidistant place, racetrack, and square and hexagonal lattices. We study spatial economic agglomerations on these platforms in economic geography (spatial economic) models. These models describe economic agglomerations in certain places through the migration of workers and firms between places due to economic incentives. This chapter introduces a multiregional formulation, the dynamics in typical economic geography models, and the associated essential concepts. A general mathematical framework for analyzing bifurcation behavior is presented. We overview group-theoretic bifurcation theory, which is a mathematical tool to tackle the bifurcation of symmetric spatial platforms.

**Keywords** Break point · Dynamics · Economic agglomeration · Economic geography model · Group-theoretic bifurcation analysis · Spatial economics · Spatial equilibrium · Stability · Sustain point

## 1.1 Introduction

*Economic geography* (*spatial economics*) aims to explain the spatial distribution of economic activities and how transportation distance between places affects the economic activities of agents. The *core–periphery model* by Krugman (1991) is the first and the most successful attempt to clarify the microeconomic underpinning of the spatial agglomeration in a full-fledged general equilibrium approach[1] (see Sect. 2.2). His study paved the way for the development of new economic geography as a mainstream field in economics. New economic geography (e.g., Fujita et al., 1999b) mostly describes for two places how economic agglomeration

[1] This statement is based on an appraisal by Fujita and Thisse (2009) in honor of Krugman's 2008 Nobel Memorial Prize in Economic Sciences.



K. Ikeda, Y. Takayama, *Bifurcation Theory of Pattern Formation in Economic Geography*, Interdisciplinary Applied Mathematics 61,
https://doi.org/10.1007/978-3-031-71686-7_1

patterns evolve due to historical increases in economic integration (i.e., decreasing worldwide transportation costs).

However, a spatial platform of *two places* is too restrictive and must be expanded to include many places. In successfully modeling many places, it is vital to appropriately model (1) economic properties and (2) the spatial platform. This book emphasizes the latter and uses typical economic geography models as examples.

In multiregional economic geography models, we investigate pattern formation in spatial agglomerations across several spatial platforms, including: two places, line segment, equidistant place, racetrack, and square and hexagonal lattices. Figure 1.1 depicts these platforms. Nodal points shown as (∘) represent places of economic activities and are connected by lines expressing roads for transportation of goods. This book devotes many chapters to the study of square and hexagonal lattices. The study of these lattices is essential, given the recent successful development of Quantitative Spatial Economics targeting a two-dimensional space (Redding & Rossi-Hansberg, 2017).

Each spatial platform has an inherent bifurcation mechanism that depends on its symmetry, which is expressed by a symmetry group. As one of the major elements of this book, we elucidate this mechanism through group-theoretic bifurcation analysis developed in nonlinear mathematics.[2] Part I presents the direct bifurcation from the uniform population distribution for these spatial platforms, whereas Part II introduces secondary and further bifurcations. This chapter introduces the multiregional formulation in typical economic geography models, such as spatial equilibrium, dynamics, and static governing equation. Essential concepts employed

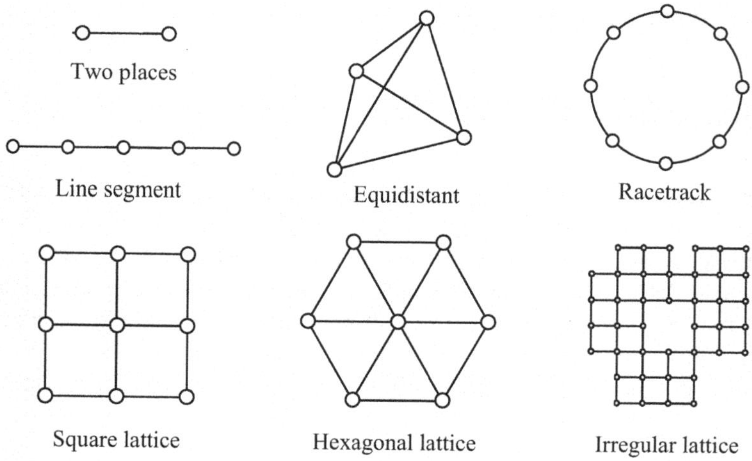


**Fig. 1.1** Spatial platforms in spatial economics

[2] See, e.g., Golubitsky et al. (1988) and Ikeda and Murota (2019).

throughout the book in the investigation of spatial economic agglomerations are explained.

We employ replicator dynamics as the standard case. This dynamics has many invariant patterns that are solutions to the governing equation for any value of the bifurcation parameter. The use of invariant patterns is helpful in the analysis of symmetric spatial platforms (see Sect. 1.7).

To elucidate the mechanism of spatial economic agglomerations, this book employs three major methodologies:

- Group-theoretic bifurcation analysis of symmetry-breaking bifurcations
- Invariant patterns in the replicator dynamics
- Global–local spatial platform for describing the hierarchical nature of spatial economic agglomerations

This chapter is organized as follows: Spatial platforms are presented in Sect. 1.2. We introduce a general mathematical framework in Sect. 1.3. In Sect. 1.4, group-theoretic bifurcation analysis is advanced. Section 1.5 presents dynamics in economic geography. In Sect. 1.6, unique behaviors in the replicator dynamics are introduced. Section 1.7 provides procedures to obtain solution curves.

## 1.2 Spatial Platforms

This book employs several spatial platforms depicted in Fig. 1.1.

**Geometric and Economic Meanings**

The spatial platforms have the following geometric and economic meanings:

- *Two-place economy* has the simplest geometrical structure and superior analytical tractability. The two places are initially identical, but an imbalance between them occurs through bifurcation.
- *Long-narrow economy*, on a line segment, accommodates a chain of places. This economy models a linear arrangement of urban areas that is widespread worldwide, as observed on Japan's Main Island and in a closed, narrow corridor between the Atlantic Ocean and the Appalachian Mountains. It can be found at transnational scales (e.g., the STRING and Golden, Blue, and Green Bananas in Europe). This economy also models the development of a chain of cities along a railway system.
- *Equidistant economy* accommodates places that are equidistant (the distance between any pair of places is identical). It is suitable as a spatial platform for the competition of airport hubs.
- *Racetrack economy* represents the geometry of a chain of identical places on the circumference of a circle. The economy does not have any borders and is an idealized counterpart of the long-narrow economy subject to the effects of borders at both ends. It is also an idealization of a cross section (line segment) of a square lattice.

- *Square and hexagonal lattices* with periodic boundary conditions express infinite two-dimensional spaces. The hexagonal lattice accommodates hexagonal patterns, racetrack patterns, and so on (see Sect. 2.3.2). These lattices with nonperiodic boundary conditions can model finite two-dimensional spaces. The hexagonal lattice is suited for the study of hexagonal patterns in central place theory and for the transportation network analysis of urban hierarchy. The square lattice models the metropolitan road network of cities, such as Chicago and Kyoto. An *irregular lattice* with an asymmetric network can express a detailed and complex geometry but relies heavily on numerical analysis.

**History of Development of Spatial Platforms**

We introduce the history of the development of spatial platforms (Fig. 1.1) in economic geography:

- *Long-narrow economy*, on a line segment or an infinite line, displays diverse agglomeration patterns ranging from continuous to discrete ones.[3]
- *Equidistant economy* is an idealized spatial platform suited for an analytical study. This economy is handled mostly using three places[4] and is studied in several settings.[5]
- *Racetrack economy* is studied extensively (Chap. 3).[6] A numerical simulation including 12 places reveals that the symmetric equilibrium can become unstable (Krugman, 1993). A local analysis (linearized eigenproblem) is conducted to identify the emergence of several spatial frequencies (Fujita et al., 1999b).
- *Square lattice* is employed in studies of two-dimensional spatial agglomerations.[7]
- *Hexagonal lattice* is used as a spatial platform of hexagonal distributions in central place theory (Christaller, 1933). Eaton and Lipsey (1975, 1982) show the existence of a hexagonal distribution of mobile production factors. Hexagonal patterns are shown to exist on a hexagonal lattice through theoretical and numerical bifurcation analyses.[8]
- *Irregular lattice* is employed for large-scale numerical simulations. An economic agglomeration pattern for cities in Europe is simulated (Stelder, 2005). The topography of trade costs, productivities, and amenities in the United States is estimated by Allen and Arkolakis (2014).

---

[3] See, e.g., Fujita and Krugman (1995), Fujita and Mori (1997), Mori (1997), Fujita et al. (1999a), and Ikeda et al. (2017a).

[4] See, e.g., Fujita et al. (1999b), Castro et al. (2012), and Commendatore et al. (2015).

[5] See, e.g., Puga (1999), Tabuchi et al. (2005), Oyama (2009), Zeng and Uchikawa (2014), and Gaspar et al. (2018, 2020).

[6] Studies on racetrack economy are given, for example, by Krugman (1993), Fujita et al. (1999b), Picard and Tabuchi (2010), Tabuchi and Thisse (2011), Akamatsu et al. (2012), Ikeda et al. (2012a, 2019), and Takayama et al. (2020).

[7] See, e.g., Clarke and Wilson (1983), Weidlich and Haag (1987), and Munz and Weidlich (1990).

[8] See, e.g., Ikeda et al. (2012b, 2014, 2017b) and Ikeda and Murota (2014).

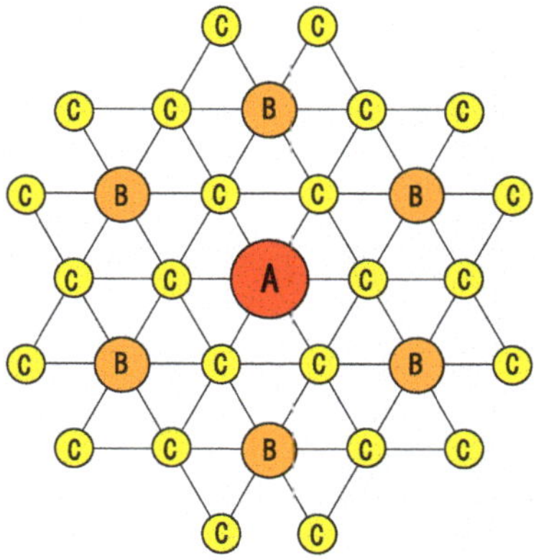


**Fig. 1.2** Christaller's $k = 3$ system (the size of a circle expresses the size of a place)

### Global–Local Spatial Platform

*Central place theory* describes a hierarchical structure of agglomerated places, such as cities, towns, villages, and so on. Figure 1.2 depicts Christaller's $k = 3$ system (1933) that expresses this hierarchical structure: The largest A-center (city) is surrounded by hexagonal B-centers (towns), and each B-center is surrounded by smaller hexagonal C-centers (villages). This book's primary target is a simulation of a hierarchical structure of urban systems in the framework of economic geography models.

It is possible to employ a hexagonal lattice with a hexagonal shape to simulate the $k = 3$ system, as demonstrated in Chap. 2. However, such a simulation relies on an excessive idealization. Quantitative Spatial Economics has been developed as a realistic spatial platform with a large number of places (Redding, 2016 and Redding & Rossi-Hansberg, 2017) but does not necessarily have insight into the hierarchical structure.

This book presents a *global–local system* as a large-scale spatial platform (see Chap. 11). This system can express a hierarchical structure of economic agglomeration and possesses insight into the bifurcation mechanism. A global network comprises a system of cities and expresses the geographical distribution and economic interaction of cities. Each city has a microstructure comprising a system of local places. The size of a local system characterizes quantitatively the population of a city, and the distribution of local places expresses local geographical properties. We can express diverse network systems of economic activities by appropriately choosing an economic geography model, a global system, and a series of local systems. The global–local system is applicable to various economic agglomeration problems, such as air networks, train networks, and a hierarchy of cities, towns, villages, and so on.

Figure 1.3 depicts a global–local system of the seven largest airport hub cities on the map of the USA (see Sect. 11.4). Local transportation networks of the cities are modeled using square lattices and two places. The local networks are connected equidistantly by an air transportation network.

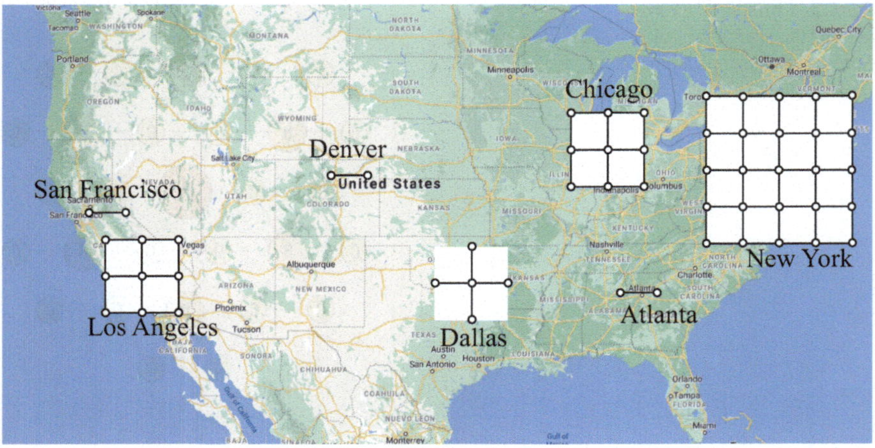


**Fig. 1.3** Global–local model of seven largest airport hub cities in the USA

## 1.3 General Mathematical Framework

We introduce a general mathematical framework based on Ikeda and Murota (2019). We consider a scalar field for a spatial platform which has a single degree of freedom in the location at each node, as is customary in most economic geography models. Depending on the case, this location can be city, town, region, place, and so on. This book mostly employs the *place* following central place theory.

### *1.3.1 Governing Equation and Bifurcation*

We consider a set of $n$ places and mobile agents migrating among these places according to an *adjustment dynamics*:

$$\frac{d\boldsymbol{\lambda}}{dt} = \boldsymbol{F}(\boldsymbol{\lambda}, \tau). \tag{1.1}$$

Here, $\boldsymbol{\lambda} = (\lambda_i \mid i = 1, \ldots, n) \in \mathbb{R}^n$ represents an *independent variable vector* for the population $\lambda_i$ in place $i$ ($\mathbb{R}$ is the set of real numbers); $\tau \in \mathbb{R}$ denotes a *bifurcation parameter* to be specified based on economic modeling; and $\boldsymbol{F} = (F_i \mid i = 1, \ldots, n)$: $\mathbb{R}^n \times \mathbb{R} \to \mathbb{R}^n$ is a sufficiently smooth nonlinear function. A typical choice of $\tau$ is the transportation cost parameter. *Stationary points* (rest points) $(\boldsymbol{\lambda}, \tau)$ in the dynamics (1.1) are defined as those satisfying the static *governing equation*:

$$\boldsymbol{F}(\boldsymbol{\lambda}, \tau) = \boldsymbol{0}. \tag{1.2}$$

**Remark 1.1** In economic geography, it is customary to use the space of the $(n-1)$-dimensional independent variable vector $\hat{\boldsymbol{\lambda}} = (\lambda_1, \ldots, \lambda_{n-1})$ with $\lambda_n = 1 - \sum_{i=1}^{n-1} \lambda_i$ (cf. (1.8)). This is an $(n-1)$-dimensional simplex and is helpful in the manipulation in spatial economic models. However, this book employs all $n$ coordinates, i.e., $\boldsymbol{\lambda} = (\lambda_1, \ldots, \lambda_n)$, because the symmetry condition and the spatial population distribution can be expressed consistently with bifurcation theory. □

In this book, the *stability* of a stationary point means its *linear stability*. We consider the *Jacobian matrix*

$$J(\boldsymbol{\lambda}, \tau) = \frac{\partial \boldsymbol{F}}{\partial \boldsymbol{\lambda}}(\boldsymbol{\lambda}, \tau) \tag{1.3}$$

of the governing equation and classify the linear stability of the points by using the eigenvalues of this matrix:

$$\begin{cases} \text{Linearly stable:} & \text{every eigenvalue has a negative real part.} \\ \text{Critical:} & \text{at least one eigenvalue is on the imaginary axis.} \\ \text{Linearly unstable:} & \text{at least one eigenvalue has a positive real part.} \end{cases}$$

When all eigenvalues of $J(\boldsymbol{\lambda}, \tau)$ are real, which is mostly the case in the analysis in this book, we can rewrite this classification as

$$\begin{cases} \text{Linearly stable:} & \text{every eigenvalue is negative.} \\ \text{Critical:} & \text{at least one eigenvalue is zero.} \\ \text{Linearly unstable:} & \text{at least one eigenvalue is positive.} \end{cases}$$

We consider a critical point $(\boldsymbol{\lambda}_{\mathrm{c}}, \tau_{\mathrm{c}})$, where $J_{\mathrm{c}} = J(\boldsymbol{\lambda}_{\mathrm{c}}, \tau_{\mathrm{c}})$ satisfies $\det J_{\mathrm{c}} = 0$. We define the *multiplicity* $M$ of the critical point by the dimension of the kernel space of $J_{\mathrm{c}}$. Let $(\boldsymbol{\eta}_i \mid i = 1, \ldots, n)$ be an orthonormal basis of $\mathbb{R}^n$ such that

$$J_{\mathrm{c}} \boldsymbol{\eta}_i = \boldsymbol{0} \quad (i = 1, \ldots, M). \tag{1.4}$$

The *critical eigenvectors* $\boldsymbol{\eta}_1, \ldots, \boldsymbol{\eta}_M$ for $M$ zero eigenvalues of $J_{\mathrm{c}}$ are related to the directions of bifurcating paths (e.g., Sect. 3.4). We further classify critical points as

$$\begin{cases} \text{Bifurcation point.} \\ \text{Stationary (limit) point of } \tau. \end{cases}$$

### 1.3.2 Simple Example of Bifurcation

As a simple example of bifurcation, we refer to a governing equation:

$$\boldsymbol{F}(\boldsymbol{\lambda}, \tau) = \begin{pmatrix} \dfrac{5}{2}\sin 2x + 2\cos x \sin y - \tau \sin x \\ 2\sin x \cos y + \sin 2y - \tau \sin y \end{pmatrix} = \boldsymbol{0}$$

with $\boldsymbol{\lambda} = (x, y)$. The Jacobian matrix is

$$J(\boldsymbol{\lambda}, \tau) = \frac{\partial \boldsymbol{F}}{\partial \boldsymbol{\lambda}} = \begin{bmatrix} 5\cos 2x - 2\sin x \sin y - \tau \cos x & 2\cos x \cos y \\ 2\cos x \cos y & -2\sin x \sin y + 2\cos 2y - \tau \cos y \end{bmatrix}.$$

There is the trivial solution $x = y = 0$, satisfying $\boldsymbol{F}(\boldsymbol{0}, \tau) = \boldsymbol{0}$ for any $\tau$. In this solution, the Jacobian matrix becomes

$$J(\boldsymbol{0}, \tau) = \begin{bmatrix} 5-\tau & 2 \\ 2 & 2-\tau \end{bmatrix}.$$

This matrix is singular at $\tau = \tau_{\rm c} = 1$ and 6. There are two simple bifurcation points ($M = 1$) at $(\boldsymbol{\lambda}_{\rm c}, \tau_{\rm c}) = (\boldsymbol{0}, 1)$ and $(\boldsymbol{0}, 6)$. The critical eigenvectors at these points for the two zero eigenvalues are given by

$$\eta_1 = \frac{1}{\sqrt{5}}\begin{pmatrix} -1 \\ 2 \end{pmatrix}, \qquad \eta_2 = \frac{1}{\sqrt{5}}\begin{pmatrix} 2 \\ 1 \end{pmatrix}.$$

Bifurcating paths at these bifurcation points are found in the directions of these eigenvectors.

## 1.4 Group-Theoretic Bifurcation Analysis

As a prelude to a general framework of group-theoretic bifurcation analysis in Sect. 3.3, we present a brief overview of this analysis using a simple example. We give a summary of the symmetry of various spatial platforms and introduce the spectrum analysis of population data.

### 1.4.1 Simple Example of Bifurcation Analysis

A bifurcating solution can be obtained as a solution to the so-called *bifurcation equation*. We demonstrate the derivation of the bifurcation equation (see Remark 1.2 below) for a simple example with the governing equation:

$$\boldsymbol{F}(x, y, \tau) = \begin{pmatrix} F_x(x, y, \tau) \\ F_y(x, y, \tau) \end{pmatrix}$$
$$= \begin{pmatrix} \left\{1 - [x^2 + (1-y)^2]^{-1/2}\right\} x + x \\ \left\{1 - [x^2 + (1-y)^2]^{-1/2}\right\} (y-1) + 2y - \tau \end{pmatrix} = \boldsymbol{0}. \tag{1.5}$$

The solution curves of this governing equation are

$$\begin{cases} x = 0, \quad \tau = 3y, & \text{pre-bifurcation curve,} \\ y = 1 - \left(1/4 - x^2\right)^{1/2}, \quad \tau = 2 - \left(1/4 - x^2\right)^{1/2}, & \text{bifurcated curve.} \end{cases}$$

The Jacobian matrix of the pre-bifurcation path is

$$J(0, y, 3y) = EA \begin{bmatrix} 2 - \frac{1}{1-y} & 0 \\ 0 & 3 \end{bmatrix}.$$

There is a bifurcation point $(x_{\mathrm{c}}, y_{\mathrm{c}}, \tau_{\mathrm{c}}) = (0, 1/2, 3/2)$, at which $\det J = 0$ holds.

**Direct Derivation of Bifurcation Equation**

In the neighborhood of this bifurcation point, we define incremental variables by

$$\tilde{x} = x, \quad \tilde{y} = y - 1/2, \quad \tilde{\tau} = \tau - 3/2$$

and expand the governing equation into a power series:

$$F_x = -4\tilde{x}\tilde{y} + 4\tilde{x}^3 + \text{higher order terms} = 0,$$
$$F_y = 3\tilde{y} - \tilde{\tau} - 2\tilde{x}^2 + \text{higher order terms} = 0.$$

The elimination of $\tilde{y}$ leads to a power series form of the so-called *bifurcation equation*:

$$\tilde{F}(\tilde{x}, \tilde{\tau}) = \frac{4}{3}\tilde{x}(\tilde{\tau} - \tilde{x}^2) + \text{higher order terms} = 0. \tag{1.6}$$

The bifurcation equation asymptotically has two solution curves: the pre-bifurcation curve $\tilde{x} = 0$ and the bifurcated curve $\tilde{\tau} - \tilde{x}^2 = 0$ intersecting at $(\tilde{x}, \tilde{\tau}) = (0, 0)$.

**Remark 1.2** The derivation of the bifurcation equation in a general setting can be conducted by a standard procedure called the *Lyapunov–Schmidt reduction*.[9] An $n$-dimensional governing equation is generically reduced, in a neighborhood of a critical point, to a system of $M$ equations, called bifurcation equation ($M$ is the multiplicity of the critical point). □

**Group-Theoretic Analysis**

Direct derivation of the bifurcation equation is very complicated in a system with a large degree of freedom. It is the basic strategy of group-theoretic bifurcation analysis to exploit the symmetry of the system to arrive at the bifurcation equation (see Sect. 4.3). This analysis does not resort to the concrete form of the governing equation, and the result of this analysis is general and is independent of the economic framework in an economic geography model.

We consider geometrical transformations $e : (x, y) \mapsto (x, y)$ and $s : (x, y) \mapsto (-x, y)$; $e$ is the identity element that leaves everything unchanged and $s$ exchanges the sign of $x$. These transformations form a dihedral group $\mathrm{D}_1 = \{e, s\}$. We introduce matrix representation:

$$T(e) = \begin{bmatrix} 1 & \\ & 1 \end{bmatrix}, \quad T(s) = \begin{bmatrix} -1 & \\ & 1 \end{bmatrix}.$$

By setting $\boldsymbol{\lambda} = (x, y)$, we get the equivariance condition

$$T(g)\boldsymbol{F}(\boldsymbol{\lambda}, \tau) = \boldsymbol{F}(T(g)\boldsymbol{\lambda}, \tau) \quad (g = e,\ s)$$

for $\boldsymbol{F} = (F_x, F_y)^\top$ in (1.5) since

$$T(s)\boldsymbol{F} = (-F_x, F_y)^\top, \quad \boldsymbol{F}(T(s)\boldsymbol{\lambda}, \tau) = \boldsymbol{F}(-x, y, \tau) = (-F_x, F_y)^\top.$$

The equivariance of $\boldsymbol{F}$ is inherited by the equivariance

$$\tilde{T}(g)\tilde{F}(\tilde{x}, \tilde{\tau}) = \tilde{F}(\tilde{T}(g)\tilde{x}, \tilde{\tau}) \quad \text{with } \tilde{T}(e) = 1,\ \tilde{T}(s) = -1$$

of the bifurcation equation. It expresses an odd function condition:

$$-\tilde{F}(\tilde{x}, \tilde{\tau}) = \tilde{F}(-\tilde{x}, \tilde{\tau}).$$

Then, together with the condition $\tilde{F}(0, 0) = \partial\tilde{F}/\partial\tilde{x}(0, 0) = 0$ of a critical point, we can arrive at a qualitative form of the bifurcation equation:

$$\tilde{F}(\tilde{x}, \tilde{\tau}) = \tilde{x}F_1(\tilde{x}^2, \tilde{\tau}) = 0$$

[9] See, e.g., Golubitsky et al. (1988) and Ikeda and Murota (2019).

for some function $F_1(\tilde{x}^2, \tilde{\tau})$ with $F_1(0, 0) = 0$. It is a general form of a pitchfork bifurcation with a pre-bifurcation curve $\tilde{x} = 0$ and a bifurcating curve $F_1(\tilde{x}^2, \tilde{\tau}) = 0$ and is qualitatively compatible with (1.6) derived directly.

### *1.4.2 Symmetry of Spatial Platforms*

In economic geography models, the mechanism of bifurcation depends on the symmetry of the spatial platform:

- The dihedral group $D_1$ expresses bilateral symmetry in *two places* (Chap. 7), *three places*, and *long-narrow economy* on a line segment (Chap. 12). All these platforms have the same bifurcation mechanism: the existence of the pitchfork bifurcation point, as explained above using the simple example.
- *Racetrack economy* with $n$ places (Chaps. 3 and 9) has the regular $n$-gonal symmetry expressed by the dihedral group $D_n$ (see (3.1) in Sect. 3.2 for the definition of $D_n$).
- *Equidistant economy* has the symmetry expressed by the symmetric group $S_n$. This economy is investigated in Chap. 14 based on mathematical studies.[10]
- *Lattice networks* of two types are employed in this book. *Square lattice* has local $D_4$-symmetry (see Chaps. 4 and 13 and Kogure et al., 2021). *Hexagonal lattice* has local $D_6$-symmetry (Chaps. 2 and 5). These lattices have an additional translational symmetry under periodic boundary conditions.

In the analysis of these spatial platforms, the bifurcation equation (cf. (1.6)) usually becomes multidimensional and possesses many bifurcating solutions.

### *1.4.3 Spectrum Analysis of Population Data*

As a systematic means to capture the pattern formation in population data, Chap. 6 advances a Fourier spectrum analysis procedure for the hexagonal lattice based on group-theoretic bifurcation theory. The population data of the Eastern USA in 2014 were taken from the City Population website (http://www.citypopulation.de/). Group-theoretic spectrum analysis of the population data was conducted to detect 18 rhombic domains accommodating core–satellite patterns and hexagonal ones. A gigantic hexagonal city network emerges there as an assemblage of hexagonal and core–satellite patterns, like a giant jigsaw puzzle with many pieces (Fig. 1.4). There is a hierarchy of subnetworks with various geographical scales that look like the overlapping hexagons (Lösch, 1940). Thus, this vast expanse of flat area

[10] See, e.g., Golubitsky and Stewart (2002), Elmhirst (2004), and Aizawa et al. (2020).

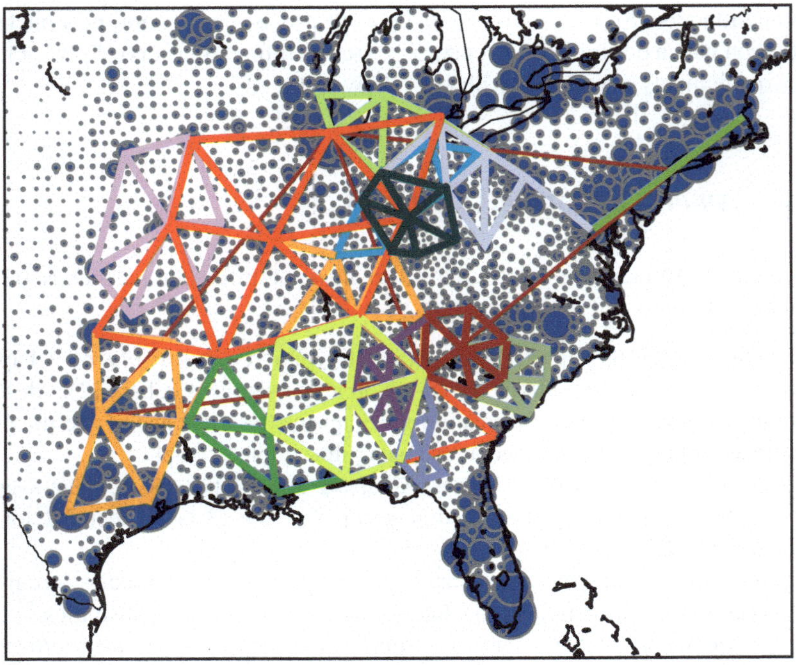

(a)

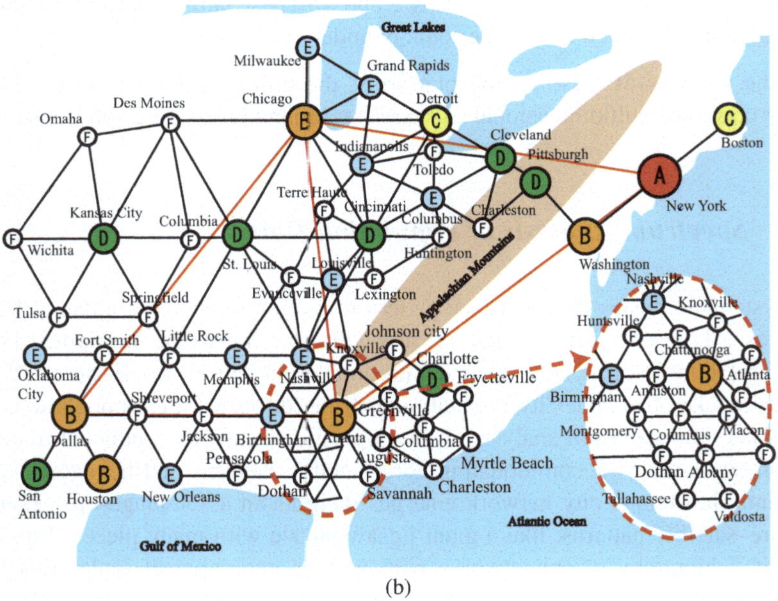


(b)

**Fig. 1.4** Assembled distribution of cities in the Eastern USA on a population map (Ikeda et al., 2018) (the area of a circle denotes the population size; lines with different colors indicate city sub-distributions in (**a**); we classify cities from A-center to F-center in decreasing order of population size in (**b**)) accommodates hierarchical hexagonal patterns, just as hexagonal patterns in physics emerge from uniformity (e.g., Bénard, 1900).

## 1.5 Dynamics in Economic Geography

There are various dynamics in spatial economics (Sandholm, 2010). Among several candidates, this book employs the replicator dynamics as the standard case, and the logit dynamics in some cases. The replicator dynamics displays a unique bifurcation mechanism that has not been given much attention in nonlinear mathematics (Sect. 1.6).

The *replicator dynamics*[11] is given as

$$F_i(\boldsymbol{\lambda}, \tau) = (v_i(\boldsymbol{\lambda}, \tau) - \bar{v}(\boldsymbol{\lambda}, \tau))\lambda_i. \tag{1.7}$$

Here, $v_i$ is *indirect utility* in the $i$th place, and $\bar{v}$ represents the *weighted average utility* defined as

$$\bar{v} = \sum_{i=1}^{n} \lambda_i v_i.$$

We introduce a *spatial equilibrium* in spatial economics, wherein the mobile agents migrate among places and choose to live in place $i$ that provides them the highest indirect utility, $v_i$. (The definition of indirect utility $v_i$ varies with economic models.) The spatial equilibrium is defined by the following problem:

$$\text{Find } (\boldsymbol{\lambda}^*, \hat{v}) \text{ satisfying: } \begin{cases} (v_i - \hat{v})\lambda_i^* = 0: & \textit{steady state}, \\ v_i - \hat{v} \leq 0: & \textit{sustainability}, \\ \lambda_i^* \geq 0: & \textit{non-negativity}, \\ \sum_{i=1}^{n} \lambda_i^* = 1: & \textit{conservation law} \end{cases} \tag{1.8}$$

(for all $i \in \{1, \ldots, n\}$). Here, $\boldsymbol{\lambda}^* = (\lambda_i^* \mid i = 1, \ldots, n)$ and $\hat{v}$ is the highest (indirect) utility of the solution to this problem. The solutions to the governing equation are not necessarily spatial equilibria as they can be unsustainable and/or violate the nonnegativity in (1.8). The problem of obtaining a set of stable spatial equilibria is related to another problem to find a set of stable *stationary points* $\boldsymbol{\lambda}^*(\tau)$ in the replicator dynamics (Hofbauer & Sigmund, 1998 and Sandholm, 2010). This dynamics has a unique feature associated with the vanishing of the population at some place(s), as highlighted in Sect. 1.6.

[11] See, e.g., Taylor and Jonker (1978) and Sandholm (2010).

The *logit dynamics* (e.g., Fudenberg & Levine, 1998) employs a particular formulation whereby mobile workers are assumed to have heterogeneous preferences regarding location choice.[12] This dynamics employs a functional form:[13]

$$\boldsymbol{F}(\boldsymbol{\lambda}, \tau) = \boldsymbol{P}(\boldsymbol{v}(\boldsymbol{\lambda}, \tau)) - \boldsymbol{\lambda}. \tag{1.9}$$

Here, $\boldsymbol{P}(\boldsymbol{v}) = (P_1, \ldots, P_n)^\top$ is the choice function vector and $\boldsymbol{v}$ is the indirect utility vector. We employ the *logit choice function*:

$$P_i(\boldsymbol{v}) = \frac{\exp(\theta v_i)}{\sum_{j=1}^{n} \exp(\theta v_j)}, \tag{1.10}$$

where $\theta \in (0, \infty)$ is a parameter for the magnitude of the heterogeneity. The spatial equilibrium of a model with heterogeneous workers is defined by the solutions to the governing equation $\boldsymbol{F}(\boldsymbol{\lambda}, \tau) = \boldsymbol{0}$ with (1.9). The assumption of the heterogeneous location choice adds a dispersion force in the model akin to congestion that becomes relatively more important as transportation costs become very low. The limit of $\theta \to \infty$ corresponds to the case without heterogeneity, and the mobile workers decide their place only by $v_i$.

The conservation law $\sum_{i=1}^{n} \lambda_i = 1$ in (1.8) is satisfied in the replicator dynamics and the logit dynamics.

## 1.6 Unique Behaviors of Replicator Dynamics

The replicator dynamics in (1.7) allows the vanishing of population $\lambda_i$ at some place(s) and has several unique features:

- Sustaining bifurcation in Sect. 1.6.2
- Invariant pattern in Sect. 1.6.3

These features demand a reorganization of the application of the conventional bifurcation theory.

---

[12] See, e.g., Tabuchi and Thisse (2002), Murata (2003), and Akamatsu et al. (2012).

[13] The indirect utility for an individual $s \in (0, h_i]$ in place $i$ is expressed as $v_i^{(s)}(\boldsymbol{\lambda}, \tau) = v_i(\boldsymbol{\lambda}, \tau) + \epsilon_i^{(s)}$, where $\epsilon_i^{(s)}$ denotes the utility representing the idiosyncratic taste for residential location. $(\epsilon_i^{(s)}, \forall s)$ is assumed to follow the Gumbel distribution, which is identical and independent across places (e.g., McFadden, 1974 and Anderson et al., 1992).

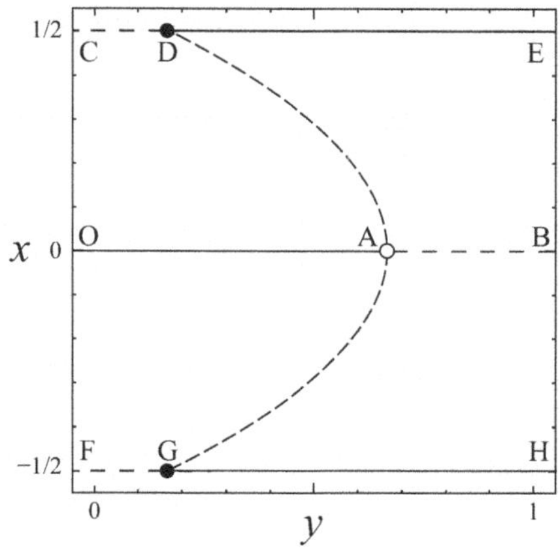


**Fig. 1.5** Pitchfork bifurcation in a two-place economy (solid curve: stable; dashed curve: unstable)

### *1.6.1 Two-Place Economy: Simple Example*

A *break point* and a *sustain point* play important roles in this book. Figure 1.5 depicts the typical behavior in the replicator dynamics in a two-place economy (cf. Fujita et al., 1999b). The vertical axis is the population difference $x = (\lambda_1 - \lambda_2)/2$ of the two places, and the horizontal axis $y \in (0, 1)$ is a trade freeness parameter (see (2.11)), which is inversely proportional to the transportation cost parameter $\tau$. The solid line denotes a stable curve and the dashed curve, an unstable one.

The governing equation has various solutions:

- The trivial solution $x = 0$ denotes the uniform state of $(\lambda_1, \lambda_2) = (1/2, 1/2)$. This solution is stable for small trade freeness $y$ and encounters an unstable *pitchfork bifurcation point* A when $y$ increases to a certain level. At this bifurcation point, there emerge two bifurcating branches AD and AG.
- Bifurcating branch AD represents a state with a larger population in place 1 ($x > 0 \mapsto \lambda_1 > \lambda_2$) and reaches the point D where the population $\lambda_2$ of place 2 vanishes.
- Bifurcating branch AG represents a state with a larger population in place 2 ($x < 0 \mapsto \lambda_1 < \lambda_2$) and reaches the point G with $\lambda_1 = 0$.
- Horizontal line CDE denotes a complete agglomeration to place 1 ($\lambda_1 = 1$) and no population in place 2 ($\lambda_2 = 0$).
- Horizontal line FGH denotes a complete agglomeration to place 2.

In economic geography, the value of $y$ at the pitchfork bifurcation point A is called *break point*, and the value of $y$ at points D and G is called *sustain point*. The state CD (excluding point D), in which $v_2 - \bar{v} = v_2 - v_1 > 0$ holds, is unsustainable and so is the state FG (excluding point G).

### 1.6.2 Corner Solution and Sustaining Bifurcation

The solutions to the governing equation in the replicator dynamics can be classified into two types:

- An *interior solution* with positive populations in all places ($\lambda_i > 0$ for all $i$)
- A *corner solution* with no population in one or more places ($\lambda_i = 0$ for some $i$).

The former solution is a spatial equilibrium, but the latter might not be a spatial equilibrium. The corner solution is absent in models with the logit dynamics in (1.9) or with housing market (e.g., Helpman, 1998 and Allen & Arkolakis, 2014).

We define sustainability and unsustainability of a corner solution:

- A corner solution is *sustainable* if $v_i - \bar{v} \leq 0$ is satisfied for any place with zero population ($\lambda_i = 0$).
- A corner solution is *unsustainable* if $v_i - \bar{v} > 0$ holds in some place.

The stability of a solution is related to the stability of a spatial equilibrium. A stable solution is sustainable (see Proposition 9.4 in Sect. 9.3 and Hofbauer & Sigmund, 1998).

The replicator dynamics has *sustaining bifurcation*, in addition to *breaking bifurcation* studied in nonlinear mathematics. One or more places with no population gain population along a bifurcating path from this bifurcation point. A *sustain point* is regarded as a special type of sustaining bifurcation point where the system's sustainability changes.

A sustaining bifurcation occurs at a critical point where $v_i - \bar{v} = 0$ and $\lambda_i = 0$ are satisfied simultaneously (see Sect. 9.3). The sustaining bifurcation arises from the factorized form $(v_i - \bar{v})\lambda_i$ of (1.7), which produces two solution curves of the governing equation (1.2) that cross at this point (Fig. 1.6). This point is a transcritical bifurcation point, and its behavior is known in the literature (e.g., Ikeda

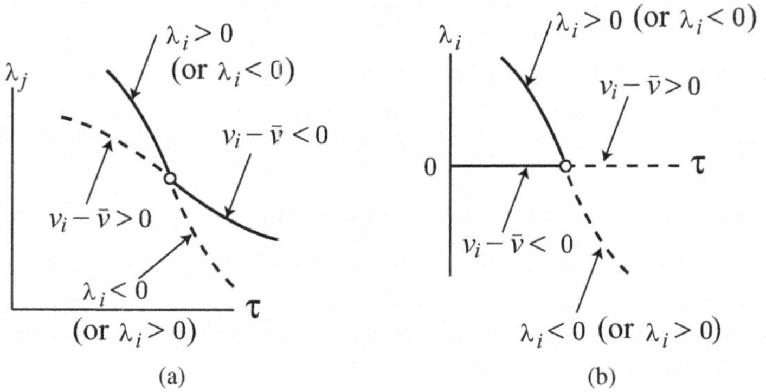


**Fig. 1.6** Sustaining bifurcation points (solid curve: stable; dashed curve: unstable). (**a**) Crossing point of two non-invariant solution curves. (**b**) Crossing point of invariant and non-invariant solution curves

& Murota, 2019, §2.5.2). Since $\lambda_i$ and $v_i - \bar{v}$ vanish simultaneously at this point, the sign of $v_i - \bar{v}$ along a (trivial) solution curve changes, as does the sign of $\lambda_i$ along the other curve. At this point, a sustainable solution ($v_i - \bar{v} < 0$, $\lambda_i = 0$) changes into an unsustainable one ($v_i - \bar{v} > 0$, $\lambda_i = 0$) along the former curve. It is Poincaré's exchange of stability. A solution curve with a kink emerges if we consider only stable equilibria. A curve with a negative population is excluded by the nonnegativity in (1.8) from stable equilibria.

The sustaining bifurcation is a unique feature of the replicator dynamics, which has a corner solution. No sustaining bifurcation exists in the logit dynamics in (1.9), which does not have a corner solution.

### *1.6.3 Invariant Pattern*

A series of solutions to the governing equation $\boldsymbol{F}(\boldsymbol{\lambda}, \tau) = \boldsymbol{0}$ in (1.2) forms a solution curve $(\boldsymbol{\lambda}(\tau), \tau)$. In general, the pattern $\boldsymbol{\lambda}$ changes along a solution curve $(\boldsymbol{\lambda}(\tau), \tau)$ parameterized by $\tau$.

There is a unique spatial pattern $\boldsymbol{\lambda} = \bar{\boldsymbol{\lambda}}$ ($\bar{\boldsymbol{\lambda}}$ is constant), for which $(\bar{\boldsymbol{\lambda}}, \tau)$ is a solution to the governing equation for any $\tau$. We call such pattern *invariant pattern*. We accordingly have the classification:

$$\begin{cases} \textit{Invariant pattern}: & \boldsymbol{\lambda}(\tau) \text{ is constant for any } \tau. \\ \textit{Non-invariant pattern}: & \boldsymbol{\lambda}(\tau) \text{ varies with } \tau. \end{cases}$$

A uniform state is an invariant pattern in many dynamics.

The replicator dynamics has many invariant patterns, in addition to the uniform state. Chapters 9 and 10 present a group-theoretic procedure to obtain invariant patterns. We present below several examples of invariant patterns in the racetrack economy with the dihedral group $\mathrm{D}_n$-symmetry.

We begin with the two-place economy in Sect. 1.6.1. In the solution curves of this economy in Fig. 1.5, we see the presence of three horizontal lines:

- Line OAB for the uniform state $(\lambda_1, \lambda_2) = (1/2, 1/2)$
- Line CDE for a full agglomeration $(\lambda_1, \lambda_2) = (1, 0)$
- Line FGH for the other full agglomeration $(\lambda_1, \lambda_2) = (0, 1)$

These three patterns $\boldsymbol{\lambda} = (1/2, 1/2)$, $(1, 0)$, and $(0, 1)$ are invariant patterns.

All invariant patterns of a racetrack economy with $n = 12$ places are depicted in Fig. 1.7. There are as many as 14 invariant patterns; the solution curves of these invariant patterns are obtained without conducting the analysis. For a large-scale analysis, the knowledge of invariant patterns is quite helpful by drastically reducing the analytical task.

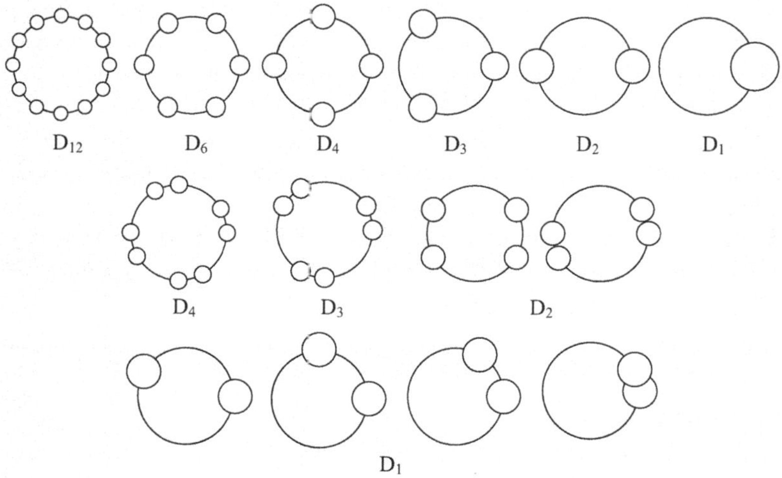


**Fig. 1.7** All invariant patterns of a racetrack economy with $n = 12$

## 1.7 Procedures to Obtain Solution Curves

Nonlinear governing equation often displays complex solution curves. It is a major ingredient of this book to propose a robust and systematic procedure to obtain such solution curves.

We search for the solutions to the governing equation in (1.2) by using the three analysis procedures:

- Direct analytical or numerical analysis[14] of the governing equation in a specific economic geography model on a spatial platform (e.g., Chap. 2)
- Group-theoretic bifurcation analysis of the spatial platform's symmetry (e.g., Chap. 4)
- The use of invariant patterns in the replicator dynamics (e.g., Chap. 10)

In the direct analysis, the concrete form of the governing equation depends on a specific economic geography model employed in the analysis and its economic parameters. Through the group-theoretic analysis, we can obtain possible direct bifurcating patterns from a uniform state. The analysis of the symmetry of the bifurcating patterns elucidates the mechanism of the secondary bifurcation. In this manner, we can obtain a tree-like structure of a hierarchical bifurcation that starts from the uniform state and ends up with a completely asymmetric state.

The direct analysis is like a car, whereas the group-theoretic analysis is like a map. The latter can advance a possible set of bifurcating patterns and the tree-like

[14] For the numerical tracing and branch switching of a solution curve, see, e.g., Crisield (1997) and Ikeda and Murota (2022).

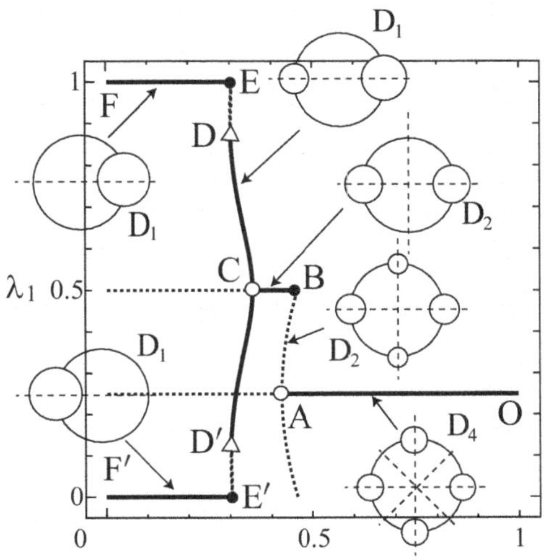


**Fig. 1.8** Solution curves of the four places ($n = 4$) (Ikeda et al., 2012a) (solid curve: stable; dashed curve: unstable; (o): simple bifurcation point; (△): limit point; (•): sustain point)

structure of the hierarchical bifurcation. The former can determine, for a particular model and its specific parameter values, the patterns that actually emerge and a course in the tree-like structure to be taken. We can employ these two types of analyses together to untangle the structure of a complex network of solution curves.

However, both of these procedures demand substantial theoretical and/or numerical tasks. We can employ the third procedure for replicator dynamics; solution curves for invariant patterns can be obtained without conducting the analysis. This procedure is like a radar, as expounded in the illustrative example below. We employ a pertinent combination of all or some of the three procedures in this book.

**Illustrative Example**

We explain the three procedures presented above using an illustrative example. Figure 1.8 shows the solution curves for a racetrack economy with four places ($n = 4$). A customary scheme, without referring to invariant patterns, obtains solution curves, bifurcation points, and sustain points in the following order:

- Horizontal line (containing stable path OA) for the uniform state ($\lambda_1 = \lambda_2 = \lambda_3 = \lambda_4 = 1/4$) and bifurcation point A
- Bifurcating curve AB from the bifurcation point A and sustain point B
- Curve BC and bifurcation point C
- Bifurcating curve CDE from the bifurcation point C and sustain point E
- Curve EF
- Curve CD′E′F′, obtained as a symmetric counterpart of path CDEF without carrying out numerical analysis

The bifurcation mechanism at each bifurcation point, obtained through group-theoretic analysis, is helpful in tracing a bifurcating solution.

An innovative scheme refers to invariant patterns and appropriately employs the other two procedures as follows:

- Obtain invariant patterns (solution curves OA, BC, and EF for the invariant patterns are obtained automatically).
- Bifurcation points A and C, and sustain points B and E, on these curves.
- Bifurcating curves AB and CDE.
- Curve CD′E′F′ as a symmetric counterpart of path CDEF.

Thus, the process of obtaining solution curves is greatly simplified.

## References


Aizawa, H., Ikeda, K., Osawa, M., & Gaspar, J. M. (2020). Breaking and sustaining bifurcations in $S_N$-invariant equidistant economy. *International Journal of Bifurcation and Chaos, 30*(16), 2050240.

Akamatsu, T., Takayama, Y., & Ikeda, K. (2012). Spatial discounting, Fourier, and racetrack economy: A recipe for the analysis of spatial agglomeration models. *Journal of Economic Dynamics and Control, 36*(11), 1729–1759.

Allen, T., & Arkolakis, C. (2014). Trade and the topography of the spatial economy. *The Quarterly Journal of Economics, 129*(3), 1085–1140.

Anderson, S. P., de Palma, A., & Thisse, J.-F. (1992). *Discrete Choice Theory of Product Differentiation*. MIT Press.

Bénard, H. (1900). Les tourbillons cellulaires dans une nappe liquide. *Revue Générale des Sciences Pures et Appliques, 11*(1261–1271), 1309–1328.

Castro, S. B. S. D., Correia-da-Silva, J., & Mossay, P. (2012). The core-periphery model with three regions and more. *Papers in Regional Science, 91*(2), 401–418.

Christaller, W. (1933). *Die zentralen Orte in Süddeutschland*. Gustav Fischer. English translation: *Central places in Southern Germany*. Prentice Hall (1966).

Clarke, M., & Wilson, A. G. (1983). The dynamics of urban spatial structure: Progress and problems. *Journal of Regional Science, 23*(1), 1–18.

Commendatore, P., Kubin, I., & Sushko, I. (2015). Typical bifurcation scenario in a three region identical new economic geography model. *Mathematics and Computers in Simulation, 108*, 63–80.

Crisield, M. A. *Non-linear finite element analysis of solids and structures* (Vol. 2). Wiley.

Eaton, B. C., & Lipsey, R. G. (1975). The principle of minimum differentiation reconsidered: Some new developments in the theory of spatial competition. *The Review of Economic Studies, 42*, 27–49.

Eaton, B. C., & Lipsey, R. G. (1982). An economic theory of central places. *The Economic Journal, 92*(365), 56–72.

Elmhirst, T. (2004). $S_N$-equivariant symmetry-breaking bifurcations. *International Journal of Bifurcation and Chaos, 14*(3), 1017–1036.

Fudenberg, D., & Levine, D. K. (1998). *The theory of learning in games*. MIT Press.

Fujita, M., & Krugman, P. (1995). When is the economy monocentric?: von Thünen and Chamberlin unified. *Regional Science and Urban Economics, 25*(4), 505–528.

Fujita, M., Krugman, P., & Mori, T. (1999a). On the evolution of hierarchical urban systems. *European Economic Review, 43*(2), 209–251.

Fujita, M., Krugman, P., & Venables, A. J. (1999b). *The spatial economy: Cities, regions, and international trade*. MIT Press.

Fujita, M., & Mori, T. (1997). Structural stability and the evolution of urban systems. *Regional Science and Urban Economics, 42,* 399–442.

Fujita, M., & Thisse, J.-F. (2009). New economic geography: An appraisal on the occasion of Paul Krugman's 2008 Nobel prize in economic sciences. *Regional Science and Urban Economics, 39*(2), 109–250.

Gaspar, J. M., Castro, S. B. S. D., & Correia-da-Silva, J. (2018). Agglomeration patterns in a multi-regional economy without income effects. *Economic Theory, 66*(4), 863–899.

Gaspar, J. M., Castro, S. B. S. D., & Correia-da-Silva, J. (2020). The footloose entrepreneur model with a finite number of equidistant regions. *International Journal of Economic Theory, 16*(4), 420–446.

Golubitsky, M., & Stewart, I. (2002). *The symmetry perspective*. Birkhäuser.

Golubitsky, M., Stewart, I., & Schaeffer, D. G. (1988). *Singularities and groups in bifurcation theory* (Vol. II). Springer.

Helpman, E. (1998). The size of regions. 33–54. In D. Pines, E. Sadka, & I. Zilcha (Eds.), *Topics in public economies: Theoretical and applied analysis*. Cambridge University Press.

Hofbauer, J., & Sigmund, K. (1998). *Evolutionary games and population dynamics*. Cambridge University Press.

Ikeda, K., Akamatsu, T., & Kono, T. (2012a). Spatial period-doubling agglomeration of a core-periphery model with a system of cities. *Journal of Economic Dynamics and Control, 36*(5), 754–778.

Ikeda, K., & Murota, K. (2014). *Bifurcation theory for hexagonal agglomeration in economic geography*. Springer.

Ikeda, K., & Murota, K. (2019). *Imperfect bifurcation in structures and materials* (3rd ed.). Springer.

Ikeda, K., & Murota, K. (2022). *Bifurcation and buckling in structures*. CRC Press.

Ikeda, K., Murota, K., & Akamatsu, T. (2012b). Self-organization of Lösch's hexagons in economic agglomeration for core-periphery models. *International Journal of Bifurcation and Chaos, 22,* 1230026-1–1230026-29.

Ikeda, K., Murota, K., Akamatsu, T., Kono, T., & Takayama, Y. (2014). Self-organization of hexagonal agglomeration patterns in new economic geography models. *Journal of Economic Behavior and Organization, 99,* 32–52.

Ikeda, K., Murota, K., Akamatsu, T., & Takayama, Y. (2017a). Agglomeration patterns in a long narrow economy of a new economic geography model: Analogy to a racetrack economy. *International Journal of Economic Theory, 13*(1), 113–145.

Ikeda, K., Murota, K., & Takayama, Y. (2017b). Stable economic agglomeration patterns in two dimensions: Beyond the scope of central place theory. *Journal of Regional Science, 57,* 132–172.

Ikeda, K., Onda, M., & Takayama, Y. (2019). Bifurcation theory of a racetrack economy in a spatial economy model. *Networks and Spatial Economics, 19*(1), 57–82.

Ikeda, K., Takayama, Y., Onda, M., & Murakami, D. (2018). Group-theoretic spectrum analysis of population distribution in Southern Germany and Eastern USA. *International Journal of Bifurcation and Chaos, 28*(14), 18300458.

Kogure, Y., Ikeda, K., & Aizawa, H. (2021). Group-theoretic bifurcation mechanism of economic agglomerations on a square lattice. *International Journal of Bifurcation and Chaos, 31*(13), 2130040.

Krugman, P. (1991). Increasing returns and economic geography. *Journal of Political Economy, 99,* 483–499.

Krugman, P. (1993). On the number and location of cities. *European Economic Review, 37,* 293–298.

Lösch, A. (1940). *Die räumliche Ordnung der Wirtschaft*. Gustav Fischer. English translation: *The Economics of Location*. Yale University Press (1954).

McFadden, D. (1974). Conditional logit analysis of qualitative choice behavior. In P. Zarembka (Ed.), *Frontiers in econometrics*. Academic.

Mori, T. (1997). A modeling of megalopolis formation: The maturing of city systems. *Journal of Urban Economics, 42,* 133–157.
Munz, M., & Weidlich, W. (1990). Settlement formation, Part II: Numerical simulation. *The Annals of Regional Science, 24,* 177–196.
Murata, Y. (2003). Product diversity, taste heterogeneity, and geographic distribution of economic activities: market vs. non-market interactions. *Journal of Urban Economics* 53(1), 126–144.
Oyama, D. (2009). Agglomeration under forward-looking expectations: Potentials and global stability. *Regional Science and Urban Economics, 39,* 696–713.
Picard, P. M., & Tabuchi, T. (2010). Self-organized agglomerations and transport costs. *Economic Theory, 42,* 565–589.
Puga, D. (1999). The rise and fall of regional inequalities. *European Economic Review, 43*(2), 303–334.
Redding, S. J. (2016). Goods trade, factor mobility and welfare. *Journal of International Economics, 101,* 148–167.
Redding, S. J., & Rossi-Hansberg, E. (2017). Quantitative spatial economics. *Annual Review of Economics, 9,* 21–58.
Sandholm, W. H. (2010). *Population games and evolution dynamics*. MIT Press.
Stelder, D. (2005). Where do cities form? A geographical agglomeration model for Europe. *Journal of Regional Science, 45,* 657–679.
Tabuchi, T., & Thisse, J.-F. (2002). Taste heterogeneity, labor mobility and economic geography. *Journal of Development Economics, 69,* 155–177.
Tabuchi, T., & Thisse, J.-F. (2011). A new economic geography model of central places. *Journal of Urban Economics, 69,* 240–252.
Tabuchi, T., Thisse, J.-F., & Zeng, D.-Z. (2005). On the number and size of cities. *Journal of Economic Geography, 5,* 423–448.
Takayama, Y., Ikeda, K., & Thisse, J.-F. (2020). Stability and sustainability of urban systems under commuting and transportation costs. *Regional Science and Urban Economics, 84,* 103553.
Taylor, P. D., & Jonker, L. B. (1978). Evolutionary stable strategies and game dynamics. *Mathematical Biosciences, 40*(1–2), 145–156.
Weidlich, W., & Haag, G. (1987). A dynamic phase transition model for spatial agglomeration processes. *Journal of Regional Science, 27*(4), 529–569.
Zeng, D.-Z., & Uchikawa, T. (2014). Ubiquitous inequality: The home market effect in a multicountry space. *Journal of Mathematical Economics, 50,* 225–233.

# Chapter 2
# Introduction to Economic Agglomeration Analysis

**Abstract** We introduce self-organizing two-dimensional agglomeration patterns in an economic geography model on hexagonal lattices with two types of boundary conditions: periodic and nonperiodic. Through a path-tracing analysis in these lattices, we obtain stable agglomeration patterns, such as hexagonal, megalopolis, and racetrack patterns. As transportation costs decrease, stable economic agglomerations progress from a uniform state to patterns with fewer and larger agglomerations. Hexagons proposed in central place theory are superior in stability compared to racetrack patterns. Despite differences in boundary conditions, a striking resemblance exists between the stable patterns on both lattices. Chapter 5 is the mathematical foundation for the bifurcation of the hexagonal lattices.

**Keywords** Bifurcation · Central place theory · Economic geography model · Hexagonal lattice · Hexagonal pattern · Logit dynamics · Megalopolis pattern · Pattern formation · Racetrack pattern · Stripe pattern

## 2.1 Introduction

Economic agglomeration displays various spatial patterns that serve as cradles of regional development and economic prosperity. Christaller (1933) developed *central place theory* through the observation of the geometrical orderliness in the spatial distribution of cities and towns in Southern Germany. Figure 2.1 shows *hexagonal patterns* called Christaller's $k = 3, 4$, and 7 systems. *Lösch's hexagons* are advanced as geometrically feasible hexagonal patterns (Lösch, 1940). The *spatial period* $L$ between spatially repeated hexagons is defined as the shortest Euclidean distance between the first-level centers with the largest population. This period takes specific values:

$$L/\tilde{L} = \sqrt{D}, \quad D = 1, 3, 4, 7, 9, 12, 13, 16, 19, 21, 25, \ldots,$$



K. Ikeda, Y. Takayama, *Bifurcation Theory of Pattern Formation in Economic Geography*, Interdisciplinary Applied Mathematics 61,
https://doi.org/10.1007/978-3-031-71686-7_2

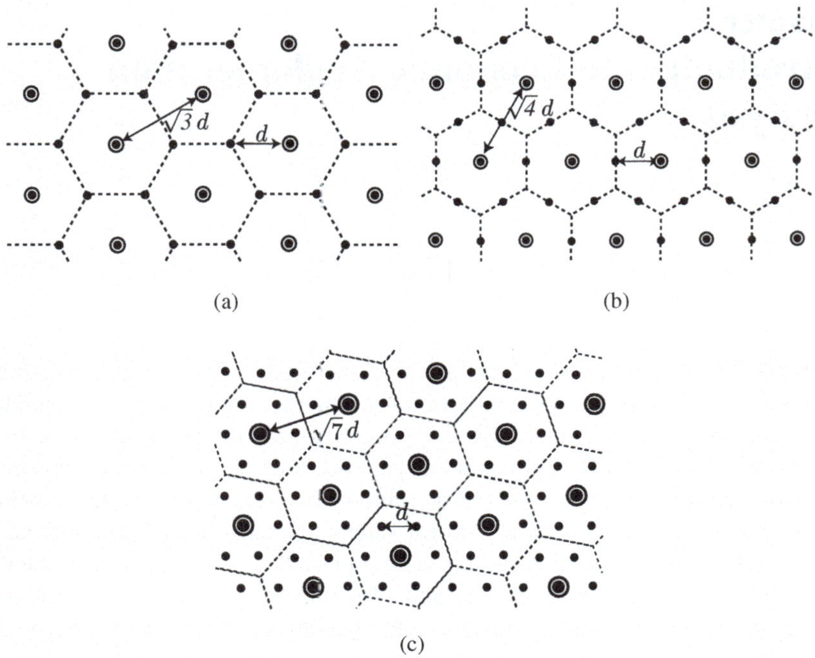


**Fig. 2.1** Three hexagonal systems predicted by Christaller (1933). (**a**) $k = 3$ system. (**b**) $k = 4$ system. (**c**) $k = 7$ system

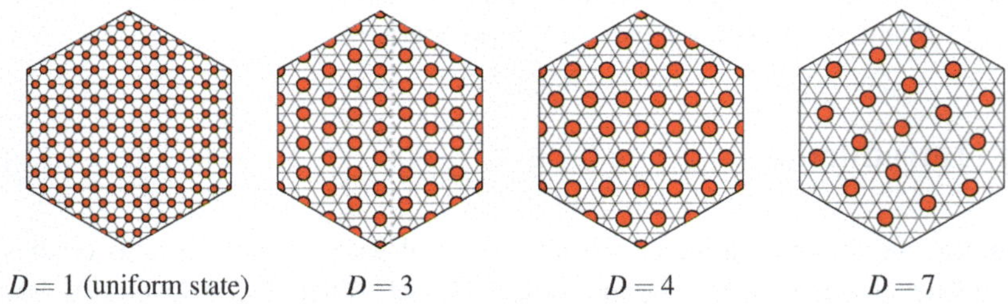


**Fig. 2.2** Lösch's hexagons on a hexagonal lattice with periodic boundary conditions

where $\tilde{L}$ is the distance between neighboring places and $D$ is an integer characterizing the size of a hexagonal pattern. Figure 2.2 depicts the four smallest hexagons. Places are located at the nodes of the hexagonal lattice. Circles denote first-level places with the same large population, whereas other places have the same small population. The smallest value $D = 1$ corresponds to a uniform state. The other three values of $D = 3$, 4, and 7 are associated with Christaller's $k = 3$, 4, and 7 systems, respectively (Fig. 2.1).

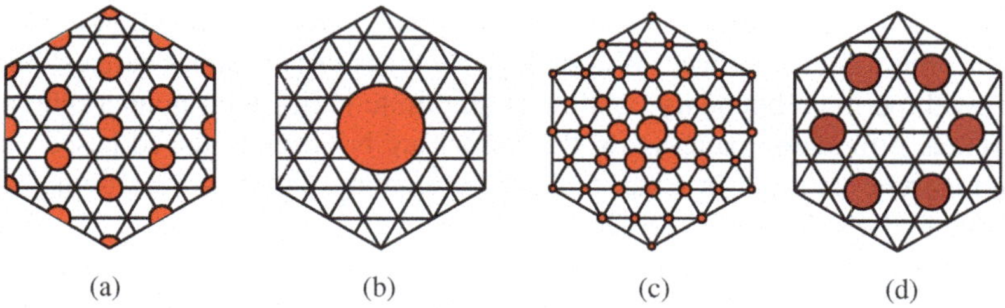


**Fig. 2.3** Agglomeration patterns on a hexagonal lattice with periodic boundary conditions (the area of a circle denotes the population size). (**a**) Hexagon. (**b**) Full agglomeration. (**c**) Megalopolis. (**d**) Racetrack

**Fig. 2.4** Agglomeration patterns observed in an irregularly shaped domain in the Forslid and Ottaviano model (2003) with logit dynamics and parameter values $(\sigma, \mu) = (5.0, 0.4)$ and $\theta = 500, 10{,}000$

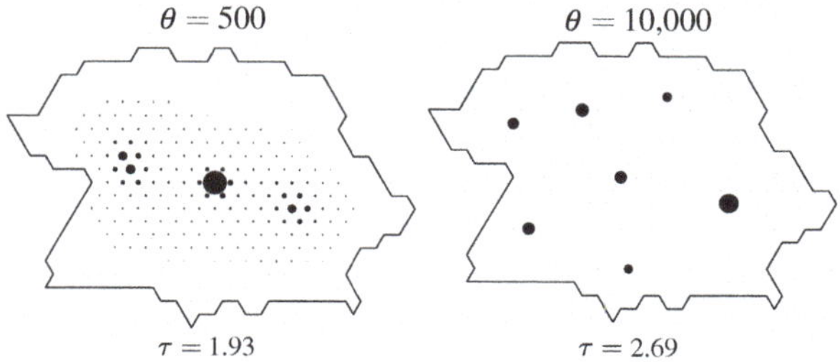


Several researchers simulated the self-organization of central places by modeling economic mechanisms.[1] There are theoretical studies on the pattern formation of hexagonal distributions and numerical analyses of their stability (see Fig. 2.3 for theoretical patterns).[2] Group-theoretic spectrum analysis of city distribution patterns in Southern Germany and the Eastern USA is conducted.[3] The geometry of cities is explained by fractal (Batty & Longley, 1994). Distributions of satellite cities, towns, and villages around Berlin and London are shown to follow a universal law (Makse et al., 1995).

As an example of an economic agglomeration analysis, we refer to the lattice model of Southern Germany on an irregularly shaped domain (*irregular lattice*) in Fig. 2.4. We conducted a numerical analysis of this lattice model in an economic geography model in Sect. 2.2. The area of the circle expresses the size of the population. We can observe distributions of agglomerated places like the $k = 3$ system. This analysis is a possible scenario in a well-known model and may change depending on the choice of model and its economic parameters. In fact, for different values of an economic parameter $\theta$ (see (1.10) for the definition of $\theta$), we see different economic agglomeration patterns in this figure.

---

[1] See, e.g., Eaton and Lipsey (1975, 1982), Clarke and Wilson (1983), Sanglier and Allen (1989), Munz and Weidlich (1990), Fujita et al. (1999), Tabuchi and Thisse (2011), and Banaszak et al. (2015).

[2] See Chap. 5 and Ikeda et al. (2012, 2014) and Ikeda and Murota (2014).

[3] See, e.g., Chap. 6 and Ikeda et al. (2018, 2022).

This chapter demonstrates the existence of stable hexagonal and non-hexagonal patterns on hexagonal lattices. We employ two types of hexagonal lattices: infinite lattice with periodic boundary conditions and finite lattice with nonperiodic ones. Theoretical bifurcating patterns for the former are briefly introduced based on the group-theoretic analysis in Chap. 5. Such bifurcating patterns are absent for the latter due to the influence of borders.

A numerical bifurcation analysis of these two lattices is presented, summarizing (Ikeda et al., 2017). In this analysis, we employ the Forslid and Ottaviano model (2003) (Sect. 2.2) with logit dynamics (see (1.9) in Sect. 1.5). We demonstrate the existence of various patterns: hexagon in Fig. 2.3a, full agglomeration (monocenter) in (b), megalopolis in (c), racetrack in (d), stripe, and deformed hexagon. In particular, the racetrack pattern expresses decentralization, in contrast to the centralization of the hexagonal patterns. As the transportation cost decreases, stable spatial patterns start from the smallest hexagonal pattern and gradually proceed to patterns with larger market areas, finally leading to the full agglomeration en route to the megalopolis pattern.

This chapter is organized as follows: An economic geography model is presented in Sect. 2.2. Pattern formation is investigated for an infinite hexagonal lattice in Sect. 2.3 and a finite lattice in Sect. 2.4.

## 2.2 Forslid and Ottaviano Model

We present the framework of an economic geography model by Forslid and Ottaviano (2003), called the FO model in this book. Among many candidates, this book employs mostly the FO model in the analysis.

**Remark 2.1** The seminal model[4] by Krugman (1991) has provided a new framework to explain interactions that occur among increasing returns at the level of firms, transportation costs, and factor mobility. This model introduces the Dixit and Stiglitz (1977) model of monopolistic competition into spatial economics, considers an economy with two sectors (agriculture and manufacturing), and assumes an upper-tier utility function of the Cobb–Douglas type with CES (constant elasticity of substitution) sub-preferences over manufacturing varieties. The Krugman model led to various variants in economic geography. The FO model is a variant that replaces the cost function of the Krugman model with that of Flam and Helpman (1987). According to the classification regarding the spatial scale of dispersion forces presented by Akamatsu et al. (2023), local dispersion forces occur within a place, whereas global dispersion forces depend on the proximity structure between places. The Krugman and the FO models belong to a class of models that exhibits only a global dispersion force leading to multi-peaked agglomerations. We see such

[4] See, e.g., Fujita et al. (1999), Baldwin et al. (2003), Brakman et al. (2009), and Combes et al. (2008) for details of this model.

agglomerations in the numerical analysis in Sects. 2.3 and 2.4. By contrast, there is a class of models (e.g., Murata & Thisse, 2005) that has only a local dispersion force and displays a relatively undiversified bifurcation behavior compared to the FO model. □

**Basic Assumptions**

There are $n$ places (labeled $i = 1, \ldots, n$), two factors of production (skilled and unskilled labor), and two sectors (manufacture, M, and agriculture, A). Both $H$ *skilled workers* and $L$ *unskilled workers* consume two types of final goods: manufacturing sector (M-sector) goods and agricultural sector (A-sector) goods. Workers supply one unit of each type of labor inelastically. Skilled workers are mobile between places, and the number of skilled workers in place $i$ is denoted by $\lambda_i$. The total number $H$ $(= \sum_{i=1}^{n} \lambda_i)$ of skilled workers is normalized as $H = 1$. Unskilled workers are immobile and equally distributed across the places.

The agricultural sector is perfectly competitive and produces homogeneous goods, whereas the M-sector provides a large variety of differentiated goods. We assume that there are a large number of potential M-sector goods, allowing us to represent the product space as being continuous. This assumption proves beneficial as it enables us to bypass integer constraints on the quantity of M-sector goods.

The utility $U$ for the M- and A-sector goods is identical across individuals and is given by a Cobb–Douglas type function:

$$U(C_i^{\mathrm{M}}, C_i^{\mathrm{A}}) = \mu \log C_i^{\mathrm{M}} + (1 - \mu) \log C_i^{\mathrm{A}} \quad (0 < \mu < 1). \tag{2.1}$$

Here, $(\cdot)_i$ denotes a variable in place $i$, $\mu$ is a constant parameter expressing the expenditure share of manufacturing sector goods, $C_i^{\mathrm{A}}$ is the consumption of the A-sector product, and $C_i^{\mathrm{M}}$ is a sub-utility function defined over a continuum of varieties of manufactured goods. This function is defined as a CES (constant elasticity of substitution) function:

$$C_i^{\mathrm{M}} \equiv \left( \sum_{j=1}^{n} \int_0^{n_j} q_{ji}(\ell)^{(\sigma-1)/\sigma} d\ell \right)^{\sigma/(\sigma-1)}, \tag{2.2}$$

where $\sigma > 1$ is the constant elasticity of substitution between any two varieties; $q_{ji}(\ell)$ is the consumption in place $i$ of a variety $\ell \in [0, n_j]$ produced in place $j$; $n_j$ is the number of produced varieties.

The *budget constraint* is given as

$$p_i^{\mathrm{A}} C_i^{\mathrm{A}} + \sum_{j=1}^{n} \int_0^{n_j} p_{ji}(\ell) q_{ji}(\ell) d\ell = Y_i, \tag{2.3}$$

where $p_i^{\mathrm{A}}$ is the price of A-sector goods, $p_{ji}(\ell)$ is the price of a variety $\ell$ in place $i$ produced in place $j$, and $Y_i$ is the income of an individual. The incomes (wages) of skilled workers and unskilled workers are represented, respectively, by $w_i$ and $w_i^{\mathrm{u}}$. That is,

$$Y_i = \begin{cases} w_i & \text{for skilled workers,} \\ w_i^{\mathrm{u}} & \text{for unskilled workers.} \end{cases}$$

An individual in place $i$ maximizes the utility in (2.1) subject to the budget constraint in (2.3). Consequently, we have the following *demand functions*:

$$C_i^{\mathrm{A}} = (1-\mu)\frac{Y_i}{p_i^{\mathrm{A}}}, \quad C_i^{\mathrm{M}} = \mu\frac{Y_i}{\rho_i}, \quad q_{ji}(\ell) = \mu\frac{\rho_i^{\sigma-1} Y_i}{p_{ji}(\ell)^{\sigma}}, \tag{2.4}$$

where $\rho_i$ denotes the *price index* of the differentiated products, which is defined as

$$\rho_i = \left( \sum_{j=1}^{n} \int_0^{n_j} p_{ji}(\ell)^{1-\sigma} d\ell \right)^{1/(1-\sigma)}. \tag{2.5}$$

The total income in place $i$ is $w_i \lambda_i + w_i^{\mathrm{u}} L/n$. Then, the total demand $Q_{ji}(\ell)$ in place $i$ for a variety $\ell$ produced in place $j$ is given by

$$Q_{ji}(\ell) = \mu\frac{\rho_i^{\sigma-1}}{p_{ji}(\ell)^{\sigma}}(w_i\lambda_i + w_i^{\mathrm{u}} L/n). \tag{2.6}$$

The A-sector is perfectly competitive and produces homogeneous goods under constant-returns-to-scale technology, which requires one unit of unskilled labor per unit output. A-sector goods are transported without a transportation cost and are chosen as the numéraire. In equilibrium, we have $p_i^{\mathrm{A}} = w_i^{\mathrm{u}} = 1$ for each $i$.

The M-sector output is subject to increasing-returns-to-scale technology and Dixit–Stiglitz monopolistic competition (1977). A firm incurs a fixed input requirement of $\alpha$ units of skilled labor and a marginal input requirement of $\beta$ units of unskilled labor. An M-sector firm produces a variety $\ell$, and a firm located in place $i$ chooses $p_{ij}(\ell)$ $(j = 1, \ldots, n)$ that maximizes its profit

$$\Pi_i(\ell) = \sum_{j=1}^{n} p_{ij}(\ell) Q_{ij}(\ell) - (\alpha w_i + \beta x_i(\ell)), \tag{2.7}$$

where $x_i(\ell)$ is the total supply of variety $\ell$ produced in place $i$ and $(\alpha w_i + \beta x_i(\ell))$ is the cost function by Flam and Helpman (1987). We have

$$x_i(\ell) = \sum_{j=1}^{n} \tau_{ij} Q_{ij}(\ell), \tag{2.8}$$

where $\tau_{ij} = \tau_{ij}(\tau)$ is the transportation cost to be defined in (2.10).

Then, the profit function of an M-sector firm, given in (2.7), can be rewritten as

$$\Pi_i(\ell) = \sum_{j=1}^{n} p_{ij}(\ell) Q_{ij}(\ell) - \left( \alpha w_i + \beta \sum_{j=1}^{n} \tau_{ij} Q_{ij}(\ell) \right),$$

which the firm maximizes. The first-order condition for this profit maximization yields

$$p_{ij}(\ell) = \frac{\sigma\beta}{\sigma - 1}\tau_{ij}. \tag{2.9}$$

Thus, $p_{ij}(\ell)$, $Q_{ij}(\ell)$, and $x_i(\ell)$ do not depend on $\ell$. Therefore, we suppress argument $\ell$ in the sequel.

**Iceberg Form of Transportation Cost**

The *transportation costs* for M-sector goods are assumed to take the *iceberg form*. That is, for each unit of M-sector goods transported from place $i$ to place $j$ ($\neq i$), only a fraction $1/\tau_{ij} < 1$ actually arrives ($\tau_{ii} = 1$). It is assumed that $\tau_{ij} = \tau_{ij}(\tau)$ is a function in a *transportation cost parameter* $\tau > 0$ as

$$\tau_{ij} = \exp(\tau L_{ij}), \tag{2.10}$$

where $L_{ij}$ is the road distance between places $i$ and $j$. We express this distance as $L_{ij} = m(i, j)\tilde{L}$ in terms of the product of the integer measure $m(i, j)$ and the distance unit $\tilde{L}$ (the distance between two neighboring places). The road resides along the circle in a racetrack economy and along the lattice in a lattice economy.

We define the *trade freeness* by

$$\phi = \exp[-\tau(\sigma - 1)\tilde{L}] \quad (0 < \phi < 1), \tag{2.11}$$

which is inversely proportional to $\tau > 0$. We introduce the *spatial discount factor*:

$$d_{ij} \equiv \tau_{ij}^{1-\sigma} = \phi^{m(i,j)}. \tag{2.12}$$

This factor represents friction between places $j$ and $i$ that decays in proportion to the road distance measure $m(i, j)$.

**Remark 2.2** We often use $\phi$ instead of $\tau$ (see Sect. 4.5) as the bifurcation parameter. The use of $\phi$ helps draw solution curves since $\phi$ is in a bounded range (0, 1), while $\tau$ goes $+\infty$. The value of $\tilde{L}$ is not required in the analysis using $\phi$ since the analysis employs only $d_{ij}$ in (2.12). □

**Market Equilibrium**

In the short run, skilled workers are immobile between places, i.e., their spatial distribution $\boldsymbol{\lambda} = (\lambda_1, \ldots, \lambda_n)$ is assumed to be given. The *market equilibrium* consists of the following three conditions: (i) The M-sector goods market clearing condition in (2.8). (ii) The zero-profit condition due to firms' free entry and exit. That is, a firm's operating profit, given in (2.7), is absorbed entirely by the wage bill of its skilled workers. This gives the *market equilibrium wage*:

$$w_i = \frac{1}{\alpha}\left(\sum_{j=1}^{n} p_{ij} Q_{ij} - \beta x_i\right). \tag{2.13}$$

(iii) The skilled labor market clearing condition:

$$\alpha n_i = \lambda_i. \tag{2.14}$$

By (2.9) and (2.14), the price index $\rho_i$ in (2.5) can be rewritten as

$$\rho_i = \frac{\sigma\beta}{\sigma-1}\left(\frac{1}{\alpha}\sum_{j=1}^{n} d_{ji}\lambda_j\right)^{1/(1-\sigma)} = \frac{\sigma\beta}{\sigma-1}\left(\frac{\Delta_i}{\alpha}\right)^{1/(1-\sigma)}. \tag{2.15}$$

Here, $\Delta_i = \sum_{j=1}^{n} d_{ji}\lambda_j$ denotes the market size of the M-sector. The *market equilibrium wage* $w_i$ in (2.13) can be represented as

$$w_i = \frac{\mu}{\sigma}\sum_{j=1}^{n}\frac{d_{ij}}{\Delta_j}(w_j\lambda_j + L/n) \tag{2.16}$$

by using (2.6), (2.8), (2.9), (2.12), and (2.15).

We evaluate $C_i^{\mathrm{A}}$ and $C_i^{\mathrm{M}}$ in (2.4) with $p_i^{\mathrm{A}} = 1$, $Y_i = w_i$, and $\rho_i$ in (2.15) and substitute them into the utility in (2.1). Omitting constant terms, we obtain the *indirect utility* $v_i$ of a skilled worker as a function of $w_i$ as

$$v_i = \frac{\mu}{\sigma-1}\log\Delta_i + \log w_i. \tag{2.17}$$

The values of some parameters, such as $\alpha$ and $\beta$ in (2.15), are not influential on $v_i$.

## 2.3 Pattern Formation on Infinite Hexagonal Lattice

This section employs a $6 \times 6$ *hexagonal lattice* with *periodic boundary conditions* (see Fig. 2.5), as a discrete mathematical counterpart of the isotropic infinite plain in central place theory. By these boundary conditions, the lattice can be repeated spatially to cover an infinite two-dimensional space by evenly spread places. Nodal

**Fig. 2.5** A $6 \times 6$ hexagonal lattice with periodic boundary conditions

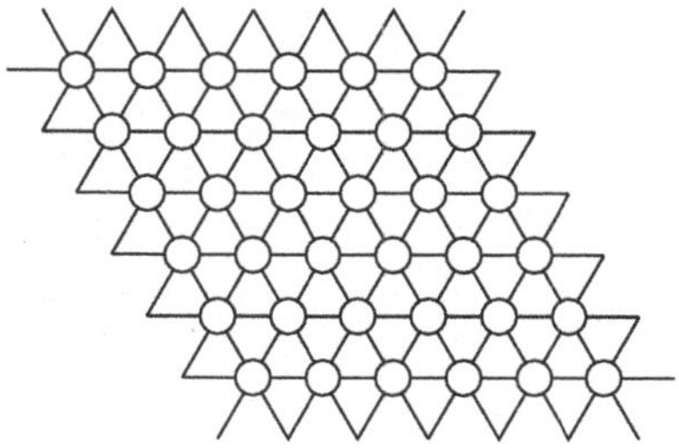

points shown as (∘) are connected by straight lines of equal length $\tilde{L}$ to form a regular triangular mesh.

In economic geography models, each nodal point accommodates a place of economic activity, and the straight lines serve as roads for the transportation of goods. We assume a spatially uniform distribution of labor, homogeneous transportation costs, and so on.

### 2.3.1 Theoretical Bifurcating Patterns

We investigate the bifurcation from the uniform state $\boldsymbol{\lambda}^* = \frac{1}{36}(1, \ldots, 1)^\top$ on the $6 \times 6$ *hexagonal lattice* by adapting the theoretical results of Chap. 5 (summary of Ikeda & Murota, 2014) to the lattice size $n = 6$.

The bifurcating hexagonal and non-hexagonal patterns of the $6 \times 6$ hexagonal lattice are described by using the eigenvectors of the Jacobian matrix of the governing equation (cf. (1.3)). Theoretically predicted hexagonal patterns are depicted in Fig. 2.6a. These patterns are obtained by spatially repeating $6 \times 6$ hexagonal lattices and cutting out a hexagonal window. The black circle denotes a positive component of the eigenvector, and the white circle expresses a negative one. In particular, the patterns with $D = 36$ are called *megalopolis patterns*; satellite places with small populations are scattered around the large place at the center of a hexagonal window to form a megalopolis-like pattern.

Figure 2.6b depicts bifurcating *racetrack patterns* in the reversed direction (reversed positive and negative components) of the hexagonal patterns. The racetrack patterns represent decentralization. In addition, there are other bifurcating patterns, such as stripe (long-narrow) and deformed hexagon (Fig. 2.7).

### 2.3.2 Numerical Simulation

A numerical analysis was conducted on the $6 \times 6$ hexagonal lattice employing the Forslid and Ottaviano model (2003) (Sect. 2.2) with the *logit dynamics* (cf. (1.9) in Sect. 1.3). We set the values of economic parameters as $(\sigma, \mu) = (5.0, 0.4)$,

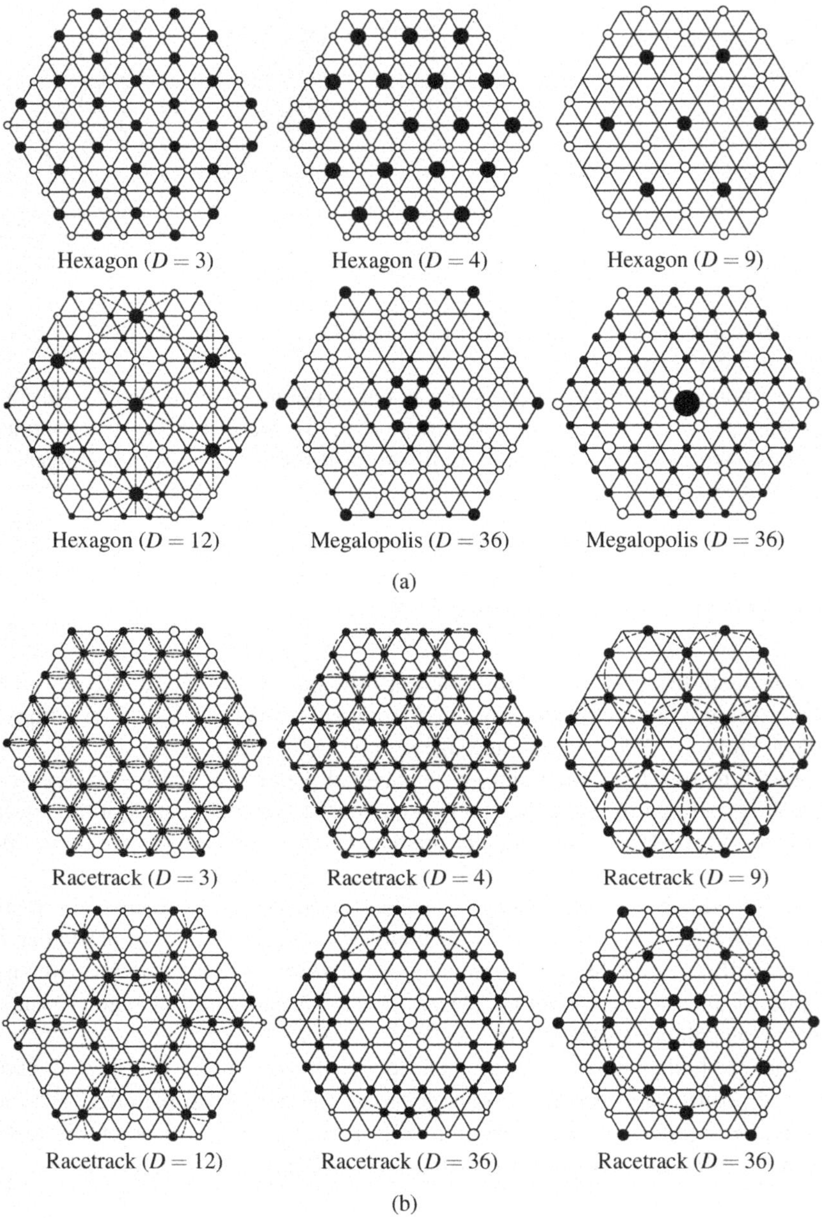


**Fig. 2.6** Bifurcating hexagonal, megalopolis, and racetrack patterns expressed by eigenvectors of the Jacobian matrix (1.3) of the governing equation (a black circle denotes a positive component; a white circle indicates a negative component; the area of a circle expresses the magnitude of the component). (**a**) Hexagonal and megalopolis patterns. (**b**) Racetrack patterns

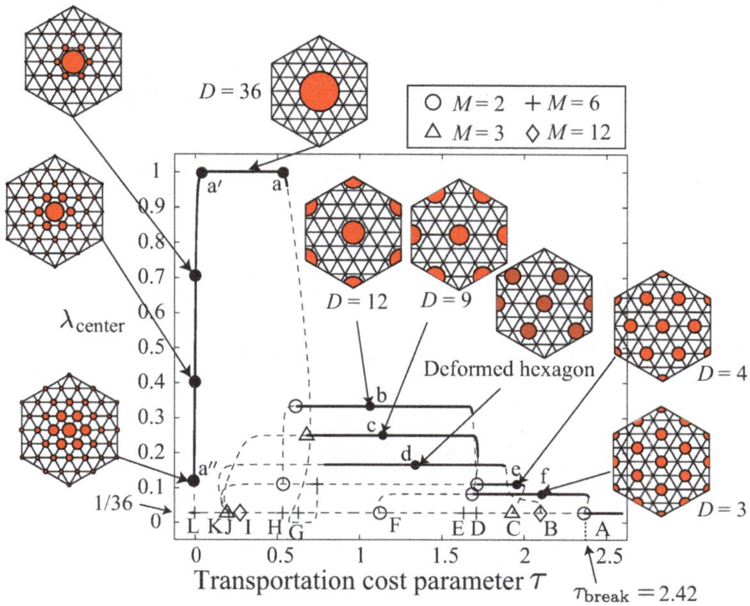


**Fig. 2.7** Solution curves for hexagonal patterns (Ikeda et al., 2017) (solid curve: stable; dashed one: unstable; $M$: multiplicity of the critical point)

in accordance with Fujita et al. (1999), as well as $\theta = 1000$ and $\tilde{L} = 1/6$. The transportation cost parameter $\tau$ in (2.10) is employed as the bifurcation parameter.

Figure 2.7 shows the solution curves, expressed as the relations between *transportation cost parameter* $\tau$ and the population $\lambda_{\text{center}}$ of the place at the center of the hexagonal window. The uniform state $\boldsymbol{\lambda}^* = \frac{1}{36}(1, \ldots, 1)^\top$ is a solution to the governing equation $\boldsymbol{F}(\boldsymbol{\lambda}, \tau) = \mathbf{0}$ for any value of $\tau$ (see Chap. 10), and its solution curve resides on the horizontal line $\lambda_{\text{center}} = 1/36$. Bifurcation points A, ..., L exist on this line. There are bifurcating curves for hexagonal patterns; solid curves are stable and dashed curves are unstable. All bifurcating curves are unstable just after bifurcation and regain stability to become stable. Population distributions on the stable curves are depicted in the figure.

Patterns that are stable for small $\tau$ are a *megalopolis pattern* (curve a′–a″ in Fig. 2.7) and a *full agglomeration* (curve a–a′).[5] A *deformed hexagonal* pattern (point d) is stable for a wide range of $\tau$.

We observe the influence of $\tau$ on stable spatial patterns. For this purpose, we refer to the ranges of $\tau$ for stable equilibria plotted against the number $N_{\text{1st}}$ of first-level centers in Fig. 2.8. When $\tau$ decreases from a large value (from right to left

[5] The megalopolis, consisting of a continuous industrial zone around the center, is also observed in the long-narrow economy (Mori, 1997). A full agglomeration is observed by Fujita and Mori (1997) and Fujita et al. (1999).

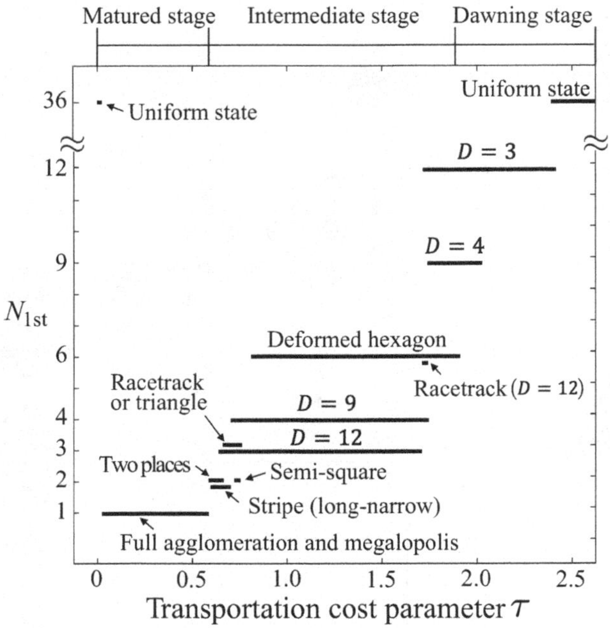


**Fig. 2.8** Ranges of the transportation cost parameter $\tau$ for stable equilibria (Ikeda et al., 2017) (the ordinate $N_{1st}$ stands for the number of the first-level places with the largest population in the hexagonal window; we count a first-level place at a vertex of the hexagonal window as 1/3 in $N_{1st}$, and that at the midpoint of a side of this window as 1/2)

**Table 2.1** Three stages of changes in stable spatial patterns for the infinite hexagonal lattice when $\tau$ decreases from a large value

| | |
|---|---|
| Dawning stage | Uniform state ⇒ Hexagon ($D = 3$) ⇒ |
| Intermediate stage | { Hexagon ($D = 4, 9, 12$)<br>Deformed hexagon<br>Racetrack<br>Triangle<br>⋮ } ⇒ |
| Matured stage | Full agglomeration ⇒ Megalopolis ⇒ Uniform state |

in this figure), $N_{1st}$ becomes smaller, thus enlarging the market area. This arises from a trade-off between transportation costs and economies of scale. Hexagons with $D = 3$, 4, 9, and 12 become stable in this order as $\tau$ decreases. We can find several stable non-hexagonal patterns: *racetrack*, semisquare, *stripe* (long-narrow), two places, and *deformed hexagon*. For example, stable racetrack patterns indicate decentralization. Changes in stable spatial patterns with decreasing $\tau$ can be classified into **dawning**, **intermediate**, and **matured stages**, as presented in Table 2.1.

The **dawning stage** for large $\tau$ aligns with central place theory. When $\tau$ decreases from a large value, the uniform state becomes unstable at the break point ($\tau \approx 2.42$) and renders the role of a unique, stable equilibrium to the hexagon with $D = 3$ ($\tau \approx 2.0$). In this stage, an increase in the market access effect enlarges an agglomeration force and, in turn, weakens the underlying predominance of the market-crowding effect. Then, firms relocate to places with greater competition, and hexagonal patterns emerge.

The **intermediate stage** accommodates multiple stable equilibria, such as hexagons with $D = 4, 9$, and 12, the racetrack patterns, and the deformed hexagonal pattern. The balance between an increasing agglomeration force and a decreasing dispersion force determines the type of emerging stable patterns.

In the **matured stage** for small $\tau$, the full agglomeration, the megalopolis pattern, and the uniform state become stable in this order. Because the transportation cost is extremely low at this stage, all places have nearly identical access to the market, and the dispersion force arising from the taste heterogeneity of workers prevails. Such re-dispersion produces a uniform state for very small $\tau$ and is a unique feature in the logit dynamics that is absent in the replicator dynamics.

The *market areas* were obtained as the Voronoi decomposition for the first-level centers with the largest population and are depicted by the bold solid lines in Fig. 2.9. These areas display hexagons, deformed hexagons, rectangles, diamonds,

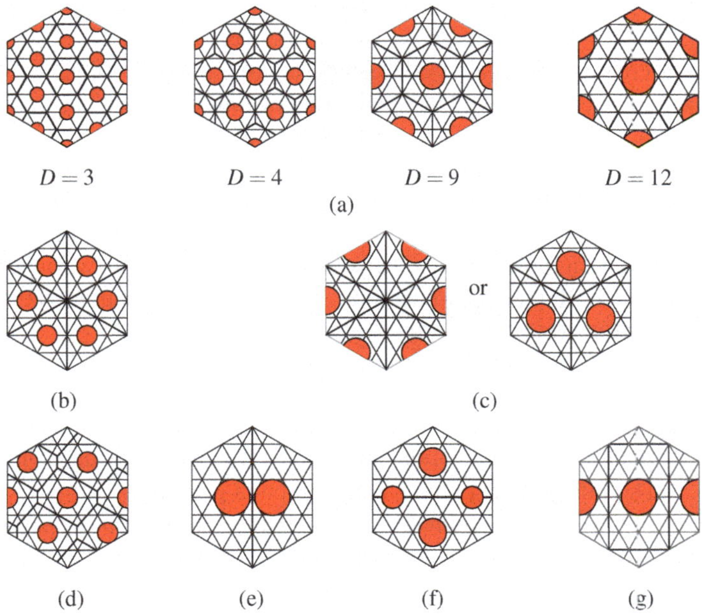


**Fig. 2.9** Stable spatial patterns and their market areas of the first-level centers observed in the numerical analysis (bold line: the market area obtained by the Voronoi decomposition for the first-level centers with the largest population). (**a**) Lösch hexagons. (**b**) Racetrack ($D = 12$). (**c**) Racetrack or triangle. (**d**) Deformed hexagon. (**e**) Twin places. (**f**) Semisquare. (**g**) Stripe (long-narrow)

and trapezoids. Lösch's hexagons in (a) have superior stability (Fig. 2.8), being stable in wide ranges of $\tau$, and have geometrically superior hexagonal shapes of market areas, highlighted in central place theory. The deformed hexagonal pattern in (d) with semihexagonal market areas also has superior stability. The racetrack pattern in (b) has regular triangular *market areas* and is stable in a very narrow range of $\tau$.

## 2.4 Pattern Formation on Finite Hexagonal Lattice

In search of realistic agglomeration patterns in a bounded domain, we resort to the *finite hexagonal lattice* with nonperiodic boundary conditions, as shown at the bottom of Fig. 2.10. This lattice lacks translational symmetry; therefore, spatially repeated patterns, such as the uniform state and hexagonal patterns, are absent in a strict sense. To compensate for this absence, spatial patterns are named by referring to theoretical spatial patterns of the infinite lattice in Fig. 2.6. We employ the same model and economic parameters as in Sect. 2.3.

**Changes in Stable Spatial Patterns**

We obtained stable solution curves and associated population distributions at points a to h in Fig. 2.10. Patterns quite similar to the hexagonal patterns are observed for the finite lattice. This similarity stems from the hexagonal shape of the lattices and the dissipating boundary effects away from the boundary.

When $\tau$ decreases from a large value, stable equilibria start with the uniform state and proceed as listed in Table 2.2. There are dawning, intermediate, and matured stages, like the infinite lattice (Table 2.1). In the dawning stage ($\tau > 2.0$), after the uniform state at point a, a hexagon with $D = 3$ appears near the center of the lattice at point b. In the intermediate stage ($1.1 < \tau < 2.0$), a hexagon with $D = 4$ (point d) emerges at $\tau \approx 2$, engendering a state of dual stable equilibria, followed by a racetrack pattern (point e) and a hexagon with $D = 9$ (point f). In the matured stage ($0 < \tau < 1.1$), the full agglomeration (point g), the megalopolis pattern (point h), and the uniform state occur in this order.

There are recurrences of decentralization (points c and e) and centralization (points f and g). At point c, the population at the center almost disappears to form inner and outer racetracks. At point e, almost the entire population concentrates on a racetrack. At point f, some population migrates to the center to produce the hexagon with $D = 9$. The full agglomeration emerges at point g.

The place at the center would recurrently undergo setbacks during short periods of the dominance by racetrack patterns. For the development of this place through investment in transportation, the present analysis implies a bumpy course undergoing several short periods of decay (stable racetrack patterns). Nonetheless, such decay would be merely transient, and the place at the center regains population en route to the development of a megalopolis if investment is maintained.

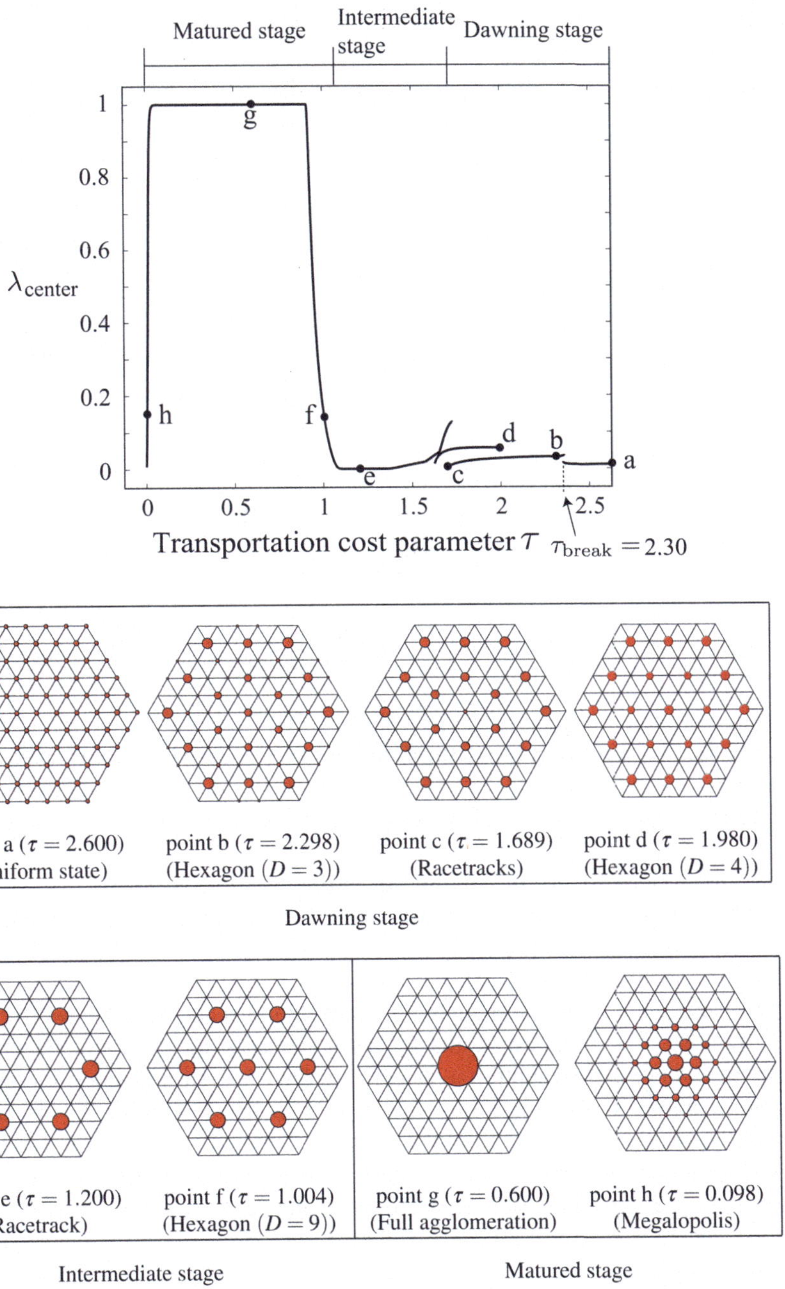


**Fig. 2.10** Stable equilibrium curves and associated population distributions in the standard case with $(\sigma, \mu) = (5.0, 0.4)$ (Ikeda et al., 2017)

**Table 2.2** Transition of stable spatial patterns for the finite hexagonal lattice when $\tau$ decreases from a large value

| | |
|---|---|
| Dawning stage | Uniform state $\Rightarrow$ Hexagon ($D = 3$) $\Rightarrow$ Hexagon ($D = 4$) $\Rightarrow$ |
| Intermediate stage | Racetrack ($D = 36$) $\Rightarrow$ Hexagon ($D = 9$) $\Rightarrow$ |
| Matured stage: | Full agglomeration $\Rightarrow$ Megalopolis $\Rightarrow$ Uniform state |

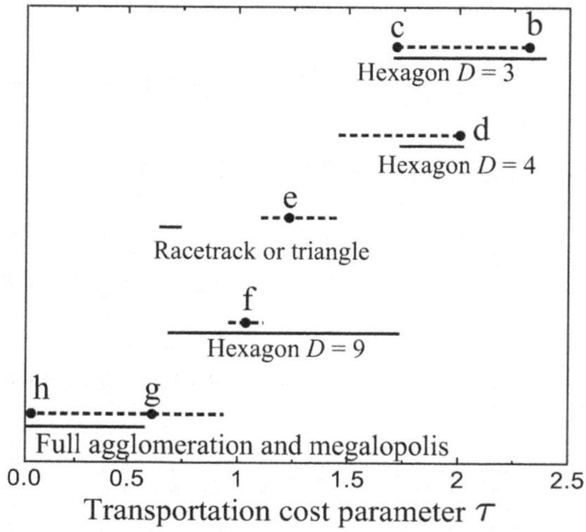


**Fig. 2.11** Comparison of the duration of $\tau$ for stable spatial patterns for the finite lattice (dashed line) and the infinite lattice (solid line) (Ikeda et al., 2017)

For the two types of lattices, the duration of $\tau$ during which some spatial pattern is stable is shown by the solid and dashed lines in Fig. 2.11. A qualitative agreement is observed in changes in stable equilibria as $\tau$ decreases.

### Parameter Dependence of Changes in Stable Spatial Patterns

We observe the dependence of spatial agglomeration on economic parameters: (1) the elasticity $\sigma$ of substitution between any two varieties and (2) the expenditure share $\mu$ of manufactured goods.

The influence of $\sigma$ on changes in stable spatial patterns was investigated for $\mu = 0.4$ and $\sigma = 3, 4, 5, 6, 8$, and 10. As $\sigma$ increases, the economic balance shifts from in favor of agglomeration to dispersion. There are strong, moderate, and weak agglomerations depending on the values of $\sigma$ as listed in Table 2.3. In all cases, spatial agglomeration starts from the uniform state and ends with the formation of a full agglomeration, en route to a megalopolis pattern, and then the uniform state. Hexagonal patterns tend to emerge for the strong agglomeration force. A racetrack pattern emerges in almost all cases and often transforms into a triangular pattern.

The influence of parameter $\mu$ is reported in Ikeda et al. (2017) for $\sigma = 5.0$. Similarly to the influence of $\sigma$, the agglomeration characteristics can be classified into strong ($0.5 \leq \mu \leq 0.8$), moderate ($0.2 \leq \mu \leq 0.4$), and weak agglomerations ($\mu = 0.1$).

**Table 2.3** Parameter dependence of changes in stable spatial patterns on $\sigma$

| |
|---|
| Strong agglomeration ($3 \leq \sigma \leq 4$): |
| Uniform state $\Rightarrow$ Hexagon ($D = 3$) $\Rightarrow$ Hexagon ($D = 4$) $\Rightarrow$ |
| $\left\{\begin{array}{l}\text{Hexagon}(D=12)\\ \text{Racetrack}\\ \text{Triangle}\end{array}\right\} \Rightarrow \left\{\begin{array}{l}\text{Hexagon}\ (D=9)\\ \text{Triangle}\end{array}\right\} \Rightarrow$ |
| Full agglomeration $\Rightarrow$ Megalopolis $\Rightarrow$ Uniform state |
| Moderate agglomeration ($5 \leq \sigma \leq 8$): |
| Uniform state $\Rightarrow$ Hexagon($D = 3$) $\Rightarrow$ Hexagon($D = 4$) $\Rightarrow$ |
| Racetrack $\Rightarrow$ Hexagon($D = 9$) $\Rightarrow$ |
| Full agglomeration $\Rightarrow$ Megalopolis $\Rightarrow$ Uniform state |
| Weak agglomeration ($\sigma = 10$): |
| Uniform state $\Rightarrow$ Racetrack $\Rightarrow$ Triangle $\Rightarrow$ Racetrack $\Rightarrow$ |
| Full agglomeration $\Rightarrow$ Megalopolis $\Rightarrow$ Uniform state |

## References


Akamatsu, T., Mori, T., Osawa, M., & Takayama, Y. (2023). Multimodal agglomeration in economic geography. *arXiv preprint* arXiv:1912.05113v5.

Baldwin, R. E., Forslid, R., Martin, P., Ottaviano, G. I. P., & Robert-Nicoud, F. (2003). *Economic geography and public policy*. Princeton University Press.

Banaszak, M., Dziecielski, M., Nijkamp, P., & Ratajczak, W. (2015). Self-organisation in spatial systems: From fractal chaos to regular patterns and vice versa. *PLOS ONE, 10,* e0136248.

Batty, M., & Longley, P. (1994). *Fractal cities*. Academic.

Brakman, S., Garretsen, H., & van Marrewijk, C. (2009). *The new introduction to geographical economics* (2nd ed.). Cambridge University Press.

Christaller, W. (1933). *Die zentralen Orte in Süddeutschland*. Gustav Fischer. English Translation: *Central places in southern Germany*. Prentice Hall (1966).

Clarke, M., & Wilson, A. G. (1983). The dynamics of urban spatial structure: Progress and problems. *Journal of Regional Science, 23,* 1–18.

Combes, P. P., Mayer, T., & Thisse, J.-F. (2008). *Economic geography: The integration of regions and nations*. Princeton University Press.

Dixit, A., & Stiglitz, J. (1977). Monopolistic competition and optimum product diversity. *American Economic Review, 67,* 297–308.

Eaton, B. C., & Lipsey, R. G. (1975). The principle of minimum differentiation reconsidered: some new developments in the theory of spatial competition. *The Review of Economic Studies, 42*(1), 27–49.

Eaton, B. C., & Lipsey, R. G. (1982). An economic theory of central places. *The Economic Journal, 92*(365), 56–72.

Flam, H., & Helpman, E. (1987). Industrial policy under monopolistic competition. *Journal of International Economics, 22,* 79–102.

Forslid, R., & Ottaviano, G. I. P. (2003). An analytically solvable core-periphery model. *Journal of Economic Geography, 3,* 229–340.

Fujita, M., Krugman, P., & Mori, T. (1999). On the evolution of hierarchical urban systems. *European Economic Review, 43*(2), 209–251.

Fujita, M., & Mori, T. (1997). Structural stability and the evolution of urban systems. *Regional Science and Urban Economics, 42,* 399–442.

Ikeda, K., & Murota, K. (2014). *Bifurcation theory for hexagonal agglomeration in economic geography*. Springer.

Ikeda, K., Murota, K., & Akamatsu, T. (2012). Self-organization of Lösch's hexagons in economic agglomeration for core-periphery models. *International Journal of Bifurcation and Chaos, 22*(08), 1230026.

Ikeda, K., Murota, K., Akamatsu, T., Kono, T., & Takayama, Y. (2014). Self-organization of hexagonal agglomeration patterns in new economic geography models. *Journal of Economic Behavior and Organization, 99,* 32–52.

Ikeda, K., Murota, K., & Takayama, Y. (2017). Stable economic agglomeration patterns in two dimensions: Beyond the scope of central place theory. *Journal of Regional Science, 57,* 132–172.

Ikeda, K., Osawa, M., & Takayama, Y. (2022). Time evolution of city distributions in Germany: Group-theoretic spectrum analysis. *Networks and Spatial Economics, 22*(1), 125–151.

Ikeda, K., Takayama, Y., Onda, M., & Murakami, D. (2018). Group-theoretic spectrum analysis of population distribution in Southern Germany and Eastern USA. *International Journal of Bifurcation and Chaos, 28*(14), 1830045.

Krugman, P. (1991). Increasing returns and economic geography. *Journal of Political Economy, 99,* 483–499.

Lösch, A. (1940). *Die räumliche Ordnung der Wirtschaft*. Gustav Fischer. English translation: *The Economics of Location*. Yale University Press (1954).

Makse, H. A., Havlin, S., & Stanley, H. E. (1995). Modelling urban growth patterns. *Nature, 377,* 608–612.

Mori, T. (1997). A modeling of megalopolis formation: The maturing of city systems. *Journal of Urban Economics, 42,* 133–157.

Munz, M., & Weidlich, W. (1990). Settlement formation, Part II: Numerical simulation. *The Annals of Regional Science, 24,* 177–196.

Murata, Y., & Thisse, J.-F. (2005). A simple model of economic geography á la Helpman-Tabuchi. *Journal of Urban Economics, 58*(1), 137–155.

Sanglier, M., & Allen, P. (1989). Evolutionary models of urban systems: An application to the Belgian provinces. *Environment Planning A, 21,* 477–498.

Tabuchi, T., & Thisse, J.-F. (2011). A new economic geography model of central places. *Journal of Urban Economics, 69,* 240–252.

# Part I
# Bifurcation from the Uniform State

Part I deals with the direct bifurcation from the uniform state of symmetric spatial platforms: racetrack economy, square lattice, and hexagonal lattice. The mechanisms of direct bifurcation are elucidated by referring to the results of the group-theoretic analysis of the groups expressing the symmetries of these platforms. Part II studies hierarchical bifurcations of these symmetric platforms that can potentially undergo repeated bifurcations until completely losing symmetry. As is customary in spatial economics, this book addresses bifurcations of steady states.

We begin with the study of the racetrack economy in Chap. 3. This economy comprises a chain of discrete places evenly distributed on the circumference of a circle. The bifurcation mechanism in this economy is studied by analyzing the dihedral group $\mathrm{D}_n$, expressing regular $n$-gonal symmetry.

We elucidate the bifurcation mechanisms of square and hexagonal lattices with periodic boundary conditions. We conduct a group-theoretic analysis of group $\mathrm{D}_4 \ltimes (\mathbb{Z}_n \times \mathbb{Z}_n)$ for the square lattice in Chap. 4 and of $\mathrm{D}_6 \ltimes (\mathbb{Z}_n \times \mathbb{Z}_n)$ for the hexagonal lattice in Chap. 5, where $\mathbb{Z}_n$ is a cyclic group and $\ltimes$ is a semidirect product. Chap. 6 applies theoretical results for the hexagonal lattice to real population data in Germany from 1987 to 2011, encompassing both pre-unification and post-unification periods. We propose a group-theoretic double Fourier spectrum as a systematic means to capture characteristic agglomeration patterns in this population data.

Individual topics of direct bifurcation are presented in Chaps. 7 and 8:

- Chapter 7 addresses a two-place economy with exogenous locational asymmetries. The imperfect pitchfork bifurcation in this economy is studied, referring to the cusp catastrophe and universal unfoldings.
- An analysis of break points for an economic geography model is conducted for square and hexagonal lattices in Chap. 8. The formulas for break points are derived and applied to the numerical analysis.

# Chapter 3
# Racetrack Economy: Introduction to Group-Theoretic Analysis

**Abstract** This chapter demonstrates how to use group-theoretic bifurcation analysis in studying pattern formation in economic agglomerations. We consider a racetrack economy, comprising identical places spread evenly on the circumference of a circle. The dihedral group describes the regular-polygon symmetry of this economy, and its subgroup expresses the symmetry of a bifurcating solution. We address the mechanism of direct bifurcation from the uniform population distribution in this economy by using the results of a group-theoretic analysis. Chapter 9 deals with the secondary and further bifurcations and presents the numerical analysis of economic geography models. This chapter is a prelude to the group-theoretic analysis of larger groups in Chap. 4 for a square lattice, Chap. 5 for a hexagonal lattice, and Chap. 14 for an equidistant economy. This chapter is a foundation for studying a long-narrow economy in Chap. 12 and a square lattice in Chap. 13 based on racetrack economy analogy.

**Keywords** Breaking bifurcation · Dihedral group · Double bifurcation · Group-theoretic bifurcation analysis · Pitchfork bifurcation · Racetrack economy · Symmetry breaking

## 3.1 Introduction

As an application of group-theoretic bifurcation analysis, we start with a racetrack economy, comprising a system of places that spread evenly on the circumference of a circle. This economy demonstrates equal competition among these places. The economy does not have any borders and is an idealized counterpart of the long-narrow economy with borders at both ends (see Chap. 12). It is also an idealization of a cross section (line segment) of a square lattice with periodic boundary conditions (see Chap. 13). This chapter describes the mechanism of direct bifurcation from the uniform state (uniform population distribution) in a racetrack economy. There are a simple bifurcation point and a series of double bifurcation points, which produce various spatial patterns.



K. Ikeda, Y. Takayama, *Bifurcation Theory of Pattern Formation in Economic Geography*, Interdisciplinary Applied Mathematics 61,
https://doi.org/10.1007/978-3-031-71686-7_3

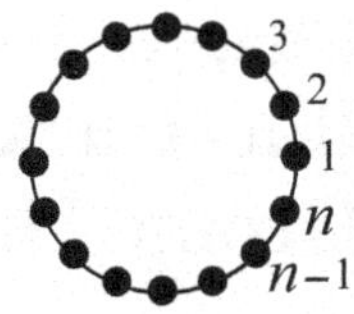


**Fig. 3.1** Racetrack economy with $n$ places ($n = 16$)

The dihedral group expresses the regular polygonal symmetry of the racetrack economy. Group-theoretic analyses of this group are provided in the literature.[1] Each bifurcation point is related to an irreducible representation of the group, and the symmetry of a bifurcating solution is expressed by its subgroup. We consider the direct bifurcation from the uniform state and do not consider sustaining bifurcations presented in Sect. 1.6.2. Chapter 9 will introduce a hierarchical bifurcation, encompassing direct, secondary, and further bifurcations, as well as sustaining bifurcations.

The literature in economic geography reports several agglomeration patterns in the racetrack economy. Krugman (1993) shows the evolution of fewer larger agglomerations from the uniform state in this economy. He regards the economy as one-dimensional and infers that it is extendable to a two-dimensional economy to engender hexagonal distributions (Krugman, 1996, p.91). Tabuchi and Thisse (2011) use a multi-industry model to show the emergence of central places: a spatial alternation of a core place with a large population and a peripheral place with a small population. A sequence of recurrent bifurcations, called the "spatial period-doubling cascade," is observed.[2] Anas (2004) demonstrates the presence of other agglomeration patterns, such as balanced agglomeration, concentrated agglomeration, and de-agglomeration.

This chapter is organized as follows: Sect. 3.2 presents the dihedral group symmetry of the racetrack economy. In Sect. 3.3, the group-theoretic analysis procedure is presented. Bifurcation from the uniform state is studied in Sect. 3.4.

## 3.2 Dihedral Group Symmetry of Racetrack Economy

We consider a *racetrack economy* with $n$ places that are evenly distributed on the circumference of a circle, as depicted in Fig. 3.1. We assume $n$ to be even in this chapter for simplicity, and an extension to the case of $n$ odd will be given in Sect. 9.4.

**Assumption 3.1 (Parity of $n$)** We set $n$ to be even.

### Dihedral Group

[1] See, e.g., Golubitsky et al. (1988) and Ikeda and Murota (2019) and (2022, Chapter 6).

[2] See, e.g., Picard and Tabuchi (2010), Ikeda et al. (2012), Akamatsu et al. (2012), and Osawa et al. (2017).

The regular $n$-gonal symmetry of the racetrack economy is denoted by the dihedral group:

$$\mathrm{D}_n = \langle r, s \rangle = \{r^i, sr^i \mid i = 0, 1, \ldots, n-1\} \tag{3.1}$$

(see, e.g., Ikeda & Murota, 2014, §3.2). Here, $\langle \cdots \rangle$ denotes a group generated by the geometrical transformations therein, and $\{\cdot\}$ denotes a group consisting of the geometrical transformations therein; $r^i$ denotes a counterclockwise *rotation* about the center of the circle at an angle of $2\pi i/n$ ($i = 0, 1, \ldots, n-1$) and $sr^i$ is the combined action of the rotation $r^i$ followed by the upside-down *reflection* $s$. Figure 3.2 displays the actions of these geometrical transformations for $n = 4$.

In our study of a system of $n$ places in the racetrack economy, each element $g$ of $\mathrm{D}_n$ permutes place numbers $\{1, \ldots, n\}$. Consequently, each *representation matrix* $T(g)$ is a *permutation matrix*, which expresses the geometrical transformation by $g$. Using the representation matrices for $r$ and $s$,

$$T(r) = \begin{pmatrix} & & & 1 \\ 1 & & & \\ & \ddots & & \\ & & 1 & \end{pmatrix}, \quad T(s) = \begin{pmatrix} 1 & & & \\ & & & 1 \\ & & \iddots & \\ & 1 & & \end{pmatrix}, \tag{3.2}$$

we can generate the representation matrices $T(g)$ ($g \in \mathrm{D}_n$):

$$T(r^i) = \{T(r)\}^i, \quad T(sr^i) = T(s)\{T(r)\}^i \quad (i = 0, 1, \ldots, n-1). \tag{3.3}$$

**Subgroups of $\mathrm{D}_n$**

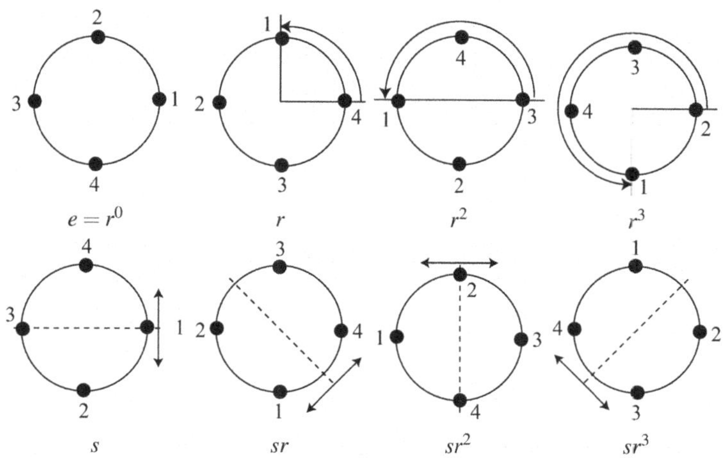


**Fig. 3.2** Actions of elements of $\mathrm{D}_4$ ($e$ is the identity element)

Bifurcated equilibria from the $D_n$-symmetric racetrack economy have partial symmetries denoted by *subgroups* of $D_n$. These subgroups are the following dihedral and *cyclic groups*

$$D_m^{k,n} = \{r^{in/m}, sr^{k-1+in/m} \mid i = 0, 1, \ldots, m-1\} \quad (k = 1, \ldots, n/m),$$

$$C_m = \{r^{in/m} \mid i = 0, 1, \ldots, m-1\}$$

(see Ikeda & Murota, 2019, §9.2). Therein, cyclic group $C_m$ denotes cyclic symmetry at an angle of $2\pi/m$; dihedral group $D_m^{k,n}$ denotes this cyclic symmetry and reflection symmetry regarding $m$-axes; the subscript $m$ ($1 \leq m \leq n/2$) is an integer that divides $n$ and the superscript $k$ ($= 1, \ldots, n/m$) expresses the directions of the reflection axes.

Since the racetrack economy has a single degree of freedom in each place, $D_{n/2}^{2,n}$-, $C_n$-, and $C_{n/2}$-symmetric modes are absent.

We consider the spatial period on the circumference of the racetrack economy. The $D_n$-invariant pattern is invariant under $2\pi/n$ rotation and has the spatial period of $T_n = 2\pi/n$. Likewise, $D_m$-, $D_m^{k,n}$- and $C_m$-invariant patterns have the spatial period of $T_m = 2\pi/m$. In general, a shorter spatial period with a larger $m$ represents a more spread spatial distribution of populations, whereas a longer spatial period with a smaller $m$ represents a more concentrated distribution.

**Example 3.1** The symmetries of agglomeration patterns for the four places ($n = 4$) are depicted in Fig. 3.3. $D_2^{2,4}$-, $C_4$-, and $C_2$-symmetric modes are absent. The patterns associated with the groups $D_1^{2,4}$ and $D_1^{4,4}$ are exchangeable by the upside-down reflection $s$; accordingly, we can identify these patterns and call them *conjugate*. Similarly, we can identify the patterns for the groups $D_1$ and $D_1^{3,4}$. □

**Irreducible Representations**

We present the *irreducible representations* of the dihedral group $D_n$ for even $n$. There are four one-dimensional and $(n/2 - 1)$ two-dimensional irreducible representations over $\mathbb{R}$. We denote these one-dimensional irreducible representations

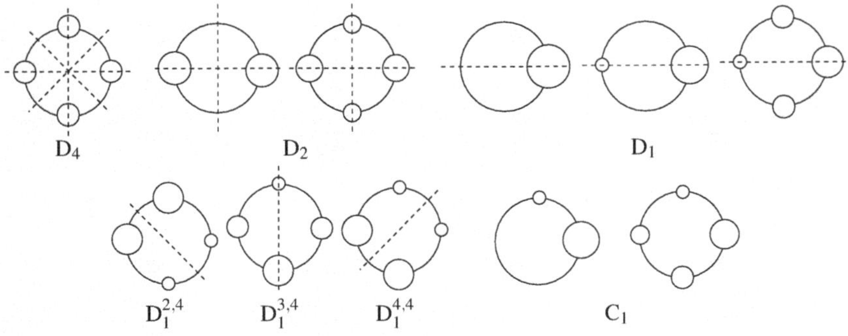


**Fig. 3.3** Symmetries of agglomeration patterns for the four places ($n = 4$; dashed line, the axis of reflection symmetry; the area of ◯ signifies the population size)

by (+, +), (+, −), (−, +), and (−, −) and two-dimensional ones by (1), (2), ..., $(n/2-1)$. The set $R(\mathrm{D}_n)$ of irreducible representations is given by

$$R(\mathrm{D}_n) = \begin{cases} \{(+,+),(+,-),(-,+),(-,-)\} & \text{for } n=2, \\ \{(+,+),(+,-),(-,+),(-,-)\} \cup \{(j) \mid j=1,\dots,n/2-1\} & \text{for } n \geq 4. \end{cases} \tag{3.4}$$

The *one-dimensional irreducible representations* are defined by

$$\begin{aligned} T^{(+,+)}(r) &= 1, & T^{(+,+)}(s) &= 1; \\ T^{(+,-)}(r) &= 1, & T^{(+,-)}(s) &= -1; \\ T^{(-,+)}(r) &= -1, & T^{(-,+)}(s) &= 1; \\ T^{(-,-)}(r) &= -1, & T^{(-,-)}(s) &= -1. \end{aligned} \tag{3.5}$$

The *two-dimensional representations* are chosen here to be unitary representations:

$$\begin{aligned} T^{(j)}(r) &= \begin{bmatrix} \cos(2\pi j/n) & -\sin(2\pi j/n) \\ \sin(2\pi j/n) & \cos(2\pi j/n) \end{bmatrix}, \\ T^{(j)}(s) &= \begin{bmatrix} 1 & 0 \\ 0 & -1 \end{bmatrix} \quad (j=1,\dots,n/2-1). \end{aligned} \tag{3.6}$$

## 3.3 Overview of Group-Theoretic Bifurcation Analysis

We overview group-theoretic bifurcation analysis (see Ikeda & Murota, 2019, Chapter 8).

We consider a nonlinear *governing equation*:

$$\boldsymbol{F}(\boldsymbol{\lambda}, \tau) = \boldsymbol{0}. \tag{3.7}$$

Here, $\boldsymbol{\lambda} = (\lambda_i \mid i = 1, \dots, N) \in \mathbb{R}^N$ represents an *independent variable vector* ($N = n$ for the racetrack economy); $\tau \in \mathbb{R}$ denotes a *bifurcation parameter*; and $\boldsymbol{F}$: $\mathbb{R}^N \times \mathbb{R} \to \mathbb{R}^N$ is a sufficiently smooth nonlinear function.

For a system with the symmetry expressed by a group $G$, the nonlinear function $\boldsymbol{F}(\boldsymbol{\lambda}, \tau)$ in (3.7) is endowed with *equivariance* to $G$:

$$T(g)\boldsymbol{F}(\boldsymbol{\lambda}, \tau) = \boldsymbol{F}(T(g)\boldsymbol{\lambda}, \tau) \quad (g \in G) \tag{3.8}$$

in terms of an $N \times N$ orthogonal matrix representation $T(g)$ of $G$ that expresses the geometrical transformation for an element $g$ of $G$ (see (3.3) for $T(g)$ of $G = \mathrm{D}_n$).

As a consequence of (3.8), the Jacobian matrix $J(\boldsymbol{\lambda}, \tau) = \frac{\partial \boldsymbol{F}}{\partial \boldsymbol{\lambda}}(\boldsymbol{\lambda}, \tau)$ is endowed with the *equivariance* (symmetry) condition

$$T(g)J(\boldsymbol{\lambda}, \tau) = J(\boldsymbol{\lambda}, \tau)T(g) \quad (g \in G) \tag{3.9}$$

if $\boldsymbol{\lambda}$ is symmetric with respect to $G$ in the sense that $T(g)\boldsymbol{\lambda} = \boldsymbol{\lambda}$ $(g \in G)$.

We consider a bifurcation from a $G$-symmetric state $\boldsymbol{\lambda} = \boldsymbol{\lambda}^*$ at a *critical point* $(\boldsymbol{\lambda}^*, \tau_c)$ of multiplicity $M$. The Jacobian matrix $J_c = \partial \boldsymbol{F}/\partial \boldsymbol{\lambda}(\boldsymbol{\lambda}^*, \tau_c)$ at this point has $M$ zero eigenvalues. Each critical point is assumed to be *group-theoretic*,[3] and, accordingly, can be associated with an *irreducible representation* $\mu$ of the group $G$. Let $(\boldsymbol{\eta}_i \mid i = 1, \ldots, N)$ be an orthonormal basis of $\mathbb{R}^N$ such that

$$J\boldsymbol{\eta}_i = \boldsymbol{0} \quad (i = 1, \ldots, M).$$

At the critical point, we define incremental variables $\boldsymbol{w} = (w_i) \in \mathbb{R}^M$ of $\boldsymbol{\lambda}$ and $\tilde{\tau}$ of $\tau$, respectively, by

$$\boldsymbol{\lambda} = \boldsymbol{\lambda}^* + \sum_{i=1}^{M} w_i \boldsymbol{\eta}_i, \qquad \tau = \tau_c + \tilde{\tau}.$$

The full system of equations $\boldsymbol{F}(\boldsymbol{\lambda}, \tau) = \boldsymbol{0}$ in (3.7) is reduced, in a neighborhood of $(\boldsymbol{\lambda}^*, \tau_c)$, to a system of $M$ equations (called *bifurcation equation*) for the incremental variables as

$$\tilde{\boldsymbol{F}}(\boldsymbol{w}, \tilde{\tau}) = \boldsymbol{0}, \tag{3.10}$$

where $\tilde{\boldsymbol{F}}\colon \mathbb{R}^M \times \mathbb{R} \to \mathbb{R}^M$ is a nonlinear function. This equation inherits the equivariance of the full system in (3.8) as

$$\tilde{T}(g)\tilde{\boldsymbol{F}}(\boldsymbol{w}, \tilde{\tau}) = \tilde{\boldsymbol{F}}(\tilde{T}(g)\boldsymbol{w}, \tilde{\tau}) \quad (g \in G), \tag{3.11}$$

where $\tilde{T}$ is the sub-representation of $T$ in the $M$-dimensional kernel space of the Jacobian matrix; $\tilde{T}$ can be related to irreducible representations $T^{(\mu)}$ in (3.5) and (3.6) for $G = \mathrm{D}_n$ in the analysis of the racetrack economy. The reduced equation (3.10) is to be solved for $\boldsymbol{w}$ as $\boldsymbol{w} = \boldsymbol{w}(\tilde{\tau})$ using the symmetry condition of $\tilde{\boldsymbol{F}}$ in (3.11). Each $\boldsymbol{w}$ uniquely determines a solution $\boldsymbol{\lambda}$ of the full system $\boldsymbol{F}(\boldsymbol{\lambda}, \tau) = \boldsymbol{0}$ in (3.7).

[3] Throughout this book, we assume that a critical point is generic (or group-theoretic) in the sense that the $M$-dimensional kernel space of the Jacobian matrix is irreducible regarding the representation $T$ (e.g., Ikeda & Murota, 2019).

## 3.4 Bifurcation from the Uniform State

We consider the $\mathrm{D}_n$-symmetric *uniform state*

$$\boldsymbol{\lambda} = \boldsymbol{\lambda}^* = \frac{1}{n}(\underbrace{1, \ldots, 1}_{n \text{ times}})^\top$$

in the racetrack economy in an economic geography model. This uniform state is a solution to the governing equation for any $\tau$.

The rule of bifurcation from the uniform state varies according to the size $n$ of this economy because the composure of $R(\mathrm{D}_n)$ in (3.4) varies with $n$. We set $n$ to be even by Assumption 3.1, and investigate its *bifurcating solutions* based on the results in the literature.[4]

In the racetrack economy with a single degree of freedom allocated to each node, the irreducible representations $(+,-)$ and $(-,-)$ are absent in (3.4) (Ikeda & Murota, 2014, §3.3.2). As a shorthand notation, we set

$$(+) = (+,+), \quad (-) = (-,+).$$

According to the condition (3.9) for $G = \mathrm{D}_n$, the Jacobian matrix $J = J^*$ in the uniform state is a symmetric circulant matrix with entries

$$J_{ij} = k_l \quad (l = \min\{|i-j|,\ n-|i-j|\}) \tag{3.12}$$

for some real value $k_l$ ($l = 0, 1, \ldots, n/2$). The *eigenvectors* of $J$ are associated with irreducible representations as (Ikeda & Murota, 2014, §3.4.1)

$$\boldsymbol{\eta}^{(+)} = \frac{1}{\sqrt{n}}\begin{pmatrix}1\\ \vdots\\ 1\end{pmatrix}, \quad \boldsymbol{\eta}^{(-)} = \frac{1}{\sqrt{n}}\begin{pmatrix}\cos\pi\cdot 0\\ \vdots\\ \cos(\pi(n-1))\end{pmatrix} = \frac{1}{\sqrt{n}}\begin{pmatrix}1\\ -1\\ \vdots\\ 1\\ -1\end{pmatrix}, \tag{3.13}$$

$$\boldsymbol{\eta}^{(j),1} = \sqrt{\frac{2}{n}}\begin{pmatrix}\cos(2\pi j\cdot 0/n)\\ \vdots\\ \cos(2\pi j(n-1)/n)\end{pmatrix}, \quad \boldsymbol{\eta}^{(j),2} = \sqrt{\frac{2}{n}}\begin{pmatrix}\sin(2\pi j\cdot 0/n)\\ \vdots\\ \sin(2\pi j(n-1)/n)\end{pmatrix}$$
$$(j = 1, \ldots, n/2-1). \tag{3.14}$$

The symmetry and the *spatial period* of these eigenvectors are:

---

[4] See Ikeda et al. (1991) and Ikeda and Murota (2019).

- $\boldsymbol{\eta}^{(+)}$ : $\mathrm{D}_n$ and $T_n = 2\pi/n$.
- $\boldsymbol{\eta}^{(-)}$ : $\mathrm{D}_{n/2}$ and $T_{n/2} = 2 \times 2\pi/n$.
- $\boldsymbol{\eta}^{(j),1}$ : $\mathrm{D}^{1,n}_{n/\hat{n}}$ and $T_{n/\hat{n}} = \hat{n} \times 2\pi/n$.
- $\boldsymbol{\eta}^{(j),2}$ : $\mathrm{D}^{1+\hat{n}/2,n}_{n/\hat{n}}$ and $T_{n/\hat{n}} = \hat{n} \times 2\pi/n$.

Here, $\hat{n}$ is an index for a double bifurcation point and is defined as

$$\hat{n} = n/\mathrm{gcd}\,(n, j) \geq 3, \tag{3.15}$$

and gcd $(n, j)$ is the greatest common divisor of the integer numbers $n$ and $j$. The integer number $\hat{n}$ in (3.15) characterizes the double bifurcation, which engenders a more localized concentration than the period-doubling bifurcation of the simple bifurcation point (see Proposition 3.2 below).

The Jacobian matrix $J$ has *eigenvalues*:

$$e^{(+)} = k_0 + k_{n/2} + 2 \sum_{l=1}^{n/2-1} k_l, \tag{3.16}$$

$$e^{(-)} = k_0 + (-1)^{n/2} k_{n/2} + 2 \sum_{l=1}^{n/2-1} (-1)^l k_l, \tag{3.17}$$

$$e^{(j)} = k_0 + \cos(\pi j)\, k_{n/2} + 2 \sum_{l=1}^{n/2-1} \cos(2\pi jl/n)\, k_l \quad (j = 1, \ldots, n/2 - 1), \tag{3.18}$$

where (3.12) gives $k_l$; $e^{(+)}$ and $e^{(-)}$ are simple eigenvalues ($M = 1$); and $e^{(j)}$ ($j = 1, \ldots, n/2 - 1$) are double eigenvalues repeated twice ($M = 2$).

Critical points at the uniform state are classified as

$$\begin{cases} e^{(+)} = 0 : & \textit{stationary (limit) point of}\ \tau\ (M = 1). \\ e^{(-)} = 0 : & \textit{simple bifurcation point}\ (M = 1). \\ e^{(j)} = 0 : & \textit{double bifurcation point}\ (M = 2). \end{cases} \tag{3.19}$$

The eigenvector(s) for a critical point is called a *critical eigenvector(s)*.

**Simple Bifurcation**

The *simple bifurcation point* with $e^{(-)} = 0$ with a $\mathrm{D}_{n/2}$-symmetric critical eigenvector $\boldsymbol{\eta}^{(-)}$ of (3.13) corresponds to the *spatial period-doubling* bifurcation, which engenders alternating large place and small place (see Proposition 3.1 below). This point is a *subcritical or supercritical pitchfork* bifurcation point, which is respectively called tomahawk or pitchfork bifurcation in economic geography.

In this case, the bifurcation equation in (3.10) becomes one-dimensional. The use of $T^{(-,+)}(r) = -1$ and $T^{(-,+)}(s) = 1$ in (3.5) in place of $\tilde{T}(g)$ in the equivariance (3.11) leads to

$$-\tilde{F}(w_1, \tilde{\tau}) = \tilde{F}(-w_1, \tilde{\tau}) = 0.$$

Thus, $\tilde{F}(w_1, \tilde{\tau})$ is an odd function in $w_1$. Then, together with the condition $\tilde{F}(0, 0) = 0$ and $\partial \tilde{F}/\partial w_1(0, 0) = 0$ of a critical point, we can arrive at an expression of the bifurcation equation:

$$\tilde{F}(w_1, \tilde{\tau}) = w_1 F_1(w_1^2, \tilde{\tau}) = 0$$

for some function $F_1(w_1^2, \tilde{\tau})$ with $F_1(0, 0) = 0$. It is a pitchfork bifurcation with a pre-bifurcation curve $w_1 = 0$ and a bifurcating curve $F_1(w_1^2, \tilde{\tau}) = 0$.

**Proposition 3.1 (Simple Bifurcation)** *At the simple bifurcation point with the critical eigenvector* $\boldsymbol{\eta}^{(-)}$ *of (3.13), we encounter a symmetry-breaking* $\mathrm{D}_n \longrightarrow \mathrm{D}_{n/2}$.

***Proof*** See Ikeda and Murota (2019, Chapter 9) for the proof. □

**Double Bifurcations**

Three or more places generally have double bifurcation points; these points are absent in the two places ($n = 2$) by (3.4). The critical eigenvector for a double bifurcation point with $e^{(j)} = 0$ is given by the superposition of the two vectors $\boldsymbol{\eta}^{(j),1}$ and $\boldsymbol{\eta}^{(j),2}$ in (3.14) as

$$\boldsymbol{\eta}(\theta) = \cos\theta \cdot \boldsymbol{\eta}^{(j),1} + \sin\theta \cdot \boldsymbol{\eta}^{(j),2}$$

for general angle $\theta$ ($0 \le \theta < 2\pi$).

The bifurcation equation in (3.10) in this case becomes two-dimensional, and $\tilde{T}(g)$ is replaced by $T^{(j)}(g)$ in (3.6). This equation is solved in the literature[5] to arrive at Proposition 3.2 below. The bifurcating paths (half-branches) do not exist in all the directions for arbitrary $\theta$, but exist in finite directions. Here, two *half-branches* are obtained by bisecting a bifurcating path at the double bifurcation point.

**Proposition 3.2 (Double Bifurcation)** *Bifurcated paths at the double bifurcation point satisfy the following properties:*

*(i) There exist* $2\hat{n}$ *half-branches* ($\hat{n}$ *bifurcated paths) in the directions of*

$$\delta\boldsymbol{\lambda} = C\boldsymbol{\eta}(\alpha_k),\ C\boldsymbol{\eta}(\alpha_k + \pi) \quad (k = 1, \ldots, \hat{n}),$$

*where* $C$ *is a scaling constant and* $\alpha_k = -\pi\ (k-1)/\hat{n}$.

[5] See Ikeda et al. (1991) and Ikeda and Murota (2019, §9.5).

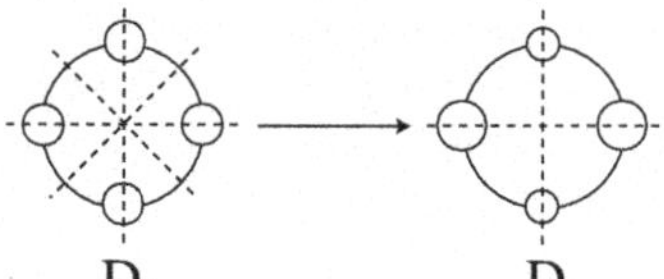


(a)

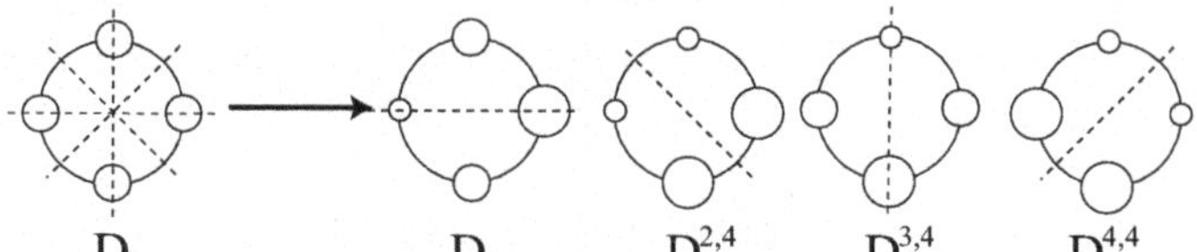


(b)

**Fig. 3.4** Direct bifurcations from the four uniform places ($n = 4$; an arrow denotes the occurrence of bifurcation). (**a**) Simple bifurcation. (**b**) Double bifurcation

*(ii) The directions* $\delta\boldsymbol{\lambda} = C\boldsymbol{\eta}(\alpha_k)$ *and* $C\boldsymbol{\eta}(\alpha_k + \pi)$ *are* $\mathrm{D}^{k,n}_{n/\hat{n}}$*-symmetric* ($k = 1, \ldots, \hat{n}$) *with the spatial period* $T_{n/\hat{n}} = \hat{n} \times 2\pi/n$.

*(iii) The* $2\hat{n}$ *half-branches can be classified into two independent ones:* $\delta\boldsymbol{\lambda} = C\boldsymbol{\eta}(\alpha_{2l-1})$ *and* $C\boldsymbol{\eta}(\alpha_{2l})$ ($l = 1, \ldots, \hat{n}$).

**Example 3.2** The change of symmetry at bifurcation points for the four places ($n = 4$) is illustrated in Fig. 3.4. At the simple bifurcation in (a), the bifurcation doubles the spatial period and triggers the concentration of the population to two places located on opposite sides of the circle, whereas the populations of the other two places decline. At the double bifurcation in (b) associated with $e^{(j)} = 0$ ($n = 4$, $j = 1, \hat{n} = 4$), the spatial period becomes four times. □

## References

Akamatsu, T., Takayama, Y., & Ikeda, K. (2012). Spatial discounting, Fourier, and racetrack economy: A recipe for the analysis of spatial agglomeration models. *Journal of Economic Dynamics and Control, 36*(11), 1729–1759.

Anas, A. (2004). Vanishing cities: What does the new economic geography imply about the efficiency of urbanization? *Journal of Economic Geography, 4*(2), 181–199.

Golubitsky, M., Stewart, I., & Schaeffer, D. G. (1988). *Singularities and groups in bifurcation theory* (Vol. II). Springer.

Ikeda, K., Akamatsu, T., & Kono, T. (2012). Spatial period-doubling agglomeration of a core-periphery model with a system of cities. *Journal of Economic Dynamics and Control, 36*(5), 754–778.

Ikeda, K., & Murota, K. (2014). *Bifurcation theory for hexagonal agglomeration in economic geography*. Springer.

Ikeda, K., & Murota, K. (2019). *Imperfect bifurcation in structures and materials* (3rd ed.) Springer.

Ikeda, K., & Murota, K. (2022). *Bifurcation and buckling in structures*. CRC Press.

Ikeda, K., Murota, K., & Fujii, H. (1991). Bifurcation hierarchy of symmetric structures. *International Journal of Solids and Structures, 27*(12), 1551–1573.

Krugman, P. (1993). On the number and location of cities. *European Economic Review, 37,* 293–298.

Krugman, P. (1996). *The self-organizing economy*. Blackwell.

Osawa, M., Akamatsu, T., & Takayama, Y. (2017). Harris and Wilson (1978) model revisited: The spatial period-doubling cascade in an urban retail model. *Journal of Regional Science, 57*(3), 442–466.

Picard, P. M., & Tabuchi, T. (2010). Self-organized agglomerations and transport costs. *Economic Theory, 42,* 565–589.

Tabuchi, T., & Thisse, J. F. (2011). A new economic geography model of central places. *Journal of Urban Economics, 69,* 240–252.

# Chapter 4
# Group-Theoretic Analysis of Square Lattice

**Abstract** After a brief introduction to group-theoretic bifurcation analysis of the racetrack economy in Chap. 3, this chapter advances a detailed group-theoretic analysis as the foundation for the whole book. We investigate the bifurcation of a scalar field on a square lattice with periodic boundary conditions through the analysis of a finite group $D_4 \ltimes (\mathbb{Z}_n \times \mathbb{Z}_n)$, where $\mathbb{Z}_n$ is a cyclic group and $\ltimes$ is a semidirect product. Bifurcating patterns from a uniform state include square, stripe, and upside-down patterns. We conduct the numerical bifurcation analysis of the square lattice for an economic geography model to observe these patterns. This chapter is a prerequisite for further studies of the square lattice in Chaps. 10 and 13.

**Keywords** Bifurcation · Group-theoretic bifurcation analysis · Pattern formation · Replicator dynamics · Square lattice · Square pattern · Stripe pattern

## 4.1 Introduction

Square road networks are widespread globally. For example, Chicago and Kyoto feature square road networks (see Fig. 4.1). A square lattice is used in several studies of spatial agglomeration[1] and quantitative spatial models.[2]

This chapter investigates the pattern formation of economic agglomerations on square lattices. We model the symmetry of a scalar field on a square lattice with periodic boundary conditions by a finite group $D_4 \ltimes (\mathbb{Z}_n \times \mathbb{Z}_n)$.[3] Through a group-theoretic bifurcation analysis of this group,[4] we present a list of bifurcating solutions from a uniform state and derive the conditions for the symmetry and asymmetry

[1] See, e.g., Clarke and Wilson (1983, 1985), Weidlich and Haag (1987), and Munz and Weidlich (1990).

[2] See Redding (2016) and Redding and Rossi-Hansberg (2017).

[3] $D_4$ is the dihedral group expressing local square symmetry, $\mathbb{Z}_n$ is a cyclic group, and $\ltimes$ is a semidirect product.

[4] This analysis is a summary of Kogure et al. (2021) and Kogure and Ikeda (2022).



K. Ikeda, Y. Takayama, *Bifurcation Theory of Pattern Formation in Economic Geography*, Interdisciplinary Applied Mathematics 61,
https://doi.org/10.1007/978-3-031-71686-7_4

(a)

(b)

**Fig. 4.1** Satellite photographs of cities provided by Google Maps displaying square road networks. (**a**) Chicago (USA). (**b**) Kyoto (Japan)

of these solutions. We conduct the numerical bifurcation analysis on an economic geography model to observe these solutions.

In nonlinear mathematics, pattern formation phenomena are modeled using partial differential equations with group equivariance on an infinite space. A system that is equivariant to an infinite group[5] $D_4 \dot{+} T^2$ has been studied for a reaction–diffusion system, Navier–Stokes flow, and the Bénard problem (e.g., Dionne et al., 1997 and Golubitsky & Stewart, 2003). By contrast, in the study of economic

[5] $T^2$ expresses the two-torus of translation symmetries and $\dot{+}$ is a semidirect product that is expressed by $\ltimes$ in this book.

geography models on a square lattice, it is essential to consider the finite group $\mathrm{D}_4 \ltimes (\mathbb{Z}_n \times \mathbb{Z}_n)$, as conducted in this chapter.

This chapter is organized as follows: Square lattice and its symmetry are described in Sect. 4.2. Section 4.3 presents a group-theoretic bifurcation analysis. Section 4.4 summarizes bifurcating solutions from the uniform state on the square lattice. In Sect. 4.5, a numerical bifurcation analysis is conducted on an economic geography model.

## 4.2 Square Lattice and Its Symmetry

We consider an $n \times n$ *square lattice* with *periodic boundary conditions* (Fig. 4.2). These boundary conditions allow the lattice to be repeated spatially, covering an infinite two-dimensional space.

**Symmetry Group**

The *symmetry group* of the $n \times n$ square lattice is

$$G = \langle r, s, p_1, p_2 \rangle = \mathrm{D}_4 \ltimes (\mathbb{Z}_n \times \mathbb{Z}_n), \tag{4.1}$$

where $\langle \cdots \rangle$ denotes a group generated by the elements therein; $\ltimes$ expresses the *semidirect product* and $\mathbb{Z}_n$ is a cyclic group of order $n$. These elements are defined as:

- $r$ : counterclockwise *rotation* about the origin at an angle of $\pi/2$
- $s$ : *reflection* $y \mapsto -y$
- $p_1$ : periodic *translation* along the vector $\boldsymbol{\ell}_1 = \tilde{L}(1,0)^\top$ by a length $\tilde{L}$
- $p_2$ : periodic *translation* along the vector $\boldsymbol{\ell}_2 = \tilde{L}(0,1)^\top$ by a length $\tilde{L}$

($\tilde{L}$ is the distance between neighboring nodes.) These elements satisfy the fundamental relations:

$$r^4 = s^2 = (rs)^2 = p_1{}^n = p_2{}^n = e, \quad p_2 p_1 = p_1 p_2,$$
$$r p_1 = p_2 r, \quad r p_2 = p_1{}^{-1} r, \quad s p_1 = p_1 s, \quad s p_2 = p_2{}^{-1} s.$$

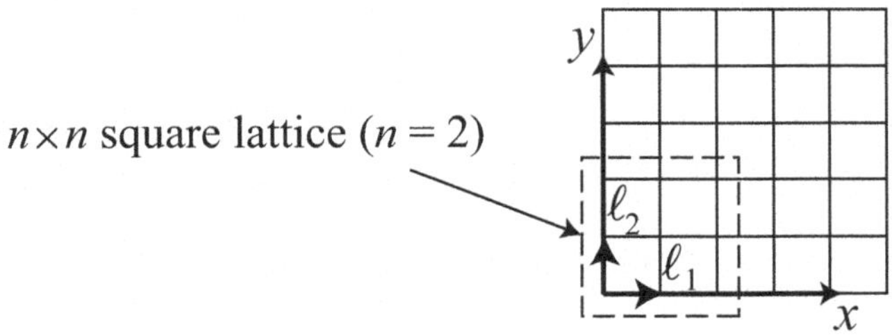


**Fig. 4.2** Square lattice

($e$ is the identity element). The group $G$ contains the dihedral group $\langle r, s\rangle \simeq \mathrm{D}_4$ and cyclic groups $\langle p_1\rangle \simeq \mathbb{Z}_n$ and $\langle p_2\rangle \simeq \mathbb{Z}_n$ as its subgroups.

**Irreducible Representations**

The irreducible representations $\mu$ of the group $G = \langle r, s, p_1, p_2\rangle$ play a pivotal role in describing and classifying bifurcating solutions. For the $n \times n$ square lattice with $n \times n$ degrees of freedom, there exists only a subset of the whole set of *irreducible representations* $\mu$ of the group $G$. This subset is[6]

$$R(G) = \{(1; +), (1; -), (2), (4; k, 0), (4; k, k), (4; n/2, \ell), (8; k, \ell)\}. \tag{4.2}$$

Therein, $\mu = (1; -)$, (2), and $(4; n/2, \ell)$ exist when the lattice size $n$ is even, and

$$1 \le k \le \left\lfloor \frac{n-1}{2} \right\rfloor \quad \text{for } (4; k, 0), (4; k, k); \quad 1 \le \ell \le \frac{n}{2} - 1 \quad \text{for } (4; n/2, \ell); \tag{4.3}$$

$$1 \le \ell \le k-1, \ 2 \le k \le \left\lfloor \frac{n-1}{2} \right\rfloor \quad \text{for } (8; k, \ell), \tag{4.4}$$

where $\lfloor x \rfloor$ is the Gauss symbol denoting the largest integer not larger than $x$. Each $\mu$ is labeled by $(M; \pm)$, $(M)$, or $(M; k, \ell)$; $M$ (= 1, 2, 4, or 8) represents the dimension of the irreducible representation $\mu$, as well as the *multiplicity* of the corresponding critical point; $k$ and $\ell$ represent the wave number of the associated eigenvector(s).

These irreducible representations are defined as follows:

$$T^{(1;+)}(r) = 1, \quad T^{(1;+)}(s) = 1, \quad T^{(1;+)}(p_1) = 1, \quad T^{(1;+)}(p_2) = 1; \tag{4.5}$$

$$T^{(1;-)}(r) = 1, \quad T^{(1;-)}(s) = 1, \quad T^{(1;-)}(p_1) = -1, \quad T^{(1;-)}(p_2) = -1; \tag{4.6}$$

$$T^{(2)}(r) = \begin{bmatrix} & 1 \\ 1 & \end{bmatrix}, \ T^{(2)}(s) = \begin{bmatrix} 1 & \\ & 1 \end{bmatrix}, \ T^{(2)}(p_1) = \begin{bmatrix} -1 & \\ & 1 \end{bmatrix}, \ T^{(2)}(p_2) = \begin{bmatrix} 1 & \\ & -1 \end{bmatrix}; \tag{4.7}$$

$$T^{(4;k,0)}(r) = \begin{bmatrix} & S \\ I & \end{bmatrix}, \quad T^{(4;k,0)}(s) = \begin{bmatrix} I & \\ & S \end{bmatrix}, \tag{4.8}$$

$$T^{(4;k,0)}(p_1) = \begin{bmatrix} R^k & \\ & I \end{bmatrix}, \quad T^{(4;k,0)}(p_2) = \begin{bmatrix} I & \\ & R^k \end{bmatrix}; \tag{4.9}$$

---

[6] Irreducible representations $(1; +, +, +)$, $(1; +, +, -)$, $(2; +, +)$, $(4; k, 0; +)$, $(4; k, k; +)$, and $(4; n/2, \ell; +)$ in (Kogure & Ikeda, 2022, §2.2) are herein simplified to $(1; +)$, $(1; -)$, $(2)$, $(4; k, 0)$, $(4; k, k)$, and $(4; n/2, \ell)$, respectively.

$$T^{(4;k,k)}(r) = \begin{bmatrix} & S \\ I & \end{bmatrix}, \quad T^{(4;k,k)}(s) = \begin{bmatrix} & S \\ S & \end{bmatrix}, \tag{4.10}$$

$$T^{(4;k,k)}(p_1) = \begin{bmatrix} R^k & \\ & R^{-k} \end{bmatrix}, \quad T^{(4;k,k)}(p_2) = \begin{bmatrix} R^k & \\ & R^k \end{bmatrix}; \tag{4.11}$$

$$T^{(4;n/2,\ell)}(r) = \begin{bmatrix} & S \\ I & \end{bmatrix}, \quad T^{(4;n/2,\ell)}(s) = \begin{bmatrix} S & \\ & I \end{bmatrix}, \tag{4.12}$$

$$T^{(4;n/2,\ell)}(p_1) = \begin{bmatrix} -I & \\ & R^{-\ell} \end{bmatrix}, \quad T^{(4;n/2,\ell)}(p_2) = \begin{bmatrix} R^\ell & \\ & -I \end{bmatrix}; \tag{4.13}$$

$$T^{(8;k,\ell)}(r) = \left[\begin{array}{cc|cc} & S & & \\ I & & & \\ \hline & & & I \\ & & S & \end{array}\right], \quad T^{(8;k,\ell)}(s) = \left[\begin{array}{cc|cc} & & I & \\ & & & I \\ \hline I & & & \\ & I & & \end{array}\right], \tag{4.14}$$

$$T^{(8;k,\ell)}(p_1) = \left[\begin{array}{cc|cc} R^k & & & \\ & R^{-\ell} & & \\ \hline & & R^k & \\ & & & R^{-\ell} \end{array}\right], \quad T^{(8;k,\ell)}(p_2) = \left[\begin{array}{cc|cc} R^\ell & & & \\ & R^k & & \\ \hline & & R^{-\ell} & \\ & & & R^{-k} \end{array}\right]. \tag{4.15}$$

Here,

$$R = \begin{bmatrix} \cos(2\pi/n) & -\sin(2\pi/n) \\ \sin(2\pi/n) & \cos(2\pi/n) \end{bmatrix}, \quad S = \begin{bmatrix} 1 & \\ & -1 \end{bmatrix}, \quad I = \begin{bmatrix} 1 & \\ & 1 \end{bmatrix}.$$

## 4.3 Group-Theoretic Bifurcation Analysis

We revisit and extend the group-theoretic bifurcation analysis procedure presented in Sect. 3.3. Through this analysis, we obtain bifurcating solutions at double and quadruple bifurcation points of the uniform state $\boldsymbol{\lambda} = \boldsymbol{\lambda}^* = \frac{1}{N}(1, \ldots, 1)^\top$.

### *4.3.1 Bifurcation Analysis Procedure*

The governing equation is

$$\boldsymbol{F}(\boldsymbol{\lambda}, \tau) = \boldsymbol{0}. \tag{4.16}$$

Here, $\boldsymbol{F}: \mathbb{R}^N \times \mathbb{R} \to \mathbb{R}^N$ is a sufficiently smooth nonlinear function in the vector $\boldsymbol{\lambda} \in \mathbb{R}^N$ and the bifurcation parameter $\tau \in \mathbb{R}$. The symmetry of this equation is described by the *equivariance* of $\boldsymbol{F}(\boldsymbol{\lambda}, \tau)$ to a group $G$ as

$$T(g)\boldsymbol{F}(\boldsymbol{\lambda}, \tau) = \boldsymbol{F}(T(g)\boldsymbol{\lambda}, \tau), \quad g \in G \tag{4.17}$$

in terms of an $N \times N$ orthogonal *matrix representation* $T(g)$ of group $G$.

**Remark 4.1** We see from (4.16) and (4.17) that if $(\boldsymbol{\lambda}, \tau)$ is a solution to $\boldsymbol{F}(\boldsymbol{\lambda}, \tau) = \boldsymbol{0}$, $(T(g)\boldsymbol{\lambda}, \tau)$ $(g \in G)$ is also a solution. Such solutions are called *conjugate*. □

We consider a bifurcation from the $G$-symmetric state $\boldsymbol{\lambda} = \boldsymbol{\lambda}^*$ at a *group-theoretic critical point* $(\boldsymbol{\lambda}^*, \tau_c)$ of multiplicity $M$ (see Footnote 3 in Sect. 3.3). The Jacobian matrix $J_c = \partial \boldsymbol{F}/\partial\boldsymbol{\lambda}(\boldsymbol{\lambda}^*, \tau_c)$ at this point has $M$ zero eigenvalues. Each critical point can be associated with an *irreducible representation* $\mu$ of the group $G$. Let $(\boldsymbol{\eta}_i \mid i = 1, \ldots, N)$ be an orthonormal basis of $\mathbb{R}^N$ such that $J\boldsymbol{\eta}_i = \boldsymbol{0}$ $(i = 1, \ldots, M)$.

At the critical point, we define incremental variables $\boldsymbol{w} = (w_i) \in \mathbb{R}^M$ of $\boldsymbol{\lambda}$ and $\tilde{\tau}$ of $\tau$, respectively, by

$$\boldsymbol{\lambda} = \boldsymbol{\lambda}^* + \sum_{i=1}^{M} w_i \boldsymbol{\eta}_i, \qquad \tau = \tau_c + \tilde{\tau}.$$

By a standard procedure called the *Lyapunov–Schmidt reduction*,[7] the full system of equations in (4.16) is reduced, in a neighborhood of $(\boldsymbol{\lambda}^*, \tau_c)$, to a system of $M$ equations (called bifurcation equation)

$$\tilde{\boldsymbol{F}}(\boldsymbol{w}, \tilde{\tau}) = \boldsymbol{0} \tag{4.18}$$

in the incremental variables, where $\tilde{\boldsymbol{F}}: \mathbb{R}^M \times \mathbb{R} \to \mathbb{R}^M$ is a nonlinear function. This equation inherits the equivariance of the full system in (4.17) as

$$\tilde{T}(g)\tilde{\boldsymbol{F}}(\boldsymbol{w}, \tilde{\tau}) = \tilde{\boldsymbol{F}}(\tilde{T}(g)\boldsymbol{w}, \tilde{\tau}) \quad (g \in G), \tag{4.19}$$

where $\tilde{T}$ is the sub-representation of $T$ in the $M$-dimensional kernel space of the Jacobian matrix and is related to (4.6)–(4.15) for the hexagonal lattice. The reduced equation (4.18) is to be solved for $\boldsymbol{w}$ as $\boldsymbol{w} = \boldsymbol{w}(\tilde{\tau})$ using the symmetry condition of $\tilde{\boldsymbol{F}}$ in (4.19). Each $\boldsymbol{w}$ uniquely determines a solution $\boldsymbol{\lambda}$ of the full system in (4.16).

A group-theoretic bifurcation analysis in this chapter proceeds as follows:

- Specify an irreducible representation $\mu$ of the group $G$.
- Derive the bifurcation equation by exploiting the equivariance in (4.17).

[7] See, e.g., Golubitsky et al. (1988) and Ikeda and Murota (2019).

- Find bifurcating solutions to the bifurcation equation by exploiting the equivariance in (4.19).
- Derive the Jacobian matrix $\tilde{J} = \partial \tilde{\boldsymbol{F}}/\partial \boldsymbol{w}$ of the bifurcation equation and investigate the stability of the bifurcating solutions.

### 4.3.2 Eigenvectors of the Jacobian Matrix

We can analytically obtain the *eigenvectors* of the Jacobian matrix $J(\boldsymbol{\lambda}^*, \tau)$. By the equivariance of $\boldsymbol{F}(\boldsymbol{\lambda}, \tau)$ in (4.17), we can construct a *transformation matrix*

$$Q = [\ldots, Q^{\mu}, \ldots]$$

made up of submatrices $Q^{\mu}$ ($\mu \in R(G)$). Each submatrix contains the eigenvectors (basis vectors) $\boldsymbol{q}_i^{\mu}$ of $J$, i.e.,

$$Q^{\mu} = (\boldsymbol{q}_i^{\mu} \mid i = 1, \ldots, M(\mu)).$$

The transformation matrix $Q$ puts the Jacobian matrix $J$ into a diagonal form:

$$Q^{\top} J Q = \mathrm{diag}(\ldots, e^{\mu} I_{M(\mu)}, \ldots), \quad \mu \in R(G),$$

where $I_{M(\mu)}$ is the $M(\mu) \times M(\mu)$ identity matrix and $\mathrm{diag}(\cdots)$ denotes a block diagonal matrix with diagonal entries therein. Bifurcating patterns from the uniform state are self-organized when some eigenvalue $e^{\mu} = (\boldsymbol{q}_i^{\mu})^{\top} J \boldsymbol{q}_i^{\mu}$ (repeated $M(\mu)$ times) becomes zero. These patterns are related to the eigenvectors $\boldsymbol{q}_i^{\mu}$.

The concrete form of $\boldsymbol{q}_i^{\mu}$ is presented using discrete trigonometric series by Kogure and Ikeda (2022, §3.4) as

$$Q^{(1;+)} = \frac{1}{n}(1, \ldots, 1)^{\top} = \langle 1 \rangle;$$

$$Q^{(1;-)} = [\langle \cos(\pi(n_1 - n_2)) \rangle] \quad \text{for } n \text{ even}; \tag{4.20}$$

$$Q^{(2)} = [\langle \cos(\pi n_1) \rangle, \ \langle \cos(\pi n_2) \rangle] \quad \text{for } n \text{ even}; \tag{4.21}$$

$$Q^{(4;k,0)} = [\langle \cos(2\pi k n_1/n) \rangle, \langle \sin(2\pi k n_1/n) \rangle, \langle \cos(2\pi k n_2/n) \rangle, \langle \sin(2\pi k n_2/n) \rangle];$$

$$\begin{aligned} Q^{(4;k,k)} = [&\langle \cos(2\pi k(n_1 + n_2)/n) \rangle, \langle \sin(2\pi k(n_1 + n_2)/n) \rangle, \\ &\langle \cos(2\pi k(-n_1 + n_2)/n) \rangle, \langle \sin(2\pi k(-n_1 + n_2)/n) \rangle]; \end{aligned}$$

$$\begin{aligned} Q^{(4;n/2,\ell)} = [&\langle \cos(\pi n_1 + 2\pi \ell n_2/n) \rangle, \ \langle \sin(\pi n_1 + 2\pi \ell n_2/n) \rangle, \\ &\langle \cos(-2\pi \ell n_1/n + \pi n_2) \rangle, \ \langle \sin(-2\pi \ell n_1/n + \pi n_2) \rangle] \quad \text{for } n \text{ even}; \end{aligned}$$

$$
\begin{aligned}
Q^{(8;k,\ell)} = [&\langle\cos(2\pi(kn_1+\ell n_2)/n)\rangle,\ \langle\sin(2\pi(kn_1+\ell n_2)/n)\rangle,\\
&\langle\cos(2\pi(-\ell n_1+kn_2)/n)\rangle,\ \langle\sin(2\pi(-\ell n_1+kn_2)/n)\rangle,\\
&\langle\cos(2\pi(kn_1-\ell n_2)/n)\rangle,\ \langle\sin(2\pi(kn_1-\ell n_2)/n)\rangle,\\
&\langle\cos(2\pi(-\ell n_1-kn_2)/n)\rangle,\ \langle\sin(2\pi(-\ell n_1-kn_2)/n)\rangle].
\end{aligned}
$$

Here, the ranges of $k$ and $\ell$ are given in (4.3) and (4.4); for a vector $(g(n_1, n_2) \mid n_1, n_2 = 0, 1, \ldots, n-1)$ on the lattice, we use the notation $\langle g(n_1, n_2)\rangle$ for its normalization ($\|\langle g(n_1, n_2)\rangle\| = 1$).

### 4.3.3 *Double Bifurcation Point*

We conduct a group-theoretic analysis of a double bifurcation point for $\mu = (2)$ $(M = 2)$. The *bifurcation equation* (4.18) for this bifurcation point is $\tilde{F}_i(\boldsymbol{w}, \tilde{\tau}) = 0$ $(i = 1, 2)$ in $\boldsymbol{w} = (w_1, w_2) \in \mathbb{R}^2$.

#### Bifurcation Analysis

In the *equivariance* condition in (4.19), we replace $\tilde{T}(g)$ by $T^{(2)}(g)$ for $\mu = (2)$ in (4.7) to obtain conditions:

$$r: \quad \tilde{F}_2(w_1, w_2) = \tilde{F}_1(w_2, w_1), \qquad \tilde{F}_1(w_1, w_2) = \tilde{F}_2(w_2, w_1), \tag{4.22}$$

$$s: \quad \tilde{F}_1(w_1, w_2) = \tilde{F}_1(w_1, w_2), \qquad \tilde{F}_2(w_1, w_2) = \tilde{F}_2(w_1, w_2), \tag{4.23}$$

$$p_1: \ -\tilde{F}_1(w_1, w_2) = \tilde{F}_1(-w_1, w_2), \qquad \tilde{F}_2(w_1, w_2) = \tilde{F}_2(-w_1, w_2), \tag{4.24}$$

$$p_2: \quad \tilde{F}_1(w_1, w_2) = \tilde{F}_1(w_1, -w_2), \qquad -\tilde{F}_2(w_1, w_2) = \tilde{F}_2(w_1, -w_2). \tag{4.25}$$

Here, we omit $\tilde{\tau}$ in the arguments of $\tilde{F}_i$ for simplicity; it suffices to consider these conditions for the generating elements $r$, $s$, $p_1$, and $p_2$ of the group $G$ in (4.1). By the first condition for $r$ in (4.22), $\tilde{F}_2$ can be obtained from $\tilde{F}_1$. The conditions (4.23) for $s$ do not impose anything.

We expand $\tilde{F}_1$ into a power series as

$$\tilde{F}_1(w_1, w_2, \tilde{\tau}) = \sum_{a=0}^{\infty}\sum_{b=0}^{\infty} A_{ab}(\tilde{\tau}) {w_1}^a {w_2}^b \tag{4.26}$$

with coefficients $A_{ab}(\tilde{\tau}) \in \mathbb{R}$. Because $(z, \overline{z}, \tilde{\tau}) = (0, 0, 0)$ corresponds to the double critical point, we have $A_{00}(0) = 0$ and $A_{10}(0) = A_{01}(0) = 0$. Therefore, we get

$$A_{10}(\tilde{\tau}) \approx A'_{10}(0)\tilde{\tau}$$

with $A'_{10}(\tilde{\tau}) = \partial A_{10}/\partial \tilde{\tau}(\tilde{\tau})$, which is generically nonzero.

By the conditions for $p_1$ and $p_2$ for $\tilde{F}_1$ in (4.24) and (4.25), we obtain

$$-\sum_{a=0}^{\infty}\sum_{b=0}^{\infty} A_{ab}(\tilde{\tau}) {w_1}^a {w_2}^b = \sum_{a=0}^{\infty}\sum_{b=0}^{\infty} A_{ab}(\tilde{\tau})(-w_1)^a {w_2}^b,$$

$$\sum_{a=0}^{\infty}\sum_{b=0}^{\infty} A_{ab}(\tilde{\tau}) {w_1}^a {w_2}^b = \sum_{a=0}^{\infty}\sum_{b=0}^{\infty} A_{ab}(\tilde{\tau}) {w_1}^a (-w_2)^b.$$

These conditions imply that $a$ is odd and $b$ is even. Then, (4.26) and the first condition in (4.22) give

$$\tilde{F}_1(w_1, w_2, \tilde{\tau}) = w_1 \sum_{a=0}^{\infty}\sum_{b=0}^{\infty} A_{2a+1,2b}(\tilde{\tau}) {w_1}^{2a} {w_2}^{2b}, \tag{4.27}$$

$$\tilde{F}_2(w_1, w_2, \tilde{\tau}) = w_2 \sum_{a=0}^{\infty}\sum_{b=0}^{\infty} A_{2a+1,2b}(\tilde{\tau}) {w_1}^{2b} {w_2}^{2a}. \tag{4.28}$$

We present Propositions 4.1 and 4.2 on the existence and symmetry of bifurcating solutions. We use the basis vectors $\boldsymbol{q}_1^{(2)}$ and $\boldsymbol{q}_2^{(2)}$ in $Q^{(2)}$ of (4.21) of the two-dimensional space for $\boldsymbol{w} = (w_1, w_2)$. These two vectors represent stripe patterns in different directions, as depicted in Fig. 4.3a. Since these two patterns are conjugate, we hereafter identify them and consider only the stripe pattern for $\boldsymbol{q}_1^{(2)}$. The sum $\boldsymbol{q}_1^{(2)} + \boldsymbol{q}_2^{(2)}$ represents the square pattern in panel (b).

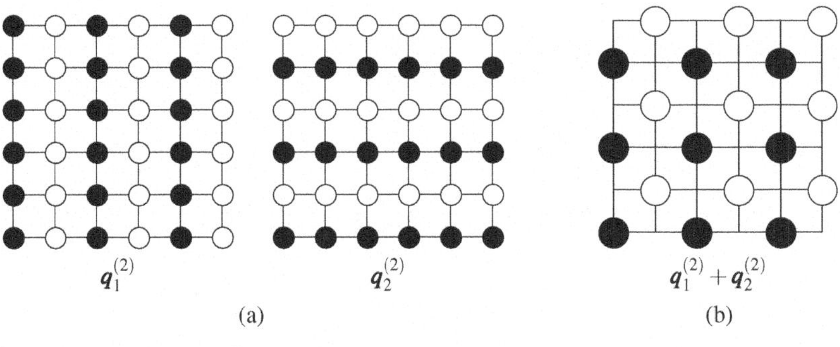


**Fig. 4.3** Patterns of the eigenvectors $\boldsymbol{q}_1^{(2)}$ and $\boldsymbol{q}_2^{(2)}$ on a $6 \times 6$ square lattice (a black circle denotes a positive component of the associated eigenvector; a white circle denotes a negative component; the size of a circle represents the magnitude of the associated component). (**a**) Stripe patterns. (**b**) Square pattern

**Proposition 4.1 (Existence)** *The double bifurcation point for $\mu = (2)$ has the following bifurcating solutions:*

*(i) Square pattern* $\boldsymbol{w}_{\text{sq}} = (w, w)$ $(w \in \mathbb{R})$
*(ii) Stripe pattern* $\boldsymbol{w}_{\text{stripe}} = (w, 0)$ $(w \in \mathbb{R})$

***Proof*** First, by (4.28), $\tilde{F}_2(w, 0, \tilde{\tau}) = 0$ is satisfied by $\boldsymbol{w}_{\text{stripe}} = (w, 0)$ for any $w$. Next, substituting $\boldsymbol{w}_{\text{stripe}} = (w, 0)$ into (4.27) leads to

$$\begin{aligned}\tilde{F}_1(w, 0, \tilde{\tau}) &= w \sum_{a=0}^{\infty} A_{2a+1,0}(\tilde{\tau}) w^{2a} \\ &= w \left\{ A_{10}'(0)\tilde{\tau} + A_{30}(0)w^2 + \text{higher order terms} \right\}\end{aligned}$$

with $A_{10}'(0) = \partial A_{10}/\partial \tilde{\tau}(0)$. Thus, $A_{10}'(0)\tilde{\tau} + A_{30}(0)w^2 \approx 0$ represents the $\tilde{\tau}$ versus $w$ relation for $\boldsymbol{w}_{\text{stripe}}$. Hence, there is a bifurcating curve satisfying $\tilde{F}_1 = \tilde{F}_2 = 0$ for $\boldsymbol{w}_{\text{stripe}}$. A similar discussion holds for $\boldsymbol{w}_{\text{sq}}$. □

We investigate the symmetry of the two *half-branches* that are obtained by bisecting a bifurcating path at the bifurcation point.

**Proposition 4.2 (Symmetry)** *The two half-branches* $(\boldsymbol{w}, \tilde{\tau})$ *and* $(-\boldsymbol{w}, \tilde{\tau})$ *are conjugate (symmetric) for* $\boldsymbol{w} = \boldsymbol{w}_{\text{sq}}$ *and* $\boldsymbol{w}_{\text{stripe}}$.

***Proof*** We note that $\tilde{F}_1(w, 0, \tilde{\tau})$ is an odd function in $w$. Then, we have $\tilde{F}_1(-w, 0, \tilde{\tau}) = -\tilde{F}_1(w, 0, \tilde{\tau})$, i.e., $\tilde{F}_1(-\boldsymbol{w}_{\text{stripe}}, \tilde{\tau}) = -\tilde{F}_1(\boldsymbol{w}_{\text{stripe}}, \tilde{\tau})$. Thus, $(\boldsymbol{w}_{\text{stripe}}, \tilde{\tau})$ and $(-\boldsymbol{w}_{\text{stripe}}, \tilde{\tau})$ are conjugate solutions for $\tilde{F}_1 = 0$ (see Remark 4.1 in Sect. 4.3.1 for conjugacy). A similar discussion holds for $\boldsymbol{w}_{\text{sq}}$. □

### Stability Analysis

The Jacobian matrix of $\tilde{\boldsymbol{F}}$ becomes

$$\tilde{J}(\boldsymbol{w}, \tilde{\tau}) \approx \begin{bmatrix} A_{10}'(0)\tilde{\tau}+3A_{30}(0){w_1}^2+A_{12}(0){w_2}^2 & 2A_{12}(0)w_1w_2 \\ 2A_{12}(0)w_1w_2 & A_{10}'(0)\tilde{\tau}+3A_{30}(0){w_2}^2+A_{12}(0){w_1}^2 \end{bmatrix}. \tag{4.29}$$

We have the following proposition on the asymptotic stability.

**Proposition 4.3 (Stability)** *At the double bifurcation point for* $\mu = (2)$, *under the assumption that the pre-bifurcation uniform state is stable, we have:*

*(i) If* $A_{30}(0) < A_{12}(0) < -A_{30}(0)$, *the square pattern* $\boldsymbol{w}_{\text{sq}}$ *is asymptotically stable.*

*(ii) If $A_{12}(0) < A_{30}(0) < 0$, the stripe pattern $\boldsymbol{w}_{\text{stripe}}$ is asymptotically stable.*
*(iii) The patterns $\boldsymbol{w}_{\text{sq}}$ and $\boldsymbol{w}_{\text{stripe}}$ are not asymptotically stable simultaneously.*

***Proof*** First, substituting the square pattern $\boldsymbol{w}_{\text{sq}} = (w, w)$ into $\tilde{F}_1$ in (4.27) and solving $\tilde{F}_1 = 0$ for $\tilde{\tau}$, we obtain $\tilde{\tau} = \tilde{\tau}_{\text{sq}} \approx -w^2 \frac{A_{30}(0)+A_{12}(0)}{A'_{10}(0)}$. Evaluating the Jacobian matrix (4.29) at $(\boldsymbol{w}_{\text{sq}}, \tilde{\tau}_{\text{sq}})$, we get $\tilde{J}(\boldsymbol{w}_{\text{sq}}, \tilde{\tau}_{\text{sq}}) \approx 2w^2 \begin{bmatrix} A_{30}(0) & A_{12}(0) \\ A_{12}(0) & A_{30}(0) \end{bmatrix}$. The eigenvalues of $\tilde{J}(\boldsymbol{w}_{\text{sq}}, \tilde{\tau}_{\text{sq}})$ are given by $e_1, e_2 \approx 2w^2(A_{30}(0) \pm A_{12}(0))$. Next, for $\boldsymbol{w}_{\text{stripe}} = (w, 0)$, we have $\tilde{\tau} = \tilde{\tau}_{\text{stripe}} \approx -w^2 \frac{A_{30}(0)}{A'_{10}(0)}$ and $\tilde{J}(\boldsymbol{w}_{\text{stripe}}, \tilde{\tau}_{\text{stripe}}) \approx w^2 \text{diag}(2A_{30}(0), A_{12}(0) - A_{30}(0))$. Its eigenvalues are given by $e_1 \approx 2w^2 A_{30}(0)$, $e_2 \approx 2w^2(A_{12}(0) - A_{30}(0))$. Then, (i) and (ii) follow. The conditions $A_{30}(0) < A_{12}(0)$ in (i) and $A_{30}(0) > A_{12}(0)$ in (ii) are contradictory, thereby proving (iii).

□

This proposition indicates possible existence of stable bifurcating solutions. Also, for other bifurcation points, the bifurcating solutions' stability depends on the case. It makes a sharp contrast with the hexagonal lattice (Sect. 5.4), for which all bifurcating paths from the uniform state are asymptotically unstable.

### 4.3.4 *Quadruple Bifurcation Point*

We consider a *quadruple bifurcation point* ($M = 4$). The *bifurcation equation* (4.18) for this point is $\tilde{\boldsymbol{F}}(\boldsymbol{w}, \tilde{\tau}) = (\tilde{F}_1, \ldots, \tilde{F}_4) = \boldsymbol{0}$ ($\boldsymbol{w} = (w_1, \ldots, w_4)$). Using $F_i = \tilde{F}_{2i-1} + \mathrm{i}\tilde{F}_{2i}$ and $z_i = w_{2i-1} + \mathrm{i}w_{2i}$ ($i = 1, 2$) ($i$ is the imaginary unit), we can rewrite this bifurcation equation into a two-dimensional complex-valued equation:

$$F_i(z_1, z_2, \overline{z}_1, \overline{z}_2, \tilde{\tau}) = 0 \quad (i = 1, 2).$$

We expand $F_1$ into a power series as

$$F_1(z_1, z_2, \overline{z}_1, \overline{z}_2, \tilde{\tau}) = \sum_{a=0}^{\infty}\sum_{b=0}^{\infty}\sum_{c=0}^{\infty}\sum_{d=0}^{\infty} A_{abcd}(\tilde{\tau}) {z_1}^a {z_2}^b\, {\overline{z}_1}^c\, {\overline{z}_2}^d \tag{4.30}$$

with expansion coefficients $A_{abcd}(\tilde{\tau})$. Since $(z_1, z_2, \overline{z}_1, \overline{z}_2, \tilde{\tau}) = (0, 0, 0, 0, 0)$ corresponds to the critical point, we have

$$A_{0000}(0) = 0, \quad A_{1000}(0) = A_{0100}(0) = A_{0010}(0) = A_{0001}(0) = 0.$$

Therefore, we get

$$A_{1000}(\tilde{\tau}) \approx A'_{1000}(0)\tilde{\tau}$$

with $A'_{1000}(\tilde{\tau}) = \partial A_{1000}/\partial\tilde{\tau}(\tilde{\tau})$, which is generically nonzero. We often omit $\overline{z}_1$, $\overline{z}_2$, and $\tilde{\tau}$ in the arguments of $F_i$ in the subsequent derivation.

This section advances the case $\mu = (4; k, 0)$, whereas the other cases $\mu = (4; k, k)$ and $(4; n/2, \ell)$ can be investigated similarly (see Kogure & Ikeda 2022, §A.6.4).

For $\mu = (4; k, 0)$, the *equivariance* condition imposes the following relations:

$$r: \quad \overline{F_2(z_1, z_2)} = F_1(\overline{z}_2, z_1), \qquad F_1(z_1, z_2) = F_2(\overline{z}_2, z_1), \tag{4.31}$$

$$s: \quad F_1(z_1, z_2) = F_1(z_1, \overline{z}_2), \qquad \overline{F_2(z_1, z_2)} = F_2(z_1, \overline{z}_2), \tag{4.32}$$

$$p_1: \omega^k F_1(z_1, z_2) = F_1(\omega^k z_1, z_2), \qquad F_2(z_1, z_2) = F_2(\omega^k z_1, z_2), \tag{4.33}$$

$$p_2: \quad F_1(z_1, z_2) = F_1(z_1, \omega^k z_2), \quad \omega^k F_2(z_1, z_2) = F_2(z_1, \omega^k z_2) \tag{4.34}$$

($\omega = \exp(\mathrm{i}2\pi/n)$). The second condition regarding $r$ in (4.31) can be rewritten as

$$F_2(z_1, z_2) = F_1(z_2, \overline{z}_1), \tag{4.35}$$

by which $F_2$ can be obtained from $F_1$. The use of (4.35) in the first of (4.31) gives $\overline{F_1(z_1, z_2)} = F_1(\overline{z}_1, \overline{z}_2)$, i.e., $A_{abcd}(\tilde{\tau}) \in \mathbb{R}$. The conditions for $s$ in (4.32) lead to $A_{abcd}(\tilde{\tau}) \in \mathbb{R}$ again and $A_{abcd}(\tilde{\tau}) = A_{adcb}(\tilde{\tau})$, which plays no role in the subsequent analysis.

The first condition for $p_1$ in (4.33) leads to

$$\omega^k A_{abcd}(\tilde{\tau}) {z_1}^a {z_2}^b {\overline{z}_1}^c {\overline{z}_2}^d = A_{abcd}(\tilde{\tau}) (\omega^k z_1)^a {z_2}^b (\omega^{-k} \overline{z}_1)^c {\overline{z}_2}^d,$$

which is equivalent to $\exp[\, \mathrm{i}2\pi k^*(a - c - 1)/\check{n}\, ] = 1$ with

$$k^* = \frac{k}{\gcd(k, n)}, \qquad \check{n} = \frac{n}{\gcd(k, n)}.$$

Here, $\gcd(\cdots)$ is the greatest common divisor of the integers therein. Then, we obtain

$$a = c + p\check{n} + 1 \geq 0, \quad p \in \mathbb{Z}. \tag{4.36}$$

Here, $\mathbb{Z}$ is a set of integers. The first condition for $p_2$ in (4.34) leads to

$$A_{abcd}(\tilde{\tau}) {z_1}^a {z_2}^b {\overline{z}_1}^c {\overline{z}_2}^d = A_{abcd}(\tilde{\tau}) {z_1}^a (\omega^k z_2)^b {\overline{z}_1}^c (\omega^{-k} \overline{z}_2)^d,$$

which is equivalent to $\exp[\, \mathrm{i}2\pi k^*(b - d)/\check{n}\, ] = 1$. Then, we get

$$b = d + q\check{n} \geq 0, \quad q \in \mathbb{Z}. \tag{4.37}$$

The second conditions for $p_1$ and $p_2$ can be rewritten using (4.35) as $F_1(z_2, \overline{z}_1) = F_1(z_2, \omega^{-k} \overline{z}_1)$ and $\omega^k F_1(z_2, \overline{z}_1) = F_1(\omega^k z_2, \overline{z}_1)$, leading to the same result as in (4.36) and (4.37). We also note, from (4.36), if $a = 0$, then $c = -p\check{n} - 1 \neq 0$. That is, terms for $A_{0bcd}$ ($c \geq 1$) do not contain $z_1$ but have $\overline{z}_1$, whereas other terms have $z_1$.

With these results, $F_1$ in (4.30) and $F_2$ in (4.35) become

$$F_1(z_1, z_2, \tilde{\tau})$$
$$= z_1 \sum_{c=0}^{\infty} \sum_{d=0}^{\infty} \sum_{p \in \mathbb{Z},\ c+p\check{n}+1>0} \sum_{q \in \mathbb{Z},\ d+q\check{n}\geq 0} A_{c+p\check{n}+1,d+q\check{n},cd}(\tilde{\tau}) z_1{}^{c+p\check{n}} z_2{}^{d+q\check{n}}\, \overline{z}_1{}^{c}\, \overline{z}_2{}^{d}$$
$$+ \overline{z}_1 \sum_{d=0}^{\infty} \sum_{p=1}^{\infty} \sum_{q \in \mathbb{Z},\ d+q\check{n}\geq 0} A_{0,d+q\check{n},p\check{n}-1,d}(\tilde{\tau}) z_2{}^{d+q\check{n}}\, \overline{z}_1{}^{p\check{n}-2}\, \overline{z}_2{}^{d}, \tag{4.38}$$

$$F_2(z_1, z_2, \tilde{\tau})$$
$$= z_2 \sum_{c=0}^{\infty} \sum_{d=0}^{\infty} \sum_{p \in \mathbb{Z},\ c+p\check{n}+1>0} \sum_{q \in \mathbb{Z},\ d+q\check{n}\geq 0} A_{c+p\check{n}+1,d+q\check{n},cd}(\tilde{\tau}) z_2{}^{c+p\check{n}} \overline{z}_1{}^{d+q\check{n}}\, \overline{z}_2{}^{c}\, z_1{}^{d}$$
$$+ \overline{z}_2 \sum_{d=0}^{\infty} \sum_{p=1}^{\infty} \sum_{q \in \mathbb{Z},\ d+q\check{n}\geq 0} A_{0,d+q\check{n},p\check{n}-1,d}(\tilde{\tau}) \overline{z}_1{}^{d+q\check{n}}\, \overline{z}_2{}^{p\check{n}-2}\, z_1{}^{d}. \tag{4.39}$$

We show the existence and symmetry of the following patterns ($w \in \mathbb{R}$):

- *Square pattern* $(z_1, z_2) = (w, w)$
- Type I *stripe pattern* $(z_1, z_2) = (w, 0)$
- Type II *stripe pattern* $(z_1, z_2) = (\mathrm{i}w, 0)$

### Square Pattern

Substituting the square pattern $(z_1, z_2) = (w, w)$ into (4.38) and (4.39), we get

$$\begin{aligned}
F_1(w, w, \tilde{\tau}) &= F_2(w, w, \tilde{\tau}) \\
&= w \sum_{c=0}^{\infty} \sum_{d=0}^{\infty} \sum_{p \in \mathbb{Z},\ c+p\check{n}+1>0} \sum_{q \in \mathbb{Z},\ d+q\check{n}\geq 0} A_{c+p\check{n}+1,d+q\check{n},cd}(\tilde{\tau}) w^{2(c+d)+(p+q)\check{n}} \\
&\quad + w \sum_{d=0}^{\infty} \sum_{p=1}^{\infty} \sum_{q \in \mathbb{Z},\ d+q\check{n}\geq 0} A_{0,d+q\check{n},p\check{n}-1,d}(\tilde{\tau}) w^{2d+(p+q)\check{n}-2} \\
&= w\{A'_{1000}(0)\tilde{\tau} + (A_{1101}(0) + A_{2010}(0))w^2 + A_{00,\check{n}-1,0}(0) w^{\check{n}-2} \\
&\qquad + \text{higher order terms}\}.
\end{aligned}$$

Thus, $F_1(w, w, \tilde{\tau}) = F_2(w, w, \tilde{\tau}) = 0$ has the trivial solution $w = 0$ and a bifurcating solution $F_1(w, w, \tilde{\tau})/w = 0$. If $\check{n}$ is even, $F_1(w, w, \tilde{\tau})$ becomes an odd function in $w$, and hence the two half-branches $(w, w, \tilde{\tau})$ and $(-w, -w, \tilde{\tau})$ are *conjugate* since $F_1(w, w, \tilde{\tau})/w = F_1(-w, -w, \tilde{\tau})/w$. If $\check{n}$ is odd, the two half-branches are *non-conjugate*.

### Stripe Patterns

Substituting Type I stripe pattern $(z_1, z_2) = (w, 0)$ into (4.38) and (4.39) and setting $A_{ac}(\tilde{\tau}) = A_{a0c0}(\tilde{\tau})$, we get

$$
\begin{aligned}
F_1(w, 0, \tilde{\tau}) &= w \sum_{c=0}^{\infty} \sum_{p \in \mathbb{Z},\ c+p\check{n}+1>0} A_{c+p\check{n}+1,c}(\tilde{\tau}) w^{2c+p\check{n}} \\
&\quad + w \sum_{p=1}^{\infty} A_{0,p\check{n}-1}(\tilde{\tau}) w^{p\check{n}-2} \\
&= w \left\{ A'_{10}(0)\tilde{\tau} + A_{21}(0) w^2 + A_{0,\check{n}-1}(0) w^{\check{n}-2} + \text{higher order terms} \right\}, \\
F_2(w, 0, \tilde{\tau}) &= 0.
\end{aligned}
$$

Thus, there are the trivial solution $w = 0$ and a bifurcating solution $F_1(w, 0, \tilde{\tau})/w = 0$. When $\check{n}$ is even, $F_1(w, 0, \tilde{\tau})/w$ is an even function in $w$, and hence the two *half-branches* for $(w, 0, \tilde{\tau})$ and $(-w, 0, \tilde{\tau})$ are *conjugate*. These half-branches are *non-conjugate* when $\check{n}$ is odd.

Substituting Type II stripe pattern $(z_1, z_2) = (\mathrm{i}w, 0)$ into (4.38) and (4.39), we have $F_2(\mathrm{i}w, 0, \tilde{\tau}) = 0$ and

$$
\begin{aligned}
& F_1(\mathrm{i}w, 0, \tilde{\tau}) \\
&= \mathrm{i}w \sum_{c=0}^{\infty} \sum_{p \in \mathbb{Z},\ c+p\check{n}+1>0} A_{c+p\check{n}+1,c}(\tilde{\tau}) (\mathrm{i}w)^{c+p\check{n}} (-\mathrm{i}w)^c \\
&\quad - \mathrm{i}w \sum_{p=1}^{\infty} A_{0,p\check{n}-1}(\tilde{\tau}) (-\mathrm{i}w)^{p\check{n}-2} \\
&= \mathrm{i}w \left\{ \sum_{c=0}^{\infty} \sum_{p \in \mathbb{Z},\ c+p\check{n}+1>0} A_{c+p\check{n}+1,c}(\tilde{\tau}) \mathrm{i}^{p\check{n}} w^{2c+p\check{n}} \right. \\
&\quad \left. - \sum_{p=1}^{\infty} A_{0,p\check{n}-1}(\tilde{\tau}) (-\mathrm{i})^{p\check{n}-2} w^{p\check{n}-2} \right\}
\end{aligned}
$$

$$= \mathrm{i}w \left\{ A'_{10}(0)\tilde{\tau} + A_{21}(0)w^2 - A_{0,\check{n}-1}(0)(-\mathrm{i})^{\check{n}-2}w^{\check{n}-2} + \text{higher order terms} \right\}.$$

When $\check{n}$ is even, $\mathrm{i}^{p\check{n}}$ and $(-\mathrm{i})^{\check{n}-2}$ become real; then, $F_1(\mathrm{i}w, 0, \tilde{\tau}) = F_2(\mathrm{i}w, 0, \tilde{\tau}) = 0$ has a bifurcating solution $F_1(w, 0, \tilde{\tau})/(\mathrm{i}w) = 0$, and the two half-branches are conjugate.

**Proposition 4.4 (Stability)** *At the critical point of multiplicity $M = 4$, under the assumption that the pre-bifurcation uniform state is stable, we have:*

*(i) For $(4; k, 0)$ and $(4; k, k)$ with $\check{n} = 3$, the square pattern $\boldsymbol{w}_{sq}$ and the Type I stripe pattern $\boldsymbol{w}_{stripe}$ are asymptotically unstable.*
*(ii) In any other cases, the asymptotic stability of bifurcating solutions $\boldsymbol{w}_{sq}$ and $\boldsymbol{w}_{stripe}$ depends on the values of $A_{abcd}(\tilde{\tau})$ in (4.30).*

***Proof*** See Kogure and Ikeda (2022) for the proof. □

## 4.4 Bifurcation from the Uniform State

We summarize the *bifurcating solutions* from the uniform state on the square lattice obtained through group-theoretic analysis of various bifurcation points: (i) $M = 1$ for a pitchfork bifurcation (Sect. 13.4.1), (ii) $M = 2$ in Sect. 4.3.3, (iii) $M = 4$ in Sect. 4.3.4, and (vi) $M = 8$ in Kogure and Ikeda (2022). Bifurcating patterns include square, stripe, and upside-down patterns. Square and stripe patterns are discretized counterparts of squares and rolls in the continuous space (e.g., Golubitsky & Stewart, 2003).

**Square Patterns**
We have bifurcating *square patterns*:[8]

$$\boldsymbol{w}_{\mathrm{sq}} = \begin{cases} w & \text{for } \mu = (1; -), \\ (w, w) & \text{for } \mu = (2), \\ (w, 0, w, 0) & \text{for } \mu = (4; k, 0), (4; k, k), (4; n/2, \ell), \\ (w, 0, w, 0, w, 0, w, 0) & \text{for } \mu = (8; k, \ell) \quad \text{(Type VM)}, \\ (w, 0, w, 0, 0, 0, 0, 0) & \text{for } \mu = (8; k, \ell) \quad \text{(Type T)} \end{cases} \tag{4.40}$$

$(w \in \mathbb{R})$ and the associated eigenvectors:

[8] Because the basis vector $\boldsymbol{q}_1^{(1;+)}$ expresses the uniform state and does not correspond to bifurcation, $\mu = (1; +)$ is not included in (4.40).

$$
\boldsymbol{q}_{\mathrm{sq}}^{(\mu)} = \begin{cases} \boldsymbol{q}_1^{(1;-)}, & \\ \boldsymbol{q}_1^{(2)} + \boldsymbol{q}_2^{(2)}, & \\ \boldsymbol{q}_1^{(4;k,\ell)} + \boldsymbol{q}_3^{(4;k,\ell)} & \text{for } (k,\ell) = (k,0),\ (k,k),\ (n/2,\ell), \\ \sum_{i=1}^{4} \boldsymbol{q}_{2i-1}^{(8;k,\ell)} & \text{for Type VM}, \\ \boldsymbol{q}_1^{(8;k,\ell)} + \boldsymbol{q}_3^{(8;k,\ell)} & \text{for Type T}. \end{cases}
$$

The existence of the square pattern for $\mu = (8; k, \ell)$ of Type T depends on the values of

$$
\hat{k} = \frac{k}{\gcd(k,\ell,n)}, \quad \hat{\ell} = \frac{\ell}{\gcd(k,\ell,n)}, \quad \hat{n} = \frac{n}{\gcd(k,\ell,n)}. \tag{4.41}
$$

Here, $\gcd(\cdots)$ is the greatest common divisor of the integers therein.

**Proposition 4.5 (Existence of Square Patterns)** *The square patterns in* (4.40) *exist in the following cases:*

*(i) Any $\mu$ and $n$, except for $\mu = (8; k, \ell)$ of Type T.*
*(ii) $\mu = (8; k, \ell)$ of Type T under the following condition:*

$$
2\gcd(\hat{k}, \hat{\ell}) \textit{ is not divisible by } \gcd(\hat{k}^2 + \hat{\ell}^2, \hat{n}). \tag{4.42}
$$

Using the corresponding eigenvectors, we depict square patterns on a $6 \times 6$ square lattice in Fig. 4.4. As depicted in Fig. 4.5a, Type T square pattern for $\mu = (8; k, \ell)$ exists, for example, for $(k, \ell) = (2, 1)$ and $n = 5$ with $(\hat{k}, \hat{\ell}, \hat{n}) = (2, 1, 5)$, for which $2\gcd(\hat{k}, \hat{\ell}) = 2$ is not divisible by $\gcd(\hat{k}^2 + \hat{\ell}^2, \hat{n}) = 5$, satisfying the condition (4.42).

**Proposition 4.6 (Symmetry of Square Patterns)** *The two half-branches for $(\boldsymbol{w}_{\mathrm{sq}}, \tilde{\tau})$ and $(-\boldsymbol{w}_{\mathrm{sq}}, \tilde{\tau})$ are conjugate in the following cases:*

*(i) $\mu = (1; -), (2), (4; n/2, \ell)$.*
*(ii) $\mu = (4; k, 0), (4; k, k)$ with $\check{n} = n/\gcd(k, n)$ even.*
*(iii) $\mu = (8; k, \ell)$ with $\hat{n} = n/\gcd(k, \ell, n)$ even.*

*They are non-conjugate in the following cases:*

*(iv) $\mu = (4; k, 0), (4; k, k)$ with $\check{n}$ odd.*
*(v) $\mu = (8; k, \ell)$ with $\hat{n}$ odd.*

For example, the two half-branches for $\boldsymbol{q}_1^{\mu} + \boldsymbol{q}_3^{\mu}$ and $-(\boldsymbol{q}_1^{\mu} + \boldsymbol{q}_3^{\mu})$ for $\mu = (4; 2, 0)$ and $(4; 2, 2)$ with odd $\check{n}$ $(= 3)$ in Fig. 4.4b represent different behaviors that are not exchangeable by any geometric transformation and, accordingly, are *non-conjugate*.

Type VM square patterns on a $10 \times 10$ square lattice are depicted in Fig. 4.6. Only the pattern for $\mu = (8; 4, 2)$ with odd $\hat{n}(= 5)$ in panel (e) is asymmetric (two half-

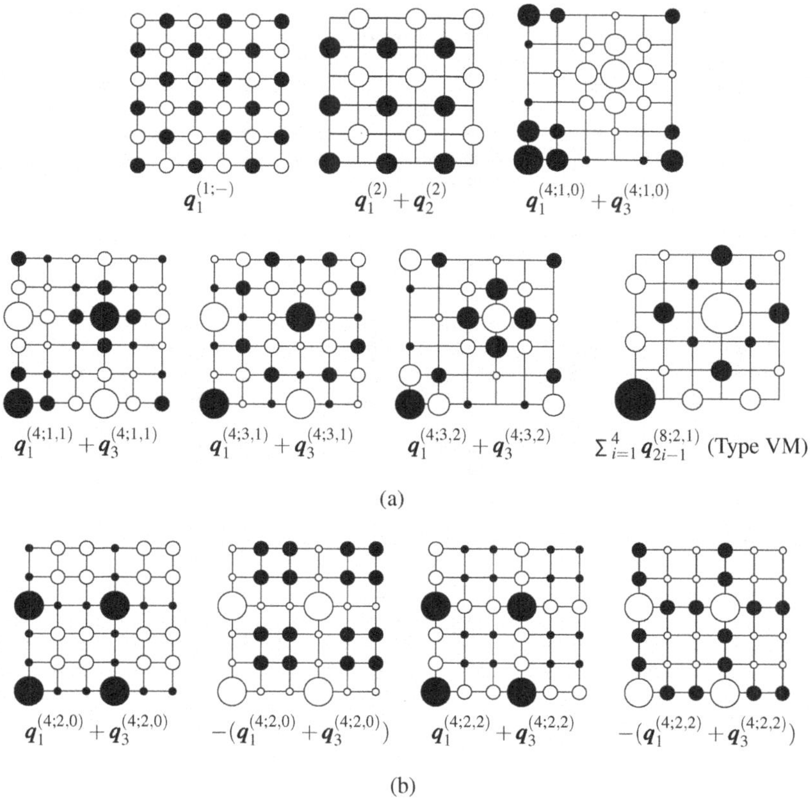


**Fig. 4.4** Square patterns on a $6 \times 6$ square lattice (black circle: positive component of the associated eigenvector; white circle: negative component; the size of a circle represents the magnitude of the associated component). (**a**) Conjugate (symmetric). (**b**) Non-conjugate (asymmetric)

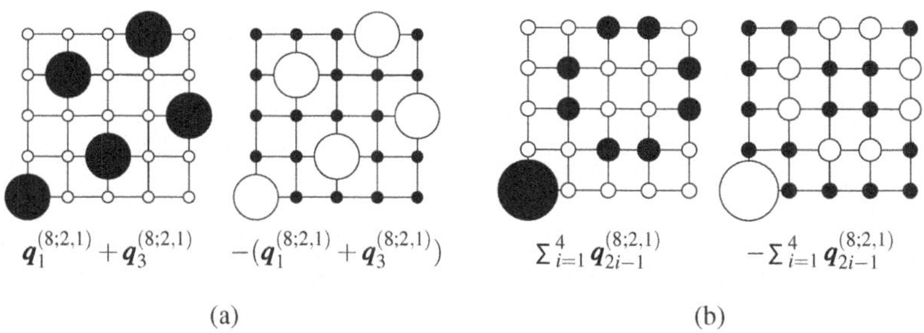


**Fig. 4.5** Non-conjugate square patterns for $\mu = (8; 2, 1)$ on a $5 \times 5$ square lattice (black circle: positive component of the associated eigenvector; white circle: negative component; the size of a circle represents the magnitude of the associated component). (**a**) Type T. (**b**) Type VM

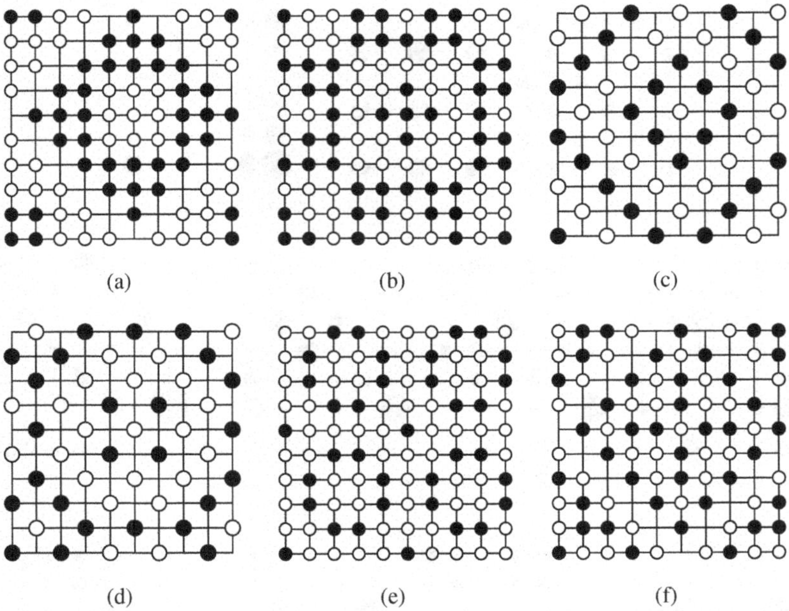

**Fig. 4.6** Type VM square patterns on a 10 × 10 square lattice (black circle: positive component of the associated eigenvector; white circle: negative component; the size of a circle represents the magnitude of the associated component). **(a)** $\sum_{i=1}^{4} \boldsymbol{q}_{2i-1}^{(8;2,1)}$. **(b)** $\sum_{i=1}^{4} \boldsymbol{q}_{2i-1}^{(8;3,1)}$. **(c)** $\sum_{i=1}^{4} \boldsymbol{q}_{2i-1}^{(8;3,2)}$. **(d)** $\sum_{i=1}^{4} \boldsymbol{q}_{2i-1}^{(8;4,1)}$. **(e)** $\sum_{i=1}^{4} \boldsymbol{q}_{2i-1}^{(8;4,2)}$. **(f)** $\sum_{i=1}^{4} \boldsymbol{q}_{2i-1}^{(8;4,3)}$

branches are non-conjugate), whereas other patterns have even $\hat{n}$ and are symmetric (two half-branches are conjugate).

**Stripe Patterns**

Bifurcating *stripe patterns* are

$$\boldsymbol{w}_{\text{stripe}} = \begin{cases} (w, 0) & \text{for } \mu = (2), \\ (w, 0, 0, 0) & \text{for } \mu = (4; k, 0), (4; k, k), (4; n/2, \ell) \quad \text{(Type I)}, \\ (0, w, 0, 0) & \text{for } \mu = (4; k, 0), (4; k, k), (4; n/2, \ell) \quad \text{(Type II)}, \\ (w, 0, 0, 0, 0, 0, 0, 0) & \text{for } \mu = (8; k, \ell) \quad \text{(Type I)}, \\ (0, w, 0, 0, 0, 0, 0, 0) & \text{for } \mu = (8; k, \ell) \quad \text{(Type II)} \end{cases} \tag{4.43}$$

($w \in \mathbb{R}$), and the conditions for their existence and symmetry are as follows:

**Proposition 4.7 (Existence of Stripe Patterns)** *The stripe patterns in* (4.43) *exist in the following cases:*

*(i)* $\mu = (2)$.
*(ii)* $\mu = (4; k, 0), (4; k, k)$ *of Type I.*
*(iii)* $\mu = (4; k, 0), (4; k, k)$ *of Type II when* $\check{n} = n/\gcd(k, n)$ *is even.*
*(iv)* $\mu = (4; n/2, \ell)$ *of Types I and II.*
*(v)* $\mu = (8; k, \ell)$ *of Type I under the following condition:*

$$\hat{k}^2 + \hat{\ell}^2,\ 2\hat{k}\hat{\ell},\ \text{and}\ \hat{k}^2 - \hat{\ell}^2 \ \text{are not divisible by}\ \hat{n}. \tag{4.44}$$

*Here,* $\hat{k}$, $\hat{\ell}$, *and* $\hat{n}$ *are defined by* (4.41).
*(vi)* $\mu = (8; k, \ell)$ *of Type II when* $\hat{n}$ *is even and the condition* (4.44) *is satisfied.*

**Proposition 4.8 (Symmetry of Stripe Patterns)** *The two half-branches* $(\boldsymbol{w}_{\text{stripe}}, \tilde{\tau})$ *and* $(-\boldsymbol{w}_{\text{stripe}}, \tilde{\tau})$ *are conjugate in the following cases:*

*(i)* $\mu = (2)$.
*(ii)* $(4; n/2, \ell)$.
*(iii)* $\mu = (4; k, 0), (4; k, k)$ *with* $\check{n} = n/\gcd(k, n)$ *even.*
*(iv)* $\mu = (8; k, \ell)$ *with* $\hat{n} = n/\gcd(k, \ell, n)$ *even.*

*They are non-conjugate in the following cases:*

*(v)* $\mu = (4; k, 0), (4; k, k)$ *with* $\check{n}$ *odd.*
*(vi)* $\mu = (8; k, \ell)$ *with* $\hat{n}$ *odd.*

Stripe patterns on the $6 \times 6$ square lattice are depicted in Fig. 4.7a. Note that the condition (4.44) is satisfied for $\mu = (8; 2, 1)$ with $\hat{n}(= 6)$ even, which guarantees the existence of Types I and II stripe patterns.[9] For $\mu = (4; 2, 0)$ and $(4; 2, 2)$, we have $\hat{n}(= 3)$ odd; accordingly, there is only Type I stripe pattern with non-conjugate half-branches.

**Upside-Down Patterns**
Bifurcating *upside-down patterns*

$$\boldsymbol{w}_{\text{upside-down}} = \begin{cases} (w, 0, 0, 0, w, 0, 0, 0)\ (w \in \mathbb{R}) & \text{for Type I,} \\ (0, w, 0, 0, 0, w, 0, 0)\ (w \in \mathbb{R}) & \text{for Type II} \end{cases}$$

conditionally exist for $\mu = (8; k, \ell)$ as follows (see Kogure & Ikeda, 2022):

**Proposition 4.9 (Existence of Upside-Down Patterns)** *For* $\mu = (8; k, \ell)$, *Type I upside-down pattern exists under the following condition:*

$$(\hat{k} + \hat{\ell}) \gcd(\hat{k} + \hat{\ell}, \hat{k} - \hat{\ell}) \ \text{is not divisible by}\ \gcd(2\hat{k}\hat{\ell}, \hat{n}). \tag{4.45}$$

[9] For $(\hat{k}, \hat{\ell}, \hat{n}) = (2, 1, 6)$, we see that $\hat{k}^2 + \hat{\ell}^2 = 5$, $2\hat{k}\hat{\ell} = 4$, and $\hat{k}^2 - \hat{\ell}^2 = 3$ are not divisible by $\hat{n} = 6$.

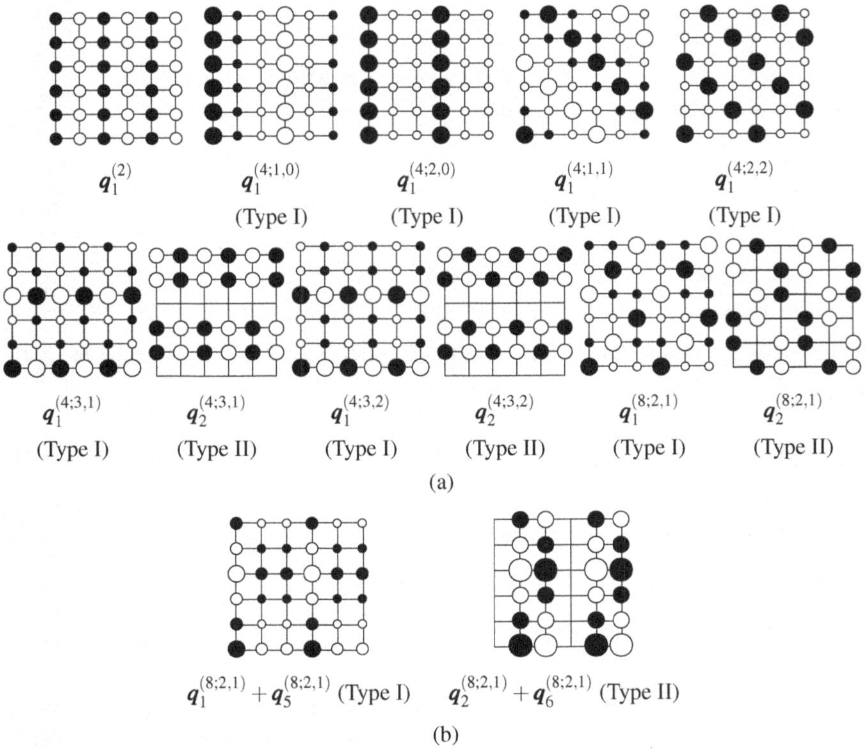


**Fig. 4.7** Stripe and upside-down patterns on the $6 \times 6$ square lattice (black circle: positive component of the associated eigenvector; white circle: negative component; the size of a circle represents the magnitude of the associated component). (**a**) Stripe patterns. (**b**) Upside-down patterns

*Type II upside-down pattern exists when $\hat{n}$ is even and* (4.45) *is satisfied.*

Upside-down patterns on the $6 \times 6$ square lattice are depicted in Fig. 4.7b. Note that the condition (4.45) is satisfied and $\hat{n}$ is even for $\mu = (8; 2, 1)$ to guarantee the existence of the associated patterns.[10]

The symmetry of the upside-down patterns depends on the parity of $\hat{n}$.

**Proposition 4.10 (Symmetry of Upside-Down Patterns)** *For $\mu = (8; k, \ell)$, the two half-branches $(\boldsymbol{w}_{\text{upside-down}}, \tilde{\tau})$ and $(-\boldsymbol{w}_{\text{upside-down}}, \tilde{\tau})$ for Types I and II are conjugate when $\hat{n}$ is even and are non-conjugate when $\hat{n}$ is odd.*

For example, Types I and II upside-down patterns for $n = 6$ in Fig. 4.7b are conjugate as $\hat{n} = 6/\gcd(2, 1, 6) = 6$ is even.

[10] For $(\hat{k}, \hat{\ell}, \hat{n}) = (2, 1, 6)$, we see that $(\hat{k} + \hat{\ell}) \gcd(\hat{k} + \hat{\ell}, \hat{k} - \hat{\ell}) = 3$ is not divisible by $\gcd(2\hat{k}\hat{\ell}, \hat{n}) = 2$, as well as $\hat{n} = 6$ is even.

## 4.5 Bifurcation Behavior in an Economic Geography Model

In the numerical analysis, we use the replicator dynamics in (1.7) and the economic geography model by Forslid and Ottaviano (2003) (Sect. 2.2). As the bifurcation parameter, we use the trade freeness $\phi \in (0, 1)$ (see Remark 2.2 in Sect. 2.2). We set $(\sigma, \mu) = (6.0, 0.4)$.

We employ the $6 \times 6$ *square lattice*, which accommodates various bifurcating patterns (see Figs. 4.4 and 4.7). The set of irreducible representations of the group $G = \mathrm{D}_4 \ltimes (\mathbb{Z}_6 \times \mathbb{Z}_6)$ is given by (4.2) for $n = 6$ as

$$R(G) = \{(1; +), (1; -), (2), (4; 1, 0), (4; 2, 0), (4; 1, 1), (4; 2, 2), (4; 3, 1), (4; 3, 2), (8; 2, 1)\}.$$

As explained in Propositions 4.5, 4.7, and 4.9, the existence of some bifurcating solutions is conditional. Bifurcating solutions satisfying the conditions of existence are listed in Table 4.1.

Figure 4.8 depicts the solution curves of the $6 \times 6$ hexagonal lattice. When the trade freeness $\phi$ increases from 0, the uniform state loses its stability at the bifurcation point A associated with $\mu = (1; -)$, and the bifurcating solution in the direction of $\boldsymbol{q}_1^{(1;-)}$ (see Fig. 4.4) is engendered. There are bifurcating curves from bifurcation points B, C, D, E, F, and G for $\mu = (4; 2, 2), (4; 3, 1), (8; 2, 1), (4; 2, 0), (4; 1, 1)$, and $(4; 1, 0)$, respectively.

**Table 4.1** List of bifurcating solutions for the $6 \times 6$ square lattice

| $\mu$ | Bifurcating solutions ($w \in \mathbb{R}$) |
|---|---|
| $(1; -)$ | $\boldsymbol{w}_{\mathrm{sq}} = w$ |
| $(2; +)$ | $\boldsymbol{w}_{\mathrm{sq}} = (w, w)$, $\boldsymbol{w}_{\mathrm{stripe}} = (w, 0)$ |
| $(4; 1, 0), (4; 2, 0), (4; 1, 1), (4; 2, 2)$ | $\boldsymbol{w}_{\mathrm{sq}} = (w, 0, w, 0)$, $\boldsymbol{w}_{\mathrm{stripeI}} = (w, 0, 0, 0)$ |
| $(4; 3, 1), (4; 3, 2)$ | $\boldsymbol{w}_{\mathrm{sq}} = (w, 0, w, 0)$, $\boldsymbol{w}_{\mathrm{stripeI}} = (w, 0, 0, 0)$, $\boldsymbol{w}_{\mathrm{stripeII}} = (0, w, 0, 0)$ |
| $(8; 2, 1)$ | $\boldsymbol{w}_{\mathrm{sqVM}} = (w, 0, w, 0, w, 0, w, 0)$, |
| | $\boldsymbol{w}_{\mathrm{stripeI}} = (w, 0, 0, 0, 0, 0, 0, 0)$, $\boldsymbol{w}_{\mathrm{stripeII}} = (0, w, 0, 0, 0, 0, 0, 0)$ |
| | $\boldsymbol{w}_{\mathrm{upside\text{-}downI}} = (w, 0, 0, 0, w, 0, 0, 0)$, |
| | $\boldsymbol{w}_{\mathrm{upside\text{-}downII}} = (0, w, 0, 0, 0, w, 0, 0)$ |

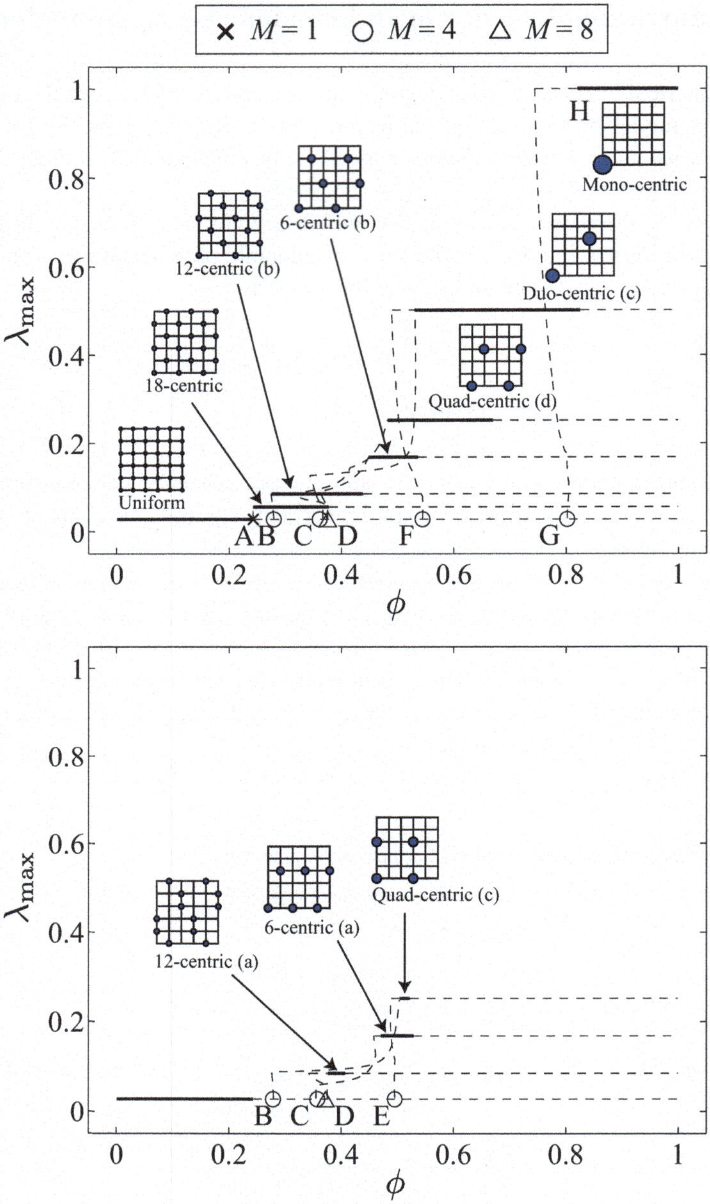


**Fig. 4.8** Direct bifurcation observed on the 6 × 6 square lattice in the FO model with $(\sigma, \mu) = (6.0, 0.4)$ (Kogure et al., 2021) (the vertical axis is $\lambda_{\max} = \max(\lambda_1, \ldots, \lambda_N)$; $M$: multiplicity of critical point; solid curve: stable; dashed one: unstable; the label of each pattern corresponds to Fig. 10.7)

# References


Clarke, M., & Wilson, A. G. (1983). The dynamics of urban spatial structure: Progress and problems. *Journal of Regional Science, 23*(1), 1–18.

Clarke, M., & Wilson, A. G. (1985). The dynamics of urban spatial structure: The progress of a research programme. *Transactions of the Institute of British Geographers, New Series, 10*(4), 427–451.

Dionne, B., Silber, M., & Skeldon, A. C. (1997) Stability results for steady, spatially periodic planforms. *Nonlinearity, 10*, 321–353.

Forslid, R., & Ottaviano, G. I. P. (2003). An analytically solvable core–periphery model. *Journal of Economic Geography, 3*, 229–240.

Golubitsky, M., & Stewart, I. (2003). *The symmetry perspective*. Birkhäuser.

Golubitsky, M., Stewart, I., & Schaeffer, D. G. (1988). *Singularities and groups in bifurcation theory* (Vol. II). Springer.

Ikeda, K., & Murota, K. (2019). *Imperfect bifurcation in structures and materials* (3rd ed.) Springer.

Kogure, Y., & Ikeda, K. (2022). Group-theoretic study of economic agglomerations on a square lattice. *MPRA Paper* 112842.

Kogure, Y., Ikeda, K., & Aizawa, H. (2021). Group-theoretic bifurcation mechanism of economic agglomerations on a square lattice. *International Journal of Bifurcation and Chaos, 31*(13), 2130040.

Munz, M., & Weidlich, W. (1990). Settlement formation, Part II: Numerical simulation. *The Annals of Regional Science, 24*, 177–196.

Redding, S. J. (2016). Goods trade, factor mobility and welfare. *Journal of International Economics, 101*, 148–167.

Redding, S. J., & Rossi-Hansberg, E. (2017). Quantitative spatial economics. *Annual Review of Economics, 9*, 21–58.

Weidlich, W., & Haag, G. (1987). A dynamic phase transition model for spatial agglomeration processes. *Journal of Regional Science, 27*(4), 529–569.

# Chapter 5
# Group-Theoretic Stability Analysis of Hexagonal Lattice

**Abstract** This chapter presents a group-theoretic bifurcation analysis of the uniform state of a hexagonal lattice with periodic boundary conditions. We conduct this analysis on a finite group $\mathrm{D}_6 \ltimes (\mathbb{Z}_n \times \mathbb{Z}_n)$ and find bifurcating hexagonal, racetrack, stripe, and upside-down patterns. We theoretically investigate the stability of these bifurcating patterns and reveal an astonishing fact: All of them are asymptotically unstable. We confirm this fact through the numerical bifurcation analysis in an economic geography model. Chapter 4 presents the basics of bifurcation analysis. This chapter provides a mathematical foundation for the numerical analysis of the hexagonal lattice in Chap. 2, the spectrum analysis of population data in Chap. 6, and the study of invariant patterns in Chap. 10.

**Keywords** Group-theoretic bifurcation analysis · Hexagonal lattice · Hexagonal pattern · Pattern formation · Stability · Stripe pattern · Upside-down pattern

## 5.1 Introduction

In central place theory, hexagonal distributions of cities and towns in Southern Germany are envisaged (Christaller, 1933), and geometrically possible hexagons are proposed (Lösch, 1940). The emergence of hexagonal distributions from a uniform state is understood as pattern formation in science. A theoretical study of the bifurcation problem equivariant to a finite group $\mathrm{D}_6 \ltimes (\mathbb{Z}_n \times \mathbb{Z}_n)$ (see (5.1)) shows the existence of hexagonal patterns on a hexagonal lattice with periodic boundary conditions (Ikeda & Murota, 2014; Ikeda et al., 2012, 2014).

In this chapter, we extend this study and investigate the asymptotic stability of these patterns, based on Ikeda et al. (2018). This chapter is motivated by the following observations:

- On the infinite hexagonal lattice in continuous space, all bifurcating solutions guaranteed by the equivariant branching lemma are, generically, unstable (Dionne et al., 1997).

 
K. Ikeda, Y. Takayama, *Bifurcation Theory of Pattern Formation in Economic Geography*, Interdisciplinary Applied Mathematics 61,
https://doi.org/10.1007/978-3-031-71686-7_5

- On the hexagonal lattice with periodic boundary conditions, all bifurcating equilibria computed numerically are unstable just after bifurcation (see Fig. 2.7 in Sect. 2.3.2 and Ikeda et al., 2017).

Through a group-theoretic analysis, we derive bifurcating hexagonal, racetrack, stripe, and upside-down patterns. The stability analysis of these bifurcating patterns reveals an astonishing fact: All of them are asymptotically unstable. We conduct a numerical bifurcation analysis of an economic geography model to observe that all bifurcating solutions are unstable just after bifurcation; however, some of these solutions recover stability and engender stable patterns.

Hexagonal patterns appear in several physical phenomena. Sattinger (1978) theoretically studied the pattern formation of the Bénard problem (1900). Buzano and Golubitsky (1983) analyzed a $D_6 \dotplus T^2$-symmetric hexagonal lattice.[1] Equivariant bifurcation analysis of six-dimensional irreducible representations of this group found hexagons, rolls, and triangles.[2] The analysis of twelve-dimensional irreducible representations obtained simple and super hexagons.[3]

This chapter is organized as follows: Sect. 5.2 presents a hexagonal lattice and its symmetry. In Sect. 5.3, we introduce bifurcating patterns on this lattice. Stability analysis of bifurcating solutions is conducted in Sect. 5.4.

## 5.2 Hexagonal Lattice and Its Symmetry

We consider a rhombic, $n \times n$ *hexagonal lattice* with periodic boundary conditions; an example of this lattice for $n = 6$ is depicted in Fig. 5.1. Nodal points, shown as (∘), represent places of economic activity. Neighboring places are connected by

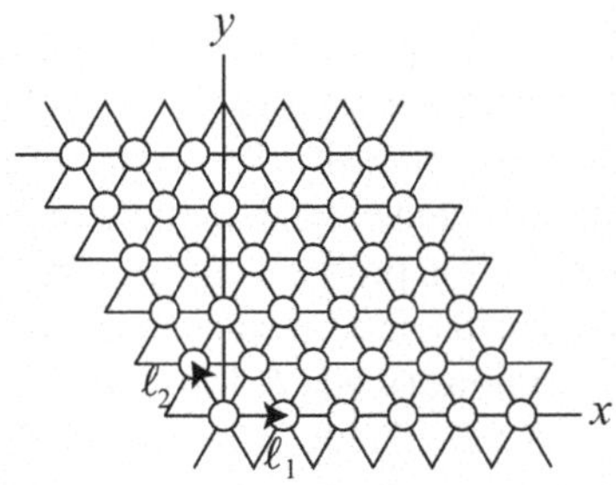


**Fig. 5.1** Hexagonal lattice ($n = 6$)

[1] $T^2$ expresses the two-torus of translation symmetries and $\dotplus$ is a semidirect product that is expressed by $\ltimes$ in this book.

[2] See, e.g., Buzano and Golubitsky (1983), Dionne and Golubitsky (1992), and Golubitsky and Stewart (2002).

[3] See, e.g., Kirchgässner (1979), Dionne et al. (1997), and Judd and Silber (2000).

straight lines of equal (unit) length $\tilde{L}$ to form a regular triangular mesh. The lattice is repeated spatially to form an infinite two-dimensional space.

The $n \times n$ hexagonal lattice has the *symmetry group*

$$G = \langle r, s, p_1, p_2 \rangle = \mathrm{D}_6 \ltimes (\mathbb{Z}_n \times \mathbb{Z}_n). \tag{5.1}$$

Here, $\langle \cdots \rangle$ denotes a group generated by the elements therein; $\ltimes$ expresses the *semidirect product*; and $\mathbb{Z}_n$ is a cyclic group of order $n$. These elements are:

- $r$ : counterclockwise *rotation* about the origin at an angle of $\pi/3$
- $s$ : *reflection* $y \mapsto -y$
- $p_1$ : periodic *translation* along the vector $\boldsymbol{\ell}_1 = \tilde{L}(1, 0)^\top$ by a length $\tilde{L}$
- $p_2$ : periodic *translation* along the vector $\boldsymbol{\ell}_2 = \tilde{L}(-1/2, \sqrt{3}/2)^\top$ by a length $\tilde{L}$

The group $G$ contains the dihedral group $\langle r, s \rangle \simeq \mathrm{D}_6$ and cyclic groups $\langle p_1 \rangle \simeq \mathbb{Z}_n$ and $\langle p_2 \rangle \simeq \mathbb{Z}_n$ as its subgroups.

In the first step of the bifurcation analysis, an *irreducible representation* $\mu$ of $G$ is specified. Only a subset of the whole set of irreducible representations $\mu$ of the group $G$ exists for the $n \times n$ hexagonal lattice with a single degree of freedom at each node. This subset is denoted as[4]

$$R(G) = \{(1), (2), (3), (6; k, 0), (6; k, k), (12; k, \ell)\}. \tag{5.2}$$

Here, $\mu = (2)$ exists when $n$ is a multiple of 3, and $\mu = (3)$ exists when $n$ is even, and

$$1 \le k \le \left\lfloor \frac{n-1}{2} \right\rfloor \quad \text{for } (6; k, 0); \quad 1 \le k \le \left\lfloor \frac{n-1}{2} \right\rfloor, \ k \ne \frac{n}{3} \quad \text{for } (6; k, k);$$
$$1 \le \ell \le k-1, \ \ 2k + \ell \le n-1 \quad \text{for } (12; k, \ell), \tag{5.3}$$

where $\lfloor x \rfloor$ denotes the largest integer not larger than $x$ for a real number $x$. For $\mu$ being expressed as $(M)$ or $(M; k, \ell)$, $M$ represents the dimension of the irreducible representation $\mu$ and as well as the *multiplicity* of the critical point ($M = 1, 2, 3, 6$, or 12); $k$ and $\ell$ represent the wave number of the associated eigenvector(s).

[4] The irreducible representations that are expressed as $(1; +, +)$, $(2; +)$, $(3; +, +)$, $(6; k, 0; +)$, and $(6; k, k; +)$ in Ikeda and Murota (2014, §6.3) are simplified herein to (1), (2), (3), $(6; k, 0)$, and $(6; k, k)$, respectively.

## 5.3 Bifurcation from the Uniform State

We present *bifurcating patterns* from the uniform state. These patterns include hexagonal, stripe, racetrack, and upside-down patterns. We recall the bifurcation equation $\tilde{\boldsymbol{F}}(\boldsymbol{w}, \tilde{\tau}) = \boldsymbol{0}$ in (4.18) and relate the components of $\boldsymbol{w}$ to the discrete trigonometric series $\boldsymbol{q}_i^{\mu}$ to be presented in Sect. 6.2.1, summarizing Ikeda and Murota (2014, §7.5.1).

We have bifurcating *hexagonal patterns*:[5]

$$\boldsymbol{w} = \boldsymbol{w}_{\text{hexa}} = \begin{cases} (w, 0) & \text{for } \mu = (2), \\ (w, w, w) & \text{for } \mu = (3), \\ (w, 0, w, 0, w, 0) & \text{for } \mu = (6; k, \ell)\ (\ell = 0,\ k), \\ (w, 0, w, 0, w, 0, w, 0, w, 0, w, 0) & \text{for } \mu = (12; k, \ell)\ \text{(Type VM)}, \\ (w, 0, w, 0, w, 0, 0, 0, 0, 0, 0, 0) & \text{for } \mu = (12; k, \ell)\ \text{(Type T)}. \end{cases}$$

The associated eigenvectors of the Jacobian matrix are

$$\boldsymbol{q} = \boldsymbol{q}_{\text{hexa}}^{(\mu)} = \begin{cases} \boldsymbol{q}_1^{(2)}, & \\ \boldsymbol{q}_1^{(3)} + \boldsymbol{q}_2^{(3)} + \boldsymbol{q}_3^{(3)}, & \\ \boldsymbol{q}_1^{(6;k,\ell)} + \boldsymbol{q}_3^{(6;k,\ell)} + \boldsymbol{q}_5^{(6;k,\ell)} & (\ell = 0,\ k), \\ \sum_{i=1}^{6} \boldsymbol{q}_{2i-1}^{(12;k,\ell)} & \text{(Type VM)}, \\ \boldsymbol{q}_1^{(12;k,\ell)} + \boldsymbol{q}_3^{(12;k,\ell)} + \boldsymbol{q}_5^{(12;k,\ell)} & \text{(Type T)}. \end{cases} \tag{5.4}$$

Type VM (non-tilted) hexagonal patterns on a $6 \times 6$ hexagonal lattice are depicted in Fig. 5.2a. The existence of the bifurcating hexagonal solution of Type T (tilted hexagon) depends on the values of $k$, $\ell$, and $n$ (Ikeda & Murota, 2014, Proposition 8.28, Page 252). For example, Type T hexagon exists for $\mu = (12; 2, 1)$ on a $7 \times 7$ hexagonal lattice (Fig. 5.2c).

A *racetrack pattern* exists in the reversed direction of a hexagonal bifurcating pattern. That is, $\boldsymbol{w}_{\text{racetrack}} = -\boldsymbol{w}_{\text{hexa}}$.

*Stripe patterns* are associated with

$$\boldsymbol{w}_{\text{stripe}} = \begin{cases} (w, 0, 0) & \text{for } \mu = (3), \\ (w, 0, 0, 0, 0, 0) & \text{for } \mu = (6; k, 0),\ (6; k, k)\ \text{(Type I)}, \\ (0, w, 0, 0, 0, 0) & \text{for } \mu = (6; k, 0),\ (6; k, k)\ \text{(Type II)}, \\ (w, 0, 0, 0, 0, 0, 0, 0, 0, 0, 0, 0) & \text{for } \mu = (12; k, \ell)\ \text{(Type I)}, \\ (0, w, 0, 0, 0, 0, 0, 0, 0, 0, 0, 0) & \text{for } \mu = (12; k, \ell)\ \text{(Type II)}. \end{cases}$$

[5] We exclude $\mu = (1)$ as it is not related to bifurcation.

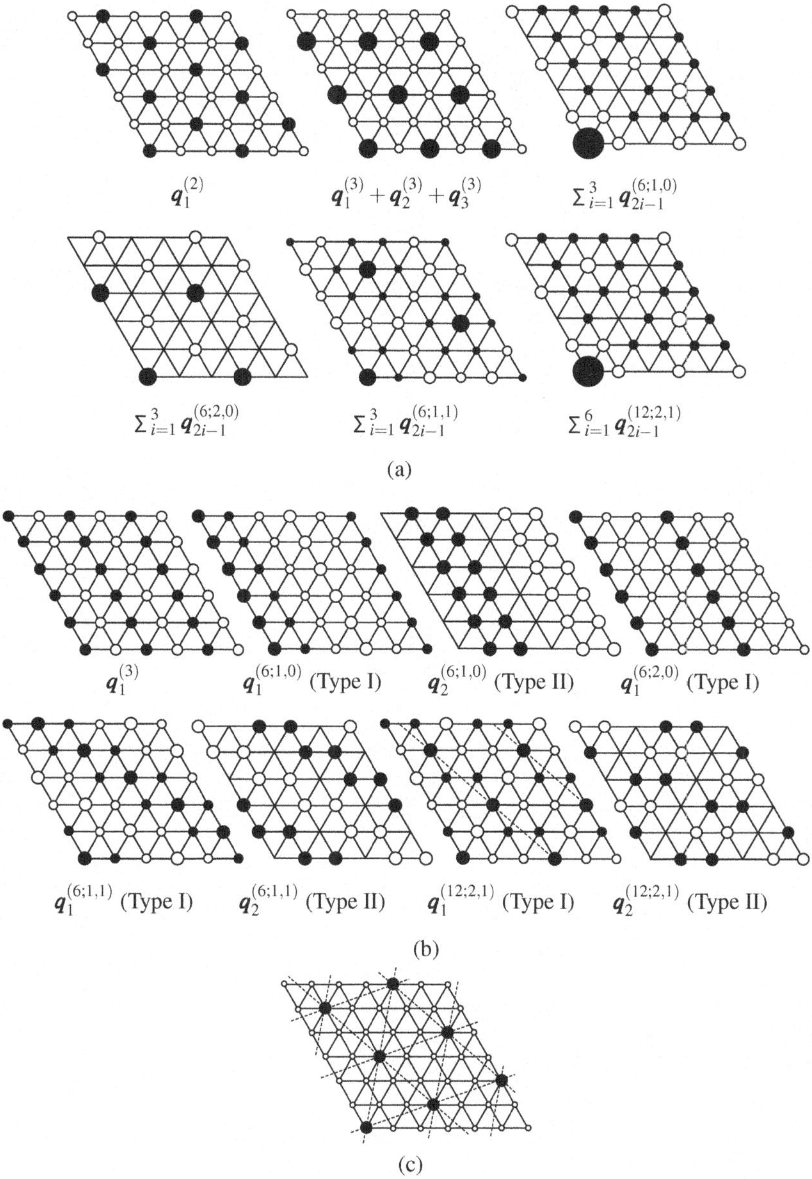


**Fig. 5.2** Bifurcating hexagonal and stripe patterns expressed by the associated eigenvectors of the Jacobian matrix of the governing equation (black circle: positive component; white circle: negative component). (**a**) Hexagonal patterns of Type VM. (**b**) Stripe patterns. (**c**) Hexagonal patterns of Type T

Stripe patterns on the $6 \times 6$ hexagonal lattice are depicted in Fig. 5.2b. As expounded in Propositions 5.1 and 5.2, the existence of some bifurcating stripe solutions depends on the values of $k$, $\ell$, and $n$.

In preparation for Proposition 5.1 for $\mu = (6; k, 0)$ and $(6; k, k)$, we define

$$\hat{n} = \frac{n}{\gcd(k, n)}, \qquad \tilde{n} = \begin{cases} \hat{n}/3 & \text{if } \hat{n} \in 3\mathbb{Z}_+, \\ \hat{n} & \text{if } \hat{n} \notin 3\mathbb{Z}_+. \end{cases} \tag{5.5}$$

Here, $\mathbb{Z}_+$ is a set of positive integers, and $\gcd(k, n)$ is the greatest common divisor of $k$ and $n$. From (5.3), we have

$$\begin{cases} \hat{n} \geq 3 & \text{for } \mu = (6; k, 0), \\ \hat{n} \geq 4 & \text{for } \mu = (6; k, k). \end{cases}$$

**Proposition 5.1** *For the bifurcation point with $M = 6$, the Type I stripe solution always exists. Type II stripe solution exists when $\hat{n}$ is even for $\mu = (6; k, 0)$ and $\tilde{n}$ is even for $\mu = (6; k, k)$.*

***Proof*** See Sect. 5.4.3 for the proof. □

In preparation for Propositions 5.2 and 5.3 for $\mu = (12; k, \ell)$, we set

$$\hat{k} = \frac{k}{\gcd(k, \ell, n)}, \qquad \hat{\ell} = \frac{\ell}{\gcd(k, \ell, n)}, \qquad \hat{n} = \frac{n}{\gcd(k, \ell, n)}. \tag{5.6}$$

**Proposition 5.2** *For $\mu = (12; k, \ell)$, Type I stripe solution exists under the following condition:*

$$\hat{k}^2 + \hat{k}\hat{\ell} + \hat{\ell}^2,\ \hat{k}^2 + 2\hat{k}\hat{\ell},\ \hat{k}^2 - \hat{\ell}^2, \text{ and } \hat{l}^2 + 2\hat{k}\hat{\ell} \textit{ are not divisible by } \hat{n}.$$

*Type II stripe solution exists when $\hat{n}$ is even and this condition is satisfied.*

***Proof*** See Ikeda et al. (2018) for the proof. □

For $\mu = (12; k, \ell)$, *upside-down patterns*

$$\boldsymbol{w}_{\text{upside-down}} = \begin{cases} (w, 0, 0, 0, 0, 0, w, 0, 0, 0, 0, 0) & \text{(Type I)}, \\ (0, w, 0, 0, 0, 0, 0, w, 0, 0, 0, 0) & \text{(Type II)} \end{cases}$$

exist conditionally depending on the values of $k$, $\ell$, and $n$ through (5.6).

**Proposition 5.3** *Type I upside-down pattern for $\mu = (12; k, \ell)$ exists under the following condition:*

$$(\hat{k} - \hat{\ell}) \gcd(\hat{k}, \hat{\ell}) \textit{ is not divisible by } \gcd(\hat{k}^2 + 2\hat{k}\hat{\ell}, \hat{n}) \textit{ and } \gcd(\hat{k}^2 + \hat{k}\hat{\ell} + \hat{\ell}^2, \hat{n}).$$

**Table 5.1** Classification of bifurcating patterns for the 6×6 hexagonal lattice (we exclude $\mu = (1)$ as it is not related to bifurcation)

| $\mu$ | Hexagon (Type VM) | Racetrack | Stripe |
|---|---|---|---|
| (2) | $(w, 0)$ | $(-w, 0)$ | Absent |
| (3) | $(w, w, w)$ | $(-w, -w, -w)$ | $(w, 0, 0)$ |
| (6; 1, 0) | $(w, 0, w, 0, w, 0)$ | $(-w, 0, -w, 0, -w, 0)$ | $(w, 0, 0, 0, 0, 0)$ (Type I) |
| | | | $(0, w, 0, 0, 0, 0)$ (Type II) |
| (6; 2, 0) | $(w, 0, w, 0, w, 0)$ | $(-w, 0, -w, 0, -w, 0)$ | $(w, 0, 0, 0, 0, 0)$ (Type I) |
| (6; 1, 1) | $(w, 0, w, 0, w, 0)$ | $(-w, 0, -w, 0, -w, 0)$ | $(w, 0, 0, 0, 0, 0)$ (Type I) |
| | | | $(0, w, 0, 0, 0, 0)$ (Type II) |
| (12; 2, 1) | $(w, 0, \ldots, w, 0)$ | $(-w, 0, \ldots, -w, 0)$ | $(w, 0, 0, \ldots, 0)$ (Type I) |
| | | | $(0, w, 0, \ldots, 0)$ (Type II) |

*Type II upside-down pattern exists when $\hat{n}$ is even and this condition is satisfied.*

***Proof*** See Ikeda et al. (2018) for the proof. □

The existence of this pattern is rare, as explained in the remark below.

**Remark 5.1** The smallest $n$ for which the upside-down pattern exists is $n = 21$ with $(k, \ell) = (9, 1)$ and $(6, 5)$. For example, for $(k, \ell) = (9, 1)$, the condition in Proposition 5.3 is satisfied since $(\hat{k} - \hat{\ell})\,\mathrm{gcd}(\hat{k}, \hat{\ell}) = 8$ is not divisible by $\mathrm{gcd}(\hat{k}^2 + 2\hat{k}\hat{\ell}, \hat{n}) = \mathrm{gcd}(99, 21) = 3$ and by $\mathrm{gcd}(\hat{k}^2 + \hat{k}\hat{\ell} + \hat{\ell}^2, \hat{n}) = \mathrm{gcd}(91, 21) = 7$. Hence, Type I upside-down pattern exists. □

**Example 5.1** We investigate the bifurcation on the $6 \times 6$ hexagonal lattice ($n = 6$). This lattice is invariant to the group $G = \mathrm{D}_6 \dot{+} (\mathbb{Z}_6 \times \mathbb{Z}_6)$, and the set of *irreducible representations* is obtained with reference to (5.2) as

$$R(G) = \{(1),\ (2),\ (3),\ (6; 1, 0),\ (6; 2, 0),\ (6; 1, 1),\ (12; 2, 1)\}. \tag{5.7}$$

The associated bifurcating patterns are listed in Table 5.1. There are Type VM hexagons, racetrack patterns and stripe patterns, whereas Type T hexagons and upside-down patterns are absent. For example, for (6; 2, 0), we have $\hat{n} = 3$ and $\tilde{n} = 1$; because $\tilde{n}$ is odd, Type II stripe patterns do not exist (Proposition 5.1); upside-down patterns for $\mu = (12; 2, 1)$ do not exist because $(\hat{k} - \hat{\ell})\,\mathrm{gcd}(\hat{k}, \hat{\ell}) = 1$ is divisible by $\mathrm{gcd}(\hat{k}^2 + \hat{k}\hat{\ell} + \hat{\ell}^2, \hat{n}) = \mathrm{gcd}(7, 6) = 1$ (Proposition 5.3). □

## 5.4 Stability Analysis of Bifurcating Solutions

We advance the stability analysis of the double, triple, and sextuple bifurcation points ($M = 2, 3, 6$, respectively) of the uniform state, whereas the analysis of the bifurcation point of multiplicity $M = 12$ is given in Ikeda et al. (2018). We refer to the concrete forms of the bifurcation equation in Ikeda and Murota (2014,

Chapter 9), obtain its Jacobian matrix, and investigate the stability of bifurcating solutions. To sum up the results of this investigation, we would like to advance a bold statement on the stability of bifurcating solutions from the uniform state.

**Proposition 5.4** *The hexagonal, racetrack, stripe, and upside-down patterns branching at the bifurcation point of the uniform state are asymptotically unstable.*

This proposition is in line with the results of the infinite hexagonal lattice in a continuous space that "all of the solution branches guaranteed by the equivariant branching lemma are, generically, unstable" (Dionne et al., 1997).

### *5.4.1 Double Bifurcation Point*

At the double bifurcation point for $\mu = (2)$, the *bifurcation equation* in (4.18) becomes two-dimensional, i.e., $\tilde{\boldsymbol{F}} = (\tilde{F}_1, \tilde{F}_2) = \boldsymbol{0}$ with a function $\tilde{F}_i = \tilde{F}_i(w_1, w_2, \tilde{\tau})$ $(i = 1, 2)$ in independent variables $w_i$ and incremental bifurcation parameter $\tilde{\tau}$.

Using $z = w_1 + \mathrm{i}w_2$ and $F = \tilde{F}_1 + \mathrm{i}\tilde{F}_2$ ($i$ is the imaginary unit), we can rewrite this bifurcation equation into a complex-valued equation: $F(z, \overline{z}, \tilde{\tau}) = 0$ and expand it into a power series as $F(z, \overline{z}, \tilde{\tau}) = \sum_{a=0}^{\infty} \sum_{b=0}^{\infty} A_{ab}(z, \overline{z}, \tilde{\tau})$. Because $(z, \overline{z}, \tilde{\tau}) = (0, 0, 0)$ corresponds to the double critical point, we have $A_{00}(0) = 0$ and $A_{10}(0) = A_{01}(0) = 0$. Therefore, we get $A_{10}(\tilde{\tau}) \approx A'_{10}(0)\tilde{\tau}$ with $A'_{10}(\tilde{\tau}) = \partial A_{10}/\partial\tilde{\tau}(\tilde{\tau})$.

The concrete form of the bifurcation equation is given as

$$\begin{aligned} F(z, \overline{z}, \tilde{\tau}) &= \sum_{a=0}^{\infty} A_{a+1,a}(\tilde{\tau}) z^{a+1}\overline{z}^{a} \\ &\quad + \sum_{p=1}^{\infty}\sum_{a=0}^{\infty} [A_{a+1+3p,a}(\tilde{\tau}) z^{a+1+3p}\overline{z}^{a} + A_{a,a-1+3p}(\tilde{\tau}) z^{a}\overline{z}^{a-1+3p}] \\ &\approx A'_{10}(0)\tilde{\tau} z + A_{02}(0)\overline{z}^{2} \end{aligned} \tag{5.8}$$

(cf. Ikeda & Murota 2014, Chapter 9). Here $A'_{10}(0)$, $A_{a+1,a}(\tilde{\tau})$, $A_{a+1+3p,a}(\tilde{\tau})$, and $A_{a,a-1+3p}(\tilde{\tau})$ are real and generically nonzero.

The Jacobian matrix of the real-valued bifurcation equation $\tilde{\boldsymbol{F}}$ is evaluated to

$$\begin{aligned} \tilde{J}(w_1, w_2, \tilde{\tau}) &= \begin{bmatrix} \frac{\partial \tilde{F}_1}{\partial w_1} & \frac{\partial \tilde{F}_1}{\partial w_2} \\ \frac{\partial \tilde{F}_2}{\partial w_1} & \frac{\partial \tilde{F}_2}{\partial w_2} \end{bmatrix} = \begin{bmatrix} \frac{\partial ReF}{\partial w_1} & \frac{\partial ReF}{\partial w_2} \\ \frac{\partial ImF}{\partial w_1} & \frac{\partial ImF}{\partial w_2} \end{bmatrix} \\ &\approx \begin{bmatrix} A'_{10}(0)\tilde{\tau} + 2A_{02}(0)w_1 & -2A_{02}(0)w_2 \\ -2A_{02}(0)w_2 & A'_{10}(0)\tilde{\tau} - 2A_{02}(0)w_1 \end{bmatrix}, \end{aligned} \tag{5.9}$$

where $Re\ F$ is the real part of $F$ and $Im(\cdot)$ is its imaginary part. The bifurcation equation $F = 0$ with (5.8) has a bifurcating hexagonal solution $z = w$, i.e., $(w_1, w_2) = (w, 0)$ $(w \in \mathbb{R})$, and its solution curve is given by $\tilde{\tau} = \tilde{\tau}_1 \approx -\frac{A_{02}(0)}{A'_{10}(0)} w$. Substitution of $(w_1, w_2, \tilde{\tau}) = (w, 0, \tilde{\tau}_1)$ into (5.9) gives $\tilde{J}(w, 0, \tilde{\tau}_1) \approx A_{02}(0)w\, \mathrm{diag}(1, -3)$. This Jacobian matrix has two eigenvalues $e_1 \approx A_{02}(0)w$ and $e_2 \approx -3A_{02}(0)w$ with opposite signs for $w \neq 0$. Hence, one of these eigenvalues is positive. The bifurcating solution, accordingly, is asymptotically unstable. Also, we can show that the racetrack pattern is asymptotically unstable by replacing $w$ by $-w$ in the discussion above.

### 5.4.2 Triple Bifurcation Point

The *bifurcation equation* $\tilde{\boldsymbol{F}} = (\tilde{F}_1, \tilde{F}_2, \tilde{F}_3)$ of the triple bifurcation point for $\mu = (3)$ is given as

$$\begin{aligned}\tilde{F}_1(w_1, w_2, w_3, \tilde{\tau}) = w_1 &\sum_{a=0}\sum_{b=0}\sum_{c=0} A_{2a,2b+1,2c}(\tilde{\tau})\, w_3^{2a} w_1^{2b} w_2^{2c} \\ + w_3 w_2 &\sum_{a=0}\sum_{b=0}\sum_{c=0} A_{2a+1,2b,2c+1}(\tilde{\tau})\, w_3^{2a} w_1^{2b} w_2^{2c},\end{aligned} \tag{5.10}$$

$$\begin{aligned}\tilde{F}_2(w_1, w_2, w_3, \tilde{\tau}) = w_2 &\sum_{a=0}\sum_{b=0}\sum_{c=0} A_{2a,2b+1,2c}(\tilde{\tau})\, w_1^{2a} w_2^{2b} w_3^{2c} \\ + w_1 w_3 &\sum_{a=0}\sum_{b=0}\sum_{c=0} A_{2a+1,2b,2c+1}(\tilde{\tau})\, w_1^{2a} w_2^{2b} w_3^{2c},\end{aligned} \tag{5.11}$$

$$\begin{aligned}\tilde{F}_3(w_1, w_2, w_3, \tilde{\tau}) = w_3 &\sum_{a=0}\sum_{b=0}\sum_{c=0} A_{2a,2b+1,2c}(\tilde{\tau})\, w_2^{2a} w_3^{2b} w_1^{2c} \\ + w_2 w_1 &\sum_{a=0}\sum_{b=0}\sum_{c=0} A_{2a+1,2b,2c+1}(\tilde{\tau})\, w_2^{2a} w_3^{2b} w_1^{2c}.\end{aligned} \tag{5.12}$$

All coefficients $A_{2a,2b+1,2c}$ and $A_{2a+1,2b,2c+1}$ are real. Since $(w_1, w_2, w_3, \tilde{\tau}) = (0, 0, 0, 0)$ corresponds to the triple critical point, we have $A_{000}(0) = 0$, $A_{100}(0) = A_{010}(0) = A_{001}(0) = 0$, and $A_{010}(\tilde{\tau}) \approx A'_{010}(0)\tilde{\tau}$, where $A'_{010}(0) = \partial A_{010}/\partial \tilde{\tau}(0)$ is generically nonzero.

The Jacobian matrix $J(w_1, w_2, w_3, \tilde{\tau})$ for $\tilde{\boldsymbol{F}} = (\tilde{F}_1, \tilde{F}_2, \tilde{F}_3)$ with (5.10)–(5.12) is evaluated to

$$\tilde{J}(w_1, w_2, w_3, \tilde{\tau}) \approx \begin{bmatrix} A'_{010}(0)\tilde{\tau} & A_{101}(0)w_3 & A_{101}(0)w_2 \\ A_{101}(0)w_3 & A'_{010}(0)\tilde{\tau} & A_{101}(0)w_1 \\ A_{101}(0)w_2 & A_{101}(0)w_1 & A'_{010}(0)\tilde{\tau} \end{bmatrix}. \quad (5.13)$$

We choose the *basis vectors* $\boldsymbol{q}_i^{(3)}$ $(i = 1, 2, 3)$ of the three-dimensional space of $\boldsymbol{w} = (w_1, w_2, w_3)$ to represent stripe patterns in three different directions, as depicted in Fig. 5.3a. Then, we note that the bifurcation equation $\tilde{\boldsymbol{F}} = \boldsymbol{0}$ has the following bifurcating solutions:

- *Hexagonal pattern* $\boldsymbol{w}_{\text{hexa}} = (w, w, w)$ (Fig. 5.3b)
- *Racetrack pattern* $\boldsymbol{w}_{\text{racetrack}} = -(w, w, w)$ (Fig. 5.3c)
- *Stripe pattern* $\boldsymbol{w}_{\text{stripe}} = (w, 0, 0)$ (Fig. 5.3a)

For the hexagonal pattern $\boldsymbol{w}_{\text{hexa}} = (w, w, w)$, we have

$$\tilde{F}_1(w, w, w, \tilde{\tau}) = \tilde{F}_2(w, w, w, \tilde{\tau}) = \tilde{F}_3(w, w, w, \tilde{\tau}) \approx A'_{010}(0)\tilde{\tau} + A_{101}(0)w.$$

Hence, the bifurcating solution takes an asymptotic form: $\tilde{\tau} = \tilde{\tau}_1(w) \approx -\frac{A_{101}(0)w}{A'_{010}(0)}$, and the Jacobian matrix in (5.13) becomes

$$\tilde{J}(\boldsymbol{w}_{\text{hexa}}, \tilde{\tau}_1) \approx A_{101}(0)w \begin{bmatrix} -1 & 1 & 1 \\ 1 & -1 & 1 \\ 1 & 1 & -1 \end{bmatrix}.$$

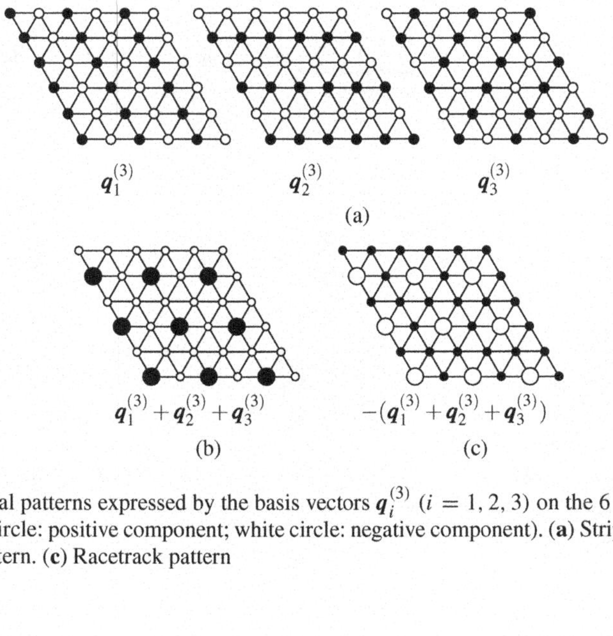


**Fig. 5.3** Spatial patterns expressed by the basis vectors $\boldsymbol{q}_i^{(3)}$ $(i = 1, 2, 3)$ on the $6 \times 6$ hexagonal lattice (black circle: positive component; white circle: negative component). (**a**) Stripe patterns. (**b**) Hexagonal pattern. (**c**) Racetrack pattern

The eigenvalues of this Jacobian matrix are $e_1 \approx A_{101}(0)w$ and $e_2 \approx -2A_{101}(0)w$ (repeated twice). Since $e_1$ and $e_2$ have opposite signs, the hexagonal pattern is asymptotically unstable. We can show that the racetrack pattern $\boldsymbol{w}_{\text{racetrack}} = -\boldsymbol{w}_{\text{hexa}}$ is asymptotically unstable by replacing $w$ with $-w$ in the discussion above.

For the stripe pattern with $\boldsymbol{w}_{\text{stripe}} = (w, 0, 0)$, we have

$$\tilde{F}_1(\boldsymbol{w}_{\text{stripe}}, \tilde{\tau}) = A'_{010}(0)\tilde{\tau}w + \sum_{b=1}^{\infty} A_{0,2b+1,0}(\tilde{\tau})w^{2b+1} \approx A'_{010}(0)\tilde{\tau}w + A_{030}(0)w^3,$$

$$\tilde{F}_2(\boldsymbol{w}_{\text{stripe}}, \tilde{\tau}) = \tilde{F}_3(\boldsymbol{w}_{\text{stripe}}, \tilde{\tau}) = 0.$$

Thus, we get $\tilde{\tau} = \tilde{\tau}_2 \approx -\frac{A_{030}(0)w^2}{A'_{010}(0)}$ from $\tilde{F}_1 = 0$, and the Jacobian matrix in (5.13) becomes

$$\tilde{J}(\boldsymbol{w}_{\text{stripe}}, \tilde{\tau}_2) \approx A_{101}(0)w \begin{bmatrix} 0 & 0 & 0 \\ 0 & 0 & 1 \\ 0 & 1 & 0 \end{bmatrix},$$

in which the diagonal terms are suppressed because they are higher order than $A_{101}(0)w$. The matrix $\tilde{J}$ has eigenvalues $e_1 \approx -A_{101}(0)w$ and $e_2 \approx A_{101}(0)w$ with opposite signs; accordingly, the bifurcating stripe pattern is asymptotically unstable.

### 5.4.3 Sextuple Bifurcation Point

We consider a *sextuple bifurcation point* ($M = 6$) for $\mu = (6; k, \ell)$ ($\ell = 0, k$). The *bifurcation equation* (4.18) for this point is $\tilde{\boldsymbol{F}}(\boldsymbol{w}, \tilde{\tau}) = (\tilde{F}_1, \ldots, \tilde{F}_6) = \boldsymbol{0}$ ($\boldsymbol{w} = (w_1, \ldots, w_6)$). Using $F_i = \tilde{F}_{2i-1} + \mathrm{i}\tilde{F}_{2i}$ and $z_i = w_{2i-1} + \mathrm{i}w_{2i}$ ($i = 1, 2, 3$) ($\mathrm{i}$ is the imaginary unit), we can rewrite this equation into a three-dimensional complex-valued equation:

$$F_i(z_1, z_2, z_3, \overline{z}_1, \overline{z}_2, \overline{z}_3, \tilde{\tau}) = 0 \quad (i = 1, 2, 3).$$

We expand $F_1$ as

$$F_1(z_1, z_2, z_3, \overline{z}_1, \overline{z}_2, \overline{z}_3, \tilde{\tau}) = \sum_{a=0}\sum_{b=0}\sum_{c=0}\sum_{d=0}\sum_{e=0}\sum_{g=0} A_{abcdeg}(\tilde{\tau}) z_1^a z_2^b z_3^c \overline{z}_1^d \overline{z}_2^e \overline{z}_3^g.$$

Since $(z_1, z_2, z_3, \overline{z}_1, \overline{z}_2, \overline{z}_3, \tilde{\tau}) = (0, \ldots, 0, 0)$ corresponds to the critical point,

$$A_{000000}(0) = 0, \quad A_{100000}(0) = A_{010000}(0) = \cdots = A_{000001}(0) = 0$$

are satisfied. Therefore, we get $A_{100000}(\tilde{\tau}) \approx a_1\tilde{\tau}$ for $a_1 = A'_{100000}(0)$. We often omit $\overline{z}_1, \overline{z}_2, \overline{z}_3$, and $\tilde{\tau}$ in the arguments of $F_i$ in the subsequent derivation.

The equivariance regarding rotation $r$ leads to $A_{abcdeg} \in \mathbb{R}$ and

$$F_2(z_1, z_2, z_3) = F_1(z_2, z_3, z_1), \quad F_3(z_1, z_2, z_3) = F_1(z_3, z_1, z_2). \tag{5.14}$$

The equivariance to $s$ gives $A_{abcdeg} = A_{acbdge}$, which plays no role in the subsequent analysis. The equivariance to $p_1$ and to $p_2$ leads to

$$a - d - c + g - 1 \equiv 0 \mod \hat{n}, \quad -b + e + c - g \equiv 0 \mod \hat{n} \text{ for } \mu = (6; k, 0), \tag{5.15}$$

$$a - d - c + g - 1 \equiv 0 \mod \tilde{n}, \quad -b + e + c - g \equiv 0 \mod \tilde{n} \text{ for } \mu = (6; k, k).$$

Recall (5.5):

$$\hat{n} = \frac{n}{\gcd(k, n)}, \quad \tilde{n} = \begin{cases} \hat{n}/3 & \text{if } \hat{n} \in 3\mathbb{Z}_+, \\ \hat{n} & \text{if } \hat{n} \notin 3\mathbb{Z}_+. \end{cases}$$

Bifurcation equation $F_1 = F_2 = F_3 = 0$ has the following bifurcating solutions:

(i) $\boldsymbol{w}_{\text{hexa}} = (w, 0, w, 0, w, 0)$, i.e., $(z_1, z_2, z_3) = (w, w, w)$: hexagonal pattern.
(ii) $\boldsymbol{w}_{\text{racetrack}} = -(w, 0, w, 0, w, 0)$ i.e., $(z_1, z_2, z_3) = -(w, w, w)$: racetrack pattern.
(iii) $\boldsymbol{w}_{\text{stripeI}} = (w, 0, 0, 0, 0, 0)$ i.e., $(z_1, z_2, z_3) = (w, 0, 0)$: Type I stripe pattern.
(iv) $\boldsymbol{w}_{\text{stripeII}} = (0, w, 0, 0, 0, 0)$, i.e., $(z_1, z_2, z_3) = (\mathrm{i}w, 0, 0)$: Type II stripe pattern.

We can express $F_1$ asymptotically as (Ikeda & Murota, 2014, Remark 9.5)

$$F_1(z_1, z_2, z_3, \tilde{\tau}) \approx \begin{cases} A_{100000}(\tilde{\tau})z_1 + A_{000011}(\tilde{\tau})\overline{z}_2\overline{z}_3 + A_{000200}(\tilde{\tau})\overline{z}_1^2 & \text{for } \hat{n} = 3 \ (\mu = (6; k, 0)), \\ A_{100000}(\tilde{\tau})z_1 + A_{000011}(\tilde{\tau})\overline{z}_2\overline{z}_3 & \text{for } \hat{n} \geq 4. \end{cases}$$

By (5.3), we have $\hat{n} \geq 4$ for $\mu = (6; k, k)$. $F_2$ and $F_3$ can be obtained from (5.14) by permuting the arguments $z_1$, $z_2$, and $z_3$ in $F_1$. We have $A_{100000}(\tilde{\tau}) \approx A'_{100000}(0)\tilde{\tau}$ and hereafter set $a_1 \equiv A'_{100000}(0)$, $a_2 \equiv A_{000011}(0)$, and $a_3 \equiv A_{000200}(0)$. Then, we obtain

$$F_1(z_1, z_2, z_3, \tilde{\tau}) \approx \begin{cases} a_1\tilde{\tau}z_1 + a_2\overline{z}_2\overline{z}_3 + a_3\overline{z}_1^2 & \text{for } \hat{n} = 3 \ (\mu = (6; k, 0)), \\ a_1\tilde{\tau}z_1 + a_2\overline{z}_2\overline{z}_3 & \text{for } \hat{n} \geq 4; \end{cases}$$

$$\tilde{J}(\boldsymbol{w}, \tilde{\tau}) \approx$$

$$\begin{bmatrix} \begin{pmatrix} a_1\tilde{\tau} \\ +2a_3w_1 \end{pmatrix} & -2a_3w_2 & a_2w_5 & -a_2w_6 & a_2w_3 & -a_2w_4 \\ -2a_3w_2 & \begin{pmatrix} a_1\tilde{\tau} \\ -2a_3w_1 \end{pmatrix} & -a_2w_6 & -a_2w_5 & -a_2w_4 & -a_2w_3 \\ a_2w_5 & -a_2w_6 & \begin{pmatrix} a_1\tilde{\tau} \\ +2a_3w_3 \end{pmatrix} & -2a_3w_4 & a_2w_1 & -a_2w_2 \\ -a_2w_6 & -a_2w_5 & -2a_3w_4 & \begin{pmatrix} a_1\tilde{\tau} \\ -2a_3w_3 \end{pmatrix} & -a_2w_2 & -a_2w_1 \\ a_2w_3 & -a_2w_4 & a_2w_1 & -a_2w_2 & \begin{pmatrix} a_1\tilde{\tau} \\ +2a_3w_5 \end{pmatrix} & -2a_3w_6 \\ -a_2w_4 & -a_2w_3 & -a_2w_2 & -a_2w_1 & -2a_3w_6 & \begin{pmatrix} a_1\tilde{\tau} \\ -2a_3w_5 \end{pmatrix} \end{bmatrix}$$

for $\hat{n} = 3$, (5.16)

$$\tilde{J}(\boldsymbol{w}, \tilde{\tau}) \approx \begin{bmatrix} a_1\tilde{\tau} & 0 & a_2w_5 & -a_2w_6 & a_2w_3 & -a_2w_4 \\ 0 & a_1\tilde{\tau} & -a_2w_6 & -a_2w_5 & -a_2w_4 & -a_2w_3 \\ a_2w_5 & -a_2w_6 & a_1\tilde{\tau} & 0 & a_2w_1 & -a_2w_2 \\ -a_2w_6 & -a_2w_5 & 0 & a_1\tilde{\tau} & -a_2w_2 & -a_2w_1 \\ a_2w_3 & -a_2w_4 & a_2w_1 & -a_2w_2 & a_1\tilde{\tau} & 0 \\ -a_2w_4 & -a_2w_3 & -a_2w_2 & -a_2w_1 & 0 & a_1\tilde{\tau} \end{bmatrix} \quad \text{for } \hat{n} \geq 4. \tag{5.17}$$

### Hexagonal and Racetrack Patterns

The bifurcating curve of the hexagonal pattern $\boldsymbol{w}_{\text{hexa}} = (w, 0, w, 0, w, 0)$ is

$$\begin{cases} a_1\tilde{\tau}_1 + (a_2 + a_3)w \approx 0 & \text{for } \hat{n} = 3, \\ a_1\tilde{\tau}_2 + a_2w \approx 0 & \text{for } \hat{n} \geq 4. \end{cases}$$

For $\hat{n} = 3$, the use of $\boldsymbol{w}_{\text{hexa}} = (w, 0, w, 0, w, 0)$ and $a_1\tilde{\tau}_1 \approx -(a_2 + a_3)w$ in (5.16) gives

$$\tilde{J}(\boldsymbol{w}_{\text{hexa}}, \tilde{\tau}_1) \approx a_2w \begin{bmatrix} C & S & S \\ S & C & S \\ S & S & C \end{bmatrix}, \quad C = \begin{bmatrix} -1+x & 0 \\ 0 & -1-3x \end{bmatrix}, \quad S = \begin{bmatrix} 1 & 0 \\ 0 & -1 \end{bmatrix}$$

with $x = a_3/a_2$. The eigenvalues of this matrix are

$$\begin{cases} e_1 \approx -3(a_2+a_3)w, & e_2 \approx (a_2+a_3)w; \\ e_3 \approx -(2a_2-a_3)w, & e_4 \approx -3a_3w \qquad \text{(repeated twice)}. \end{cases}$$

Since the eigenvalues $e_1$ and $e_2$ have opposite signs, the associated bifurcating solution is asymptotically unstable.

For $\hat{n} \geq 4$, the use of $\boldsymbol{w}_{\text{hexa}} = (w, 0, w, 0, w, 0)$ and $a_1\tilde{\tau}_2 \approx -a_2w$ in (5.17) gives

$$\tilde{J}(\boldsymbol{w}_{\text{hexa}}, \tilde{\tau}_2) \approx a_2w \begin{bmatrix} -I & S & S \\ S & -I & S \\ S & S & -I \end{bmatrix}, \quad I = \begin{bmatrix} 1 & 0 \\ 0 & 1 \end{bmatrix}.$$

The eigenvalues of $\tilde{J}(\boldsymbol{w}_{\text{hexa}}, \tilde{\tau}_2)$ are $e_1 \approx -3a_2w$, $e_2 \approx a_2w$, $e_3 \approx -2a_2w$ (repeated twice), and $e_4 \approx 0$ (repeated twice). Since $e_1$ and $e_2$ have opposite signs, the bifurcating hexagonal solution is asymptotically unstable. The bifurcating racetrack solution can be shown to be also asymptotically unstable by replacing $w$ with $-w$.

### Type I Stripe Pattern

The bifurcating solution curve of the Type I stripe pattern is given by

$$\begin{cases} a_1\tilde{\tau}_3 + a_3w \approx 0 & \text{for } \hat{n} = 3, \\ a_1\tilde{\tau}_4 + a_4w^2 \approx 0 & \text{for } \hat{n} \geq 4 \end{cases} \tag{5.18}$$

with $a_4 = A_{000300}(0) + A_{200100}(0)$ for $\hat{n} = 4$ and $a_4 = A_{200100}(0)$ for $\hat{n} \geq 5$.

For $\hat{n} = 3$, the use of $\boldsymbol{w}_{\text{stripeI}} = (w, 0, 0, 0, 0, 0)$ and $a_1\tilde{\tau}_3 \approx -a_3w$ in (5.16) yields

$$\tilde{J}(\boldsymbol{w}_{\text{stripeI}}, \tilde{\tau}_3) \approx a_3w \begin{bmatrix} D & O & O \\ O & -I & \frac{a_2}{a_3}S \\ O & \frac{a_2}{a_3}S & -I \end{bmatrix} \quad \text{with} \quad D = \begin{bmatrix} 1 & 0 \\ 0 & -3 \end{bmatrix}.$$

The eigenvalues of this Jacobian matrix are $e_1 \approx -3a_3w$, $e_2 \approx a_3w$, $e_3 \approx -(a_2 + a_3)w$ (repeated twice), and $e_4 \approx (a_2 - a_3)w$ (repeated twice). Since the eigenvalues $e_1$ and $e_2$ have opposite signs, the bifurcating solution is asymptotically unstable.

For $\hat{n} \geq 4$, the use of $\boldsymbol{w}_{\text{stripeI}} = (w, 0, 0, 0, 0, 0)$ and $a_1\tilde{\tau}_4 \approx -a_4w^2$ (obtained from (5.18)) in (5.17) leads to

$$\tilde{J}(\boldsymbol{w}_{\text{stripeI}}, \tilde{\tau}_4) \approx a_2w \begin{bmatrix} O & O & O \\ O & O & S \\ O & S & O \end{bmatrix},$$

and the diagonal terms $-\frac{a_4 w^2}{a_2}$ are suppressed because they are higher order terms than $a_2 w$. Since the matrix $\tilde{J}(\boldsymbol{w}_{\text{stripeI}}, \tilde{\tau}_4)$ has eigenvalues $e_1 \approx -a_2 w$ and $e_2 \approx a_2 w$ with opposite signs, the bifurcating solution is asymptotically unstable.

### Type II Stripe Pattern

We search for a solution of the form $\boldsymbol{w}_{\text{stripeII}} = (0, w, 0, 0, 0, 0)$, i.e., $(z_1, z_2, z_3) = (\mathrm{i}w, 0, 0)$. We consider $\mu = (6; k, 0)$, whereas $\mu = (6; k, k)$ can be dealt with similarly (Ikeda et al. 2018). From (5.15), we get $a = d + 1 + p\hat{n}$ $(p \in \mathbb{Z})$. This solution satisfies $F_2 = F_3 = 0$, and $F_1$ becomes

$$F_1(\mathrm{i}w, 0, 0, -\mathrm{i}w, 0, 0)/\mathrm{i}$$
$$= \sum_{d+1+p\hat{n} \geq 0,\ d \geq 0} A_{d+1+p\hat{n},00d00}(\tilde{\tau})(-1)^d\, \mathrm{i}^{2d+p\hat{n}}\, w^{1+2d+p\hat{n}}.$$

Then, we see that $F_1(\mathrm{i}w, 0, 0, -\mathrm{i}w, 0, 0) = 0$ has a solution if $\hat{n}$ is even, for which $\mathrm{i}^{2d+p\hat{n}} = (-1)^{d+p\hat{n}/2}$ is real. Hence, $\hat{n} \neq 3$.

The use of $\boldsymbol{w}_{\text{stripeII}} = (0, w, 0, 0, 0, 0)$ and its solution curve $a_1\tilde{\tau}_4 \approx -a_4 w^2$ (obtained from (5.18)) in (5.17) for $\hat{n} \geq 4$ leads to

$$\tilde{J}(\boldsymbol{w}_{\text{stripeII}}, \tilde{\tau}_4) \approx -a_2 w \begin{bmatrix} O & O & O \\ O & O & \hat{S} \\ O & \hat{S} & O \end{bmatrix}, \quad \hat{S} = \begin{bmatrix} 0 & 1 \\ 1 & 0 \end{bmatrix}.$$

Since the matrix $\tilde{J}(\boldsymbol{w}_{\text{stripeII}}, \tilde{\tau}_4)$ has eigenvalues $e_1 \approx -a_2 w$ and $e_2 \approx a_2 w$ with opposite signs, the bifurcating solution is asymptotically unstable.

## References


Bénard, H. (1900). Les tourbillons cellulaires dans une nappe liquide. *Revue Générale des Sciences Pures et Appliquées, 11*, 1261–1271, 1309–1328.

Buzano, E., & Golubitsky, M. (1983). Bifurcation on the hexagonal lattice and the planar Bénard problem. *Philosophical Transactions of the Royal Society of London, A 308*, 617–667.

Christaller, W. (1933). *Die zentralen Orte in Süddeutschland*. Gustav Fischer. English translation: *Central places in Southern Germany*. Prentice Hall (1966).

Dionne, B., & Golubitsky, M. (1992). Planforms in two and three dimensions. *Zeitschrift für angewandte Mathematik und Physik, 43*, 36–62.

Dionne, B., Silber, M., & Skeldon, A. C. (1997). Stability results for steady, spatially, periodic planforms. *Nonlinearity, 10*, 321–353.

Golubitsky, M., & Stewart, I. (2002). *The symmetry perspective*. Birkhäuser.

Ikeda, K., Aizawa, H., & Kogure, Y. (2018). Stability of bifurcating patterns of spatial economy models on a hexagonal lattice. *International Journal of Bifurcation and Chaos, 28*(11), 1850138.

Ikeda, K., & Murota, K. (2014). *Bifurcation theory for hexagonal agglomeration in economic geography*. Springer.

Ikeda, K., Murota, K., & Akamatsu, T. (2012). Self-organization of Lösch's hexagons in economic agglomeration for core–periphery models. *International Journal of Bifurcation and Chaos, 22*, 1230026-1–1230026-29.

Ikeda, K., Murota, K., Akamatsu, T., Kono, T., & Takayama, Y. (2014). Self-organization of hexagonal agglomeration patterns in new economic geography models. *Journal of Economic Behavior and Organization, 99*, 32–52.

Ikeda, K., Murota, K., & Takayama, Y. (2017). Stable economic agglomeration patterns in two dimensions: Beyond the scope of central place theory. *Journal of Regional Science, 57*, 132–172.

Judd, S. L., & Silber, M. (2000). Simple and superlattice turning patterns in reaction-diffusion systems: Bifurcation, bistability, and parameter collapse. *Physica D, 136*, 45–65.

Kirchgässner, K. (1979). Exotische Lösungen Bénardschen Problems. *Mathematical Methods in the Applied Sciences, 1*, 453–467.

Lösch, A. (1940). *Die räumliche Ordnung der Wirtschaft*. Gustav Fischer. English translation: *The Economics of Location*. Yale University Press (1954).

Sattinger, D. H. (1978). Group representation theory, bifurcation theory and pattern formation. *Journal of Functional Analysis, 28*, 58–101.

# Chapter 6
# Spectrum Analysis of Population Data in German Reunification

**Abstract** We propose a group-theoretic double Fourier spectrum for the hexagonal lattice as a systematic means of capturing pattern formation in real population data. Among many patterns that can be self-organized from a uniform state, we focus on megalopolis, rhombic, and core–satellite patterns (a downtown surrounded by hexagonal satellite cities). We introduce a *principal vector* as a superposition of a few predominant patterns to grasp the multi-scale nature of population agglomeration patterns. Benchmark spectra for these patterns are advanced and are found in Germany's population data. Incremental population data of Germany are processed to detect a shift in predominant population increase/decrease patterns from 1987 to 2011 in the pre- and post-unification periods. Chapter 2 advances typical patterns on a hexagonal lattice. Chapter 5 is a theoretical foundation of the spectrum analysis.

**Keywords** Central place theory · City distribution · Core–satellite pattern · Fourier spectrum analysis · German reunification · Hexagonal pattern · Megalopolis pattern · Pattern formation

## 6.1 Introduction

Chapter 5 introduced a group-theoretic bifurcation analysis of a hexagonal lattice. Through this analysis, we have demonstrated the existence of hexagonal distributions on this lattice in the framework of economic geography. This existence underpins the hexagonal distributions of the cities and towns in Southern Germany envisaged in central place theory (Christaller, 1933).

This chapter searches for the existence of spatial patterns in large-scale population data that have come to be stored worldwide. We propose a systematic methodology to capture characteristic spatial patterns in population data. Measuring spatial agglomeration patterns in actual population data can be challenging. As described in central place theory (Christaller, 1933 and Lösch, 1940), in the real world, spatial agglomeration patterns form a hierarchical structure of centers and



K. Ikeda, Y. Takayama, *Bifurcation Theory of Pattern Formation in Economic Geography*, Interdisciplinary Applied Mathematics 61,
https://doi.org/10.1007/978-3-031-71686-7_6

subcenters. As a result, measurement becomes scale-dependent; data at a given scale allow us to identify some centers but not some lower-level subcenters. We aim to handle the multi-scale nature of spatial patterns and capture characteristic agglomeration patterns in population data with significant noise. For this purpose, we assemble the Fourier basis into several groups that express self-organized hexagonal modes from a uniform state (cf. (5.4) in Sect. 5.3 and Ikeda et al., 2018). By contrast, a conventional double Fourier series cannot express such self-organized hexagons.

First, we introduce a *principal vector* as a superposition of a few Fourier modes to grasp the multi-scale nature of agglomerations. Fourier modes with the first few largest eigenvalues of the adjacency matrix are used in the principal vector not only to filter noise but also to capture important agglomeration patterns. Owing to the symmetry of the hexagonal lattice, the basis vectors of the group-theoretic Fourier series coincide with the eigenvectors of the adjacency matrix in spatial statistics, which capture distinctive spatial patterns with associated spatial autocorrelation distances (Tiefelsdorf & Griffith, 2007).

Next, as the targets of the spectrum analysis, we advance benchmark spectra for several important patterns: megalopolis, rhombic, and core–satellite patterns (a downtown surrounded by hexagonal satellite cities), shown in Fig. 6.1. These patterns are observed in the numerical analysis (Sect. 2.3). We find such targeted patterns in the population data of Germany in 2011 (Sect. 6.4), and agglomeration patterns are successfully grasped by the principal vector with only four terms.

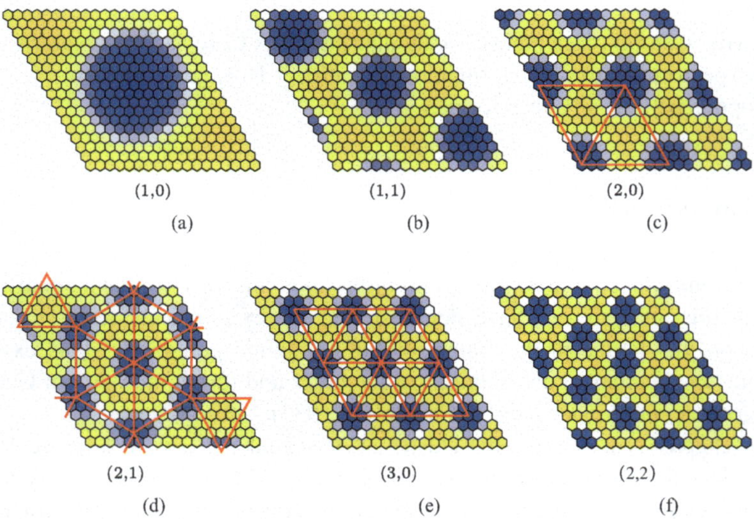


**Fig. 6.1** Megalopolis, hexagonal, and core–satellite patterns (shifted to the center of the domain) on an $18 \times 18$ hexagonal lattice (a blue hexagonal area denotes a positive component; a yellow one indicates a negative one; the magnitude of the component increases as the color becomes darker; a red line clarifies spatial patterns). (**a**) $\boldsymbol{q}_{\mathrm{hexa}}^{(1,0)}$ (megalopolis). (**b**) $\boldsymbol{q}_{\mathrm{hexa}}^{(1,1)}$ (three hexagons). (**c**) $\boldsymbol{q}_{\mathrm{hexa}}^{(2,0)}$ (rhombic). (**d**) $\boldsymbol{q}_{\mathrm{hexa}}^{(2,1)}$ (core–satellite pattern). (**e**) $\boldsymbol{q}_{\mathrm{hexa}}^{(3,0)}$ (nine hexagons). (**f**) $\boldsymbol{q}_{\mathrm{hexa}}^{(2,2)}$ (twelve hexagons)

(a) (b) (c) (d)

**Fig. 6.2** Bifurcating patterns from the uniform distribution on an 18 × 18 square lattice (a blue (yellow) circle at each node denotes a positive (negative) component of the associated eigenvector; the size of a circle at each node represents the magnitude of the associated component). (**a**) Megalopolis. (**b**) Quad-centric. (**c**) Nine-centric. (**d**) Core–satellite pattern

Last, incremental populations during the pre-reunification and post-reunification periods (from 1987 to 2011) are investigated. We detect a shift in the predominance of a megalopolis pattern around Frankfurt to that of a core–satellite pattern for several large cities.

Although we employ the hexagonal lattice in this chapter, a square lattice is another solid candidate. Square road networks prosper worldwide, e.g., in Chicago (USA) and Kyoto (Japan) (see Fig. 4.1). Figure 6.2 depicts bifurcating solutions on an 18 × 18 square lattice.

There are several related studies. Bridging empirics and theory is an important topic (e.g., Stelder, 2005 and Bosker et al., 2010). There are several studies on Germany: German division and reunification by Redding and Sturm (2008), and the city-size distribution of the West Germany by Bosker et al. (2008) and Findeisena and Südekum (2008). The reemergence of central place theory in regional science has come to be acknowledged (Mulligan et al., 2012). Hexagonal patterns on a hexagonal lattice are proved to exist (Ikeda & Murota, 2014 and Ikeda et al., 2014, 2017). The evolution of network analysis in geography is reviewed by Ducruet and Beauguitte (2014). Role of spatial topology in the core–periphery model is studied by Barbero and Zofío (2016).

This chapter is organized as follows: In Sect. 6.2, the group-theoretic Fourier spectrum is presented. Section 6.3 advances benchmark spectra for important spatial patterns. Hexagonal distributions of cities are detected in the population data of Germany in Sect. 6.4. In Sect. 6.5, spectrum analysis in the Eastern USA is overviewed.

## 6.2 Group-Theoretic Fourier Spectrum Analysis

We introduce the bifurcation mechanism of the self-organizing hexagons on an $n \times n$ hexagonal lattice with periodic boundary conditions and an oblique discrete Fourier series on this lattice, based on Chap. 5 and Ikeda and Murota (2014). We construct this Fourier series compatibly with the bifurcation mechanism of self-organizing

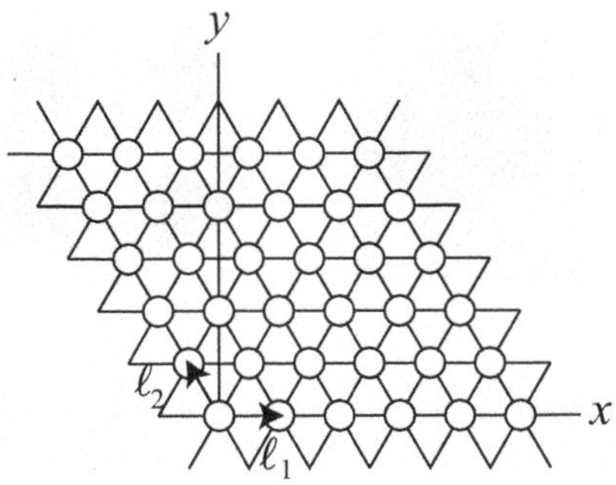


**Fig. 6.3** A 6 × 6 hexagonal lattice with periodic boundary conditions

hexagonal patterns and present a spectrum analysis procedure that can filter noise and detect spatial agglomeration patterns.

### 6.2.1 *Group-Theoretic Fourier Series and Self-Organizing Patterns*

We consider a rhombic $n \times n$ *hexagonal lattice* with periodic boundary conditions (see Fig. 6.3 for an example of $n = 6$). We index evenly distributed $n \times n$ discrete places on this lattice by $(n_1, n_2)$ and define the population distribution vector:

$$\boldsymbol{\lambda} = (\lambda_{n_1 n_2} \mid n_1, n_2 = 0, \ldots, n-1).$$

Using the group-theoretic, oblique, *discrete Fourier series* by Ikeda and Murota (2014), we expand the population distribution vector $\boldsymbol{\lambda}$ on the $n \times n$ hexagonal lattice (where $n$ is a multiple of 6) as

$$\boldsymbol{\lambda} = \sum_{m=1}^{3}\sum_{i=1}^{m} c_i^{(m)} \boldsymbol{q}_i^{(m)} + \sum_{k=1}^{n/2-1}\sum_{i=1}^{6} c_i^{(k,0)} \boldsymbol{q}_i^{(k,0)} + \sum_{k=1}^{n/2-1,\ k\neq n/3}\ \sum_{i=1}^{6} c_i^{(k,k)} \boldsymbol{q}_i^{(k,k)} + \sum_{1\le \ell \le k-1,\ 2k+\ell\le n-1}\ \sum_{i=1}^{12} c_i^{(k,\ell)} \boldsymbol{q}_i^{(k,\ell)} \qquad (6.1)$$

with Fourier coefficients $c_i^{(m)}$ and $c_i^{(k,\ell)}$. This series is meticulously rearranged and regrouped based on the bifurcation mechanism. In (6.1), $(m)$ and $(k, \ell)$ denote the irreducible representations $\mu$; $(6; k, \ell)$ and $(12; k, \ell)$ in (5.4) are expressed as $(k, \ell)$ for simplicity.

The associated *basis vectors* are given as (Ikeda & Murota, 2014)

$$\boldsymbol{q}_1^{(1)} = \frac{1}{n}(1, \ldots, 1)^\top = \langle 1 \rangle,$$

$$\left[\boldsymbol{q}_1^{(2)}, \boldsymbol{q}_2^{(2)}\right] = [\ \langle \cos(2\pi(n_1 - 2n_2)/3) \rangle,\ \langle \sin(2\pi(n_1 - 2n_2)/3) \rangle\ ],$$

$$\left[\boldsymbol{q}_1^{(3)}, \boldsymbol{q}_2^{(3)}, \boldsymbol{q}_3^{(3)}\right] = [\ \langle \cos(\pi n_1) \rangle,\ \langle \cos(\pi n_2) \rangle,\ \langle \cos(\pi(n_1 - n_2)) \rangle\ ],$$

$$\begin{aligned}\left[\boldsymbol{q}_1^{(k,0)}, \ldots, \boldsymbol{q}_6^{(k,0)}\right] = [\ &\langle \cos(2\pi k\, n_1/n) \rangle,\ \langle \sin(2\pi k\, n_1/n) \rangle,\ \langle \cos(2\pi k(-n_2)/n) \rangle,\\ &\langle \sin(2\pi k(-n_2)/n) \rangle,\ \langle \cos(2\pi k(-n_1 + n_2)/n) \rangle,\\ &\langle \sin(2\pi k(-n_1 + n_2)/n) \rangle\ ] \qquad \text{for } 1 \le k \le \frac{n}{2} - 1,\end{aligned}$$

$$\begin{aligned}&\left[\boldsymbol{q}_1^{(k,k)}, \ldots, \boldsymbol{q}_6^{(k,k)}\right]\\ &= [\ \langle \cos(2\pi k(n_1 + n_2)/n) \rangle,\ \langle \sin(2\pi k(n_1 + n_2)/n) \rangle,\ \langle \cos(2\pi k(n_1 - 2n_2)/n) \rangle,\\ &\quad \langle \sin(2\pi k(n_1 - 2n_2)/n) \rangle,\ \langle \cos(2\pi k(-2n_1 + n_2)/n) \rangle,\\ &\quad \langle \sin(2\pi k(-2n_1 + n_2)/n) \rangle\ ] \qquad \text{for } 1 \le k \le \frac{n}{2} - 1,\ k \ne \frac{n}{3},\end{aligned}$$

$$\begin{aligned}\left[\boldsymbol{q}_1^{(k,\ell)}, \ldots, \boldsymbol{q}_{12}^{(k,\ell)}\right] = [\ &\langle \cos(2\pi(kn_1 + \ell n_2)/n) \rangle,\ \langle \sin(2\pi(kn_1 + \ell n_2)/n) \rangle,\\ &\langle \cos(2\pi(\ell n_1 - (k + \ell)n_2)/n) \rangle,\ \langle \sin(2\pi(\ell n_1 - (k + \ell)n_2)/n) \rangle,\\ &\langle \cos(2\pi(-(k + \ell)n_1 + kn_2)/n) \rangle, \langle \sin(2\pi(-(k + \ell)n_1 + kn_2)/n) \rangle,\\ &\langle \cos(2\pi(kn_1 - (k + \ell)n_2)/n) \rangle,\ \langle \sin(2\pi(kn_1 - (k + \ell)n_2)/n) \rangle,\\ &\langle \cos(2\pi(\ell n_1 + kn_2)/n) \rangle,\ \langle \sin(2\pi(\ell n_1 + kn_2)/n) \rangle,\\ &\langle \cos(2\pi(-(k + \ell)n_1 + \ell n_2)/n) \rangle, \langle \sin(2\pi(-(k + \ell)n_1 + \ell n_2)/n) \rangle\ ]\\ &\qquad \text{for } 1 \le \ell \le k - 1,\ 2k + \ell \le n - 1.\end{aligned}$$

Here, for a vector $(g(n_1, n_2) \mid n_1, n_2 = 0, 1, \ldots, n - 1)$, we use the notation $\langle g(n_1, n_2) \rangle$ for its normalization ($\|\langle g(n_1, n_2) \rangle\| = 1$).

Among self-organizing bifurcating patterns from the uniform state ($\boldsymbol{q}_1^{(1)} \equiv \langle 1 \rangle$), we are particularly interested in a *core–satellite pattern* and a series of *hexagonal patterns* in central place theory. These patterns are given by special combinations of the basis vectors in (5.4):

$$
\boldsymbol{q}^{\mu}_{\text{hexa}} = \begin{cases} \boldsymbol{q}^{(2)}, \\ \boldsymbol{q}^{(3)}_1 + \boldsymbol{q}^{(3)}_2 + \boldsymbol{q}^{(3)}_3, \\ \boldsymbol{q}^{(k,\ell)}_1 + \boldsymbol{q}^{(k,\ell)}_3 + \boldsymbol{q}^{(k,\ell)}_5 \quad (\ell = 0 \text{ or } k), \\ \sum_{i=1}^{6} \boldsymbol{q}^{(k,\ell)}_{2i-1}. \end{cases} \tag{6.2}
$$

Figure 6.1 shows these patterns. The *core–satellite pattern* in (d) for $\boldsymbol{q}^{(2,1)}_{\text{hexa}}$ plays a vital role in the search for distributions of cities. This pattern represents a large circle (core place) surrounded by six small ellipses (peripheral places).

### 6.2.2 Spectrum Analysis Procedure

To capture characteristic agglomeration patterns in the population data (Sect. 6.4), we reassemble the double Fourier components in (6.1) into the so-called *isotypic components* (cf. Golubitsky et al., 1988) as

$$
\boldsymbol{q}^{\mu} = \sum_{i=1}^{M(\mu)} c^{\mu}_i \boldsymbol{q}^{\mu}_i \quad (\mu \in R).
$$

Here, $\mu$ is either (1), (2), (3), or $(k, \ell)$; $R$ is the whole set of $\mu$; and $M(\mu)$ (= $1, 2, 3, 6$, or 12) is the number of basis vectors $\boldsymbol{q}^{\mu}_i$ for $\mu$. This assemblage is compatible with the bifurcation mechanism (Sect. 6.2.1), and the vector $\boldsymbol{q}^{\mu}$ for appropriately chosen $c^{\mu}_i$ can represent hexagonal patterns $\boldsymbol{q}^{\mu}_{\text{hexa}}$ in (6.2). We rewrite the double Fourier series in (6.1) as

$$
\boldsymbol{\lambda} = \sum_{\mu \in R} \boldsymbol{q}^{\mu}. \tag{6.3}
$$

The component vectors $\boldsymbol{q}^{\mu}$ of (6.3) depicted in Fig. 6.1 do indeed look like agglomeration patterns. However, some vectors with high spatial frequencies are noise-like patterns (e.g., Fig 6.4). To filter such noise, we use an adjacency matrix in spatial statistics (Dray et al., 2006 and Murakami & Griffith, 2015). Eigenvectors of this matrix corresponding to large eigenvalues are known to capture clusters or agglomeration effects (Tiefelsdorf & Griffith, 2007), which central place theory aims to describe. In our analysis, these eigenvectors, called spatial eigenvectors, are used in the selection of the principal components that can express agglomeration effects.

The *adjacency matrix* $A = \{a_{ij} \mid i, j = 1, \ldots, n\}$ of the $n \times n$ hexagonal lattice network is defined such that $a_{ij} = 1$ if $i$ and $j$ are connected and $a_{ij} = 0$ otherwise.

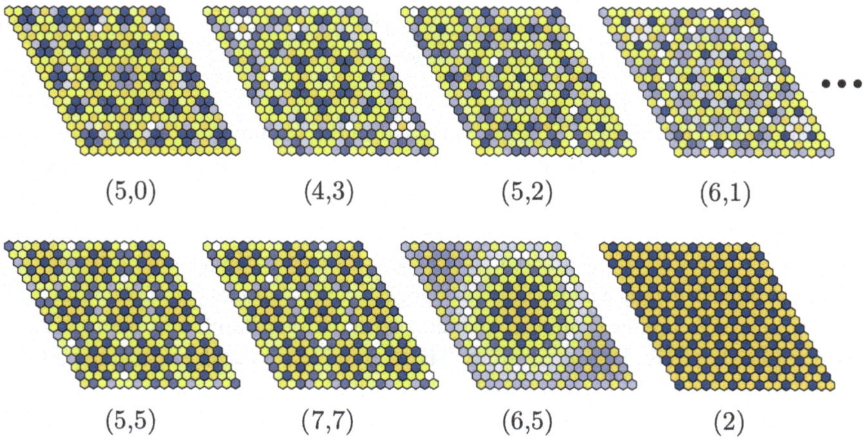


**Fig. 6.4** Hexagonal patterns expressed by eigenvectors with high frequencies on an $18 \times 18$ hexagonal lattice (a blue hexagonal area denotes a positive component; a yellow one indicates a negative one; the magnitude of the component increases as the color becomes darker)

This matrix takes the form:

$$A = \begin{bmatrix} B & C & & C^\top \\ C^\top & B & \ddots & \\ & \ddots & \ddots & C \\ C & & C^\top & B \end{bmatrix}; \quad \text{with} \quad B = \begin{bmatrix} & 1 & & 1 \\ 1 & & \ddots & \\ & \ddots & & 1 \\ 1 & & 1 & \end{bmatrix}, \quad C = \begin{bmatrix} 1 & 1 & & \\ & \ddots & \ddots & \\ & & 1 & 1 \\ 1 & & & 1 \end{bmatrix}.$$

It is noteworthy that the eigenvectors $\boldsymbol{q}_i^\mu$ of $A$ are also the *eigenvectors* of the Jacobian matrix $J$ of the governing equation, thereby related to self-organizing patterns.[1]

Using the eigenvalues $e^\mu = (\boldsymbol{q}_i^\mu)^\top A \boldsymbol{q}_i^\mu$ of the adjacency matrix $A$, we would like to depict a subset of $\mu$ for large eigenvalues, that is,

$$R^* = \{\mu \mid e^\mu \geq e^* \text{ and } \mu \neq (1)\} \tag{6.4}$$

as *principal components* for some threshold value $e^*$; we exclude $\mu = (1)$ associated with the uniform distribution here and in the remainder of this chapter since we target not uniformity but pattern formation. Then, we define the *principal vector* for these components as

[1] There exists the same set of eigenvectors for any matrix $Q$ satisfying the commutativity $T(g)Q = QT(g)$ $(g \in G)$ (Ikeda & Murota, 2014, §13.1). The matrices $J$ and $A$ satisfy this commutativity.

$$\boldsymbol{q}_m^* = \sum_{\mu \in R^*} \boldsymbol{q}^{\mu}, \tag{6.5}$$

where $m$ denotes the number of terms involved in the summation on the right-hand side of this equation. In application, it is vital to choose the number $m$ of the terms appropriately as $\boldsymbol{q}_m^*$ with too small $m$ may fail to capture spatial distribution property and $\boldsymbol{q}_m^*$ with too large $m$ may be subject to noise.

The eigenvalues $e^{\mu}$ of the adjacency matrix $A$ for an $18 \times 18$ *hexagonal lattice* are listed in Table 6.1 in decreasing order. The eigenvectors with larger wave numbers $k$ and $\ell$, which have higher spatial frequencies with a larger number of agglomerated zones, tend to have smaller eigenvalues. These eigenvectors with noise-like patterns (Fig. 6.4) are systematically excluded by using the principal components in (6.4). The hexagonal and core–satellite patterns in Fig. 6.1a–d correspond to the first to the fourth largest positive eigenvalues and play an important role in the description of the real population data (Sect. 6.4).

**Table 6.1** Order of the eigenvalues $e^{\mu}$ of the $18 \times 18$ adjacency matrix $A$

| Order | $\mu$ | Eigenvalue | Name | Corresponding figure |
|---|---|---|---|---|
| 1 | (1,0) | 5.76 | Megalopolis | Fig. 6.1a |
| 2 | (1,1) | 5.29 | Three hexagons | b |
| 3 | (2,0) | 5.06 | Rhombic | c |
| 4 | (2,1) | 4.41 | Core–satellite | d |
| 5 | (3,0) | 4.00 | Nine hexagons | e |
| 6 | (2,2) | 3.41 | Twelve hexagons | f |

| Order | $\mu$ | Eigenvalue |
|---|---|---|
| 7 | (3,1) | 3.23 |
| 8 | (4,0) | 2.69 |
| 9 | (3,2) | 2.18 |
| 10 | (4,1) | 1.88 |
| 11 | (5,0) | 1.31 |
| 12 | (3,3) | 1.00 |
| 13 | (4,2) | 0.88 |
| 14 | (5,1) | 0.53 |
| 15 | (6,0) | 0.00 |
| 16 | (4,3) | −0.18 |
| 17 | (5,2) | −0.35 |
| 18 | (6,1) | −0.65 |
| 19 | (7,0) | −1.06 |
| 20 | (4,4) | −1.18 |

| Order | $\mu$ | Eigenvalue |
|---|---|---|
| 21 | (5,3) | −1.23 |
| 22 | (6,2) | −1.35 |
| 23 | (7,1) | −1.53 |
| 24 | (8,0) | −1.76 |
| 25–29 | (5,4), (7,2), (6,3), (8,1) | −2.00 |
| | (3): $k = 4$ system | −2.00 |
| 30 | (8,8) | −2.23 |
| 31 | (7,3) | −2.41 |
| 32 | (6,4) | −2.53 |
| 33 | (5,5) | −2.57 |
| 34 | (7,7) | −2.72 |
| 35 | (6,5) | −2.88 |
| 36 | (2): $k = 3$ system | −3.00 |

## 6.3 Benchmark Spectra for Important Spatial Patterns

In preparation for the spectrum analysis of population data, we advance benchmark spectra for important patterns: *megalopolis*, *rhombic*, and *core–satellite* patterns in Fig. 6.1. These patterns are the target of the spectrum analysis. In particular, we would like to determine a sufficient number $m$ of terms included in the principal vector $\boldsymbol{q}_m^* = \sum_{\mu \in R^*} \boldsymbol{q}^\mu$ in (6.5) to express these patterns unambiguously.

Megalopolis patterns are shown on the left of Fig. 6.5a and b, whereas their *group-theoretic Fourier spectra* are shown on the right,[2] which plots the squared magnitudes $||\boldsymbol{q}^\mu||^2$ ($\mu \in R$) of the assembled Fourier terms in (6.3); the red bars express the spectra for the first to the fourth largest eigenvalues $e^\mu$.

When the patterns become more diffused from (a) to (b), we can see the predominance of the spectrum for $(k, \ell) = (1, 0)$ accompanied by that for $(1, 1)$ as the second largest one. Thus, the spectrum of $(1, 0)$ signifies the presence of a *megalopolis pattern*, whereas the spectrum of $(1, 1)$ is a byproduct.

A *core–satellite pattern* and its spectrum are shown in Fig. 6.5c. We can see the predominance of the spectrum for $(k, \ell) = (2, 1)$, whereas even the second and the third predominant ones for $(1, 1)$ and $(4, 0)$ have small magnitudes. By contrast, real data often feature two predominant spectra for $(k, \ell) = (2, 1)$ and $(1, 0)$ (Fig. 6.9c). As a possible scenario for such predominance, we consider the development of an industrial belt between the core place and a satellite place shown in Fig. 6.5d. In association with the formation of an industrial belt from (c) to (d), the spectra for $(1, 0)$ and $(1, 1)$ increase.

For the *rhombic pattern* (four hexagons) shown in Fig. 6.6a, $(k, \ell) = (2, 0)$ has the largest spectrum. In the pattern in Fig. 6.6b, for which one city has a larger population in comparison with the other three cities, the spectrum for $(1, 0)$ increases and for $(1, 1)$ emerges as a byproduct.

Thus, the spectra for $(k, \ell) = (1, 0)$, $(1, 1)$, $(2, 0)$, and $(2, 1)$ characterize the existence of megalopolis, core–satellite, and rhombic patterns. Moreover, these four spectra correspond to the first to the fourth largest eigenvalues of the adjacency matrix $A$ (Table 6.1). Accordingly, we choose the *principal vector* in (6.5) as

$$\boldsymbol{q}_4^* = \sum_{\mu \in R^*} \boldsymbol{q}^\mu, \qquad R^* = \{(1, 0), (1, 1), (2, 0), (2, 1)\}. \tag{6.6}$$

The principal vectors $\boldsymbol{q}_4^*$ for the megalopolis, core–satellite, and rhombic patterns depicted in Fig. 6.7a–c are capable of capturing their agglomeration properties.

---

[2] In this figure and the remainder of this chapter, the squared magnitude $||\boldsymbol{q}^{(1)}||^2$ for the uniform distribution is suppressed since such a distribution is not of interest in the present study.

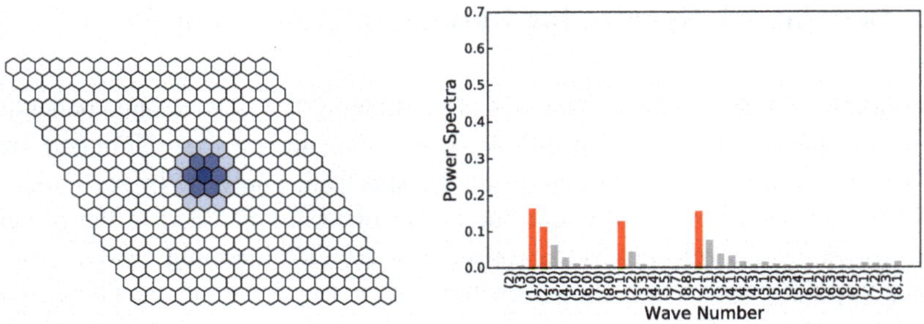


(a)

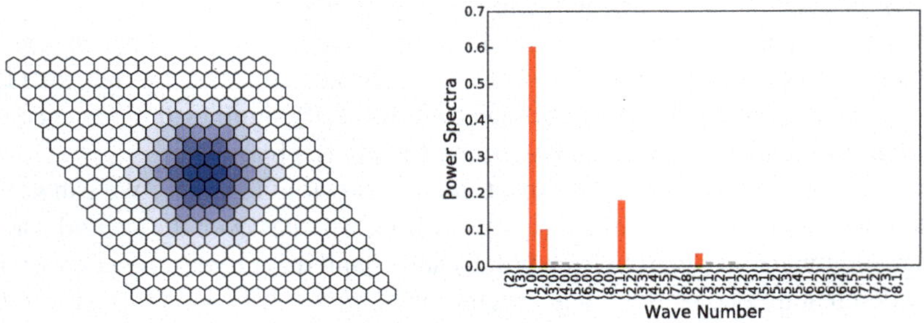


(b)

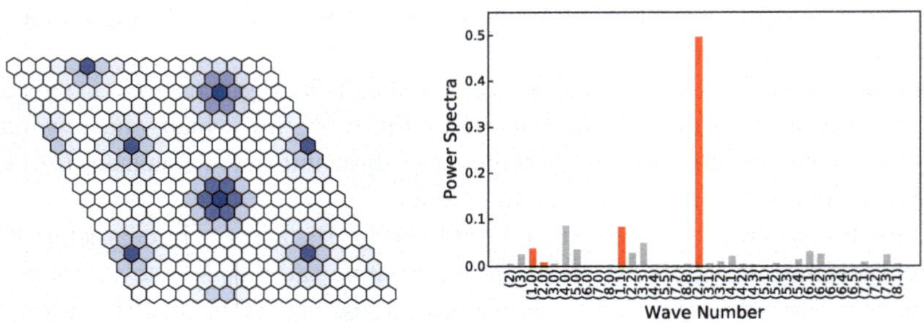


(c)

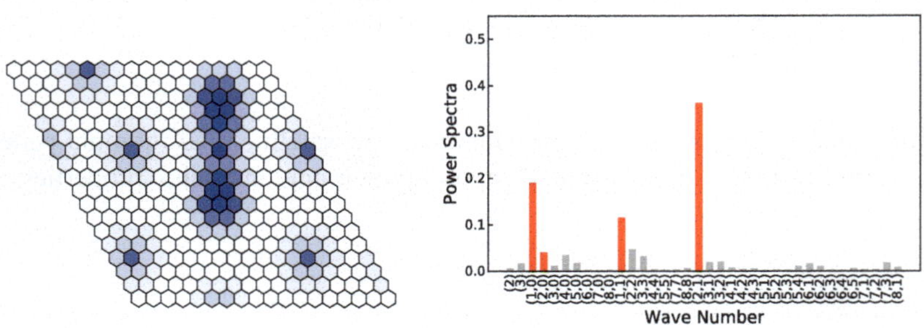


(d)

**Fig. 6.5** Megalopolis and core–satellite patterns and their spectra (on the left, the magnitude of the component increases as the blue color becomes darker; on the right, the red bars correspond to the spectra with the first to the fourth largest eigenvalues $e^{\mu}$ and the brown bars to others). (**a**) Megalopolis pattern (concentrated). (**b**) Megalopolis pattern (diffused). (**c**) Core–satellite pattern. (**d**) Emergence of an industrial belt

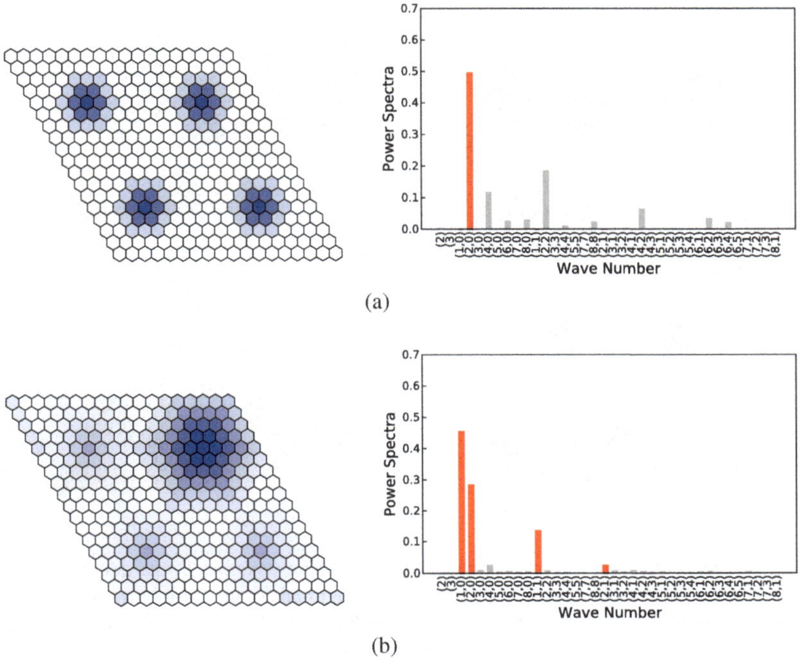


**Fig. 6.6** Rhombic patterns and their spectra (on the left, the magnitude of the component increases as the blue color becomes darker; on the right, the red bars correspond to the spectra with the first to the fourth largest eigenvalues $e^{\mu}$; the brown bars to others). (**a**) Four identical cities. (**b**) One large and three small cities

## 6.4 Spectrum Analysis of Germany

The existence of hexagonal distributions of cities and towns in Southern Germany, envisaged by Christaller (1933), remains to be verified scientifically. As a step toward this verification, the principal vector $\boldsymbol{q}_4^*$ ($m = 4$) in (6.6) is used to capture spatial patterns based on the benchmark spectra analysis in Sect. 6.3.

### *6.4.1 Setting of Spectrum Analysis*

We employ the *population data map* on the Mercator projection shown in Fig. 6.8. This map contains Germany and neighboring countries. Panel (a) denotes the population in 2011, panel (b) the increment of the population from 1987 to 2000,[3]

[3] Although the population map of East Germany is not fully covered due to the lack of data, it does not affect the discussion in this section.

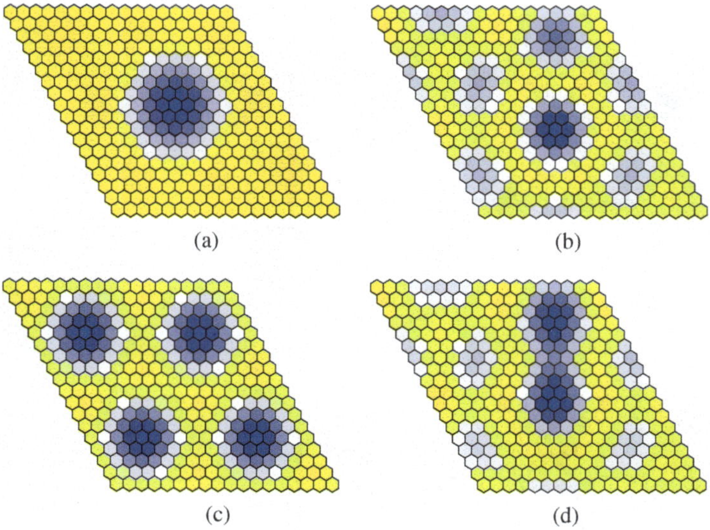

**Fig. 6.7** The principal vector $q_4^*$ for spatial patterns (a blue hexagonal area denotes a positive component; a yellow one indicates a negative one; the magnitude of the component increases as the color becomes darker). (**a**) Megalopolis pattern (Fig. 6.5b). (**b**) Core–satellite pattern (Fig. 6.5c). (**c**) Rhombic pattern (Fig. 6.6a). (**d**) Industrial belt pattern (Fig. 6.5d)

and panel (c) from 2000 to 2011. Germany experienced fluctuations in population from 77.7 million in 1987 to 82.1 million in 2000, and to 80.2 million in 2011. We observe an overall increase in the population, indicated by blue circles, in West Germany from 1987 to 2000. However, from 2000 to 2011, there is a mixture of an increase and a decrease, shown by blue and yellow circles, respectively.

We acquired the population data from the City Population website (http://www.citypopulation.de/) based on the sources listed in Table 6.2. The latitude and longitude of a location were acquired by GoogleMap and Nominatim of OpenStreetMap (https://nominatim.openstreetmap.org/).

On the map in Fig. 6.8a, two rhombic regions were chosen based on preliminary analyses: Region 1 in Southern Germany and Region 2 in Middle Germany. These regions are overlaid by a rhombic 18 × 18 *hexagonal lattice* (regular triangular mesh), and the population is allocated to the nearest node to arrive at the discretized population distribution (e.g., Fig. 6.9b for Region 1).

## 6.4.2 Southern Germany

### Population Data in 2011

The spectrum in Fig. 6.9c is obtained through the spectrum analysis of the discrete population data in Fig. 6.9b of Southern Germany (Region 1) in 2011. There are two

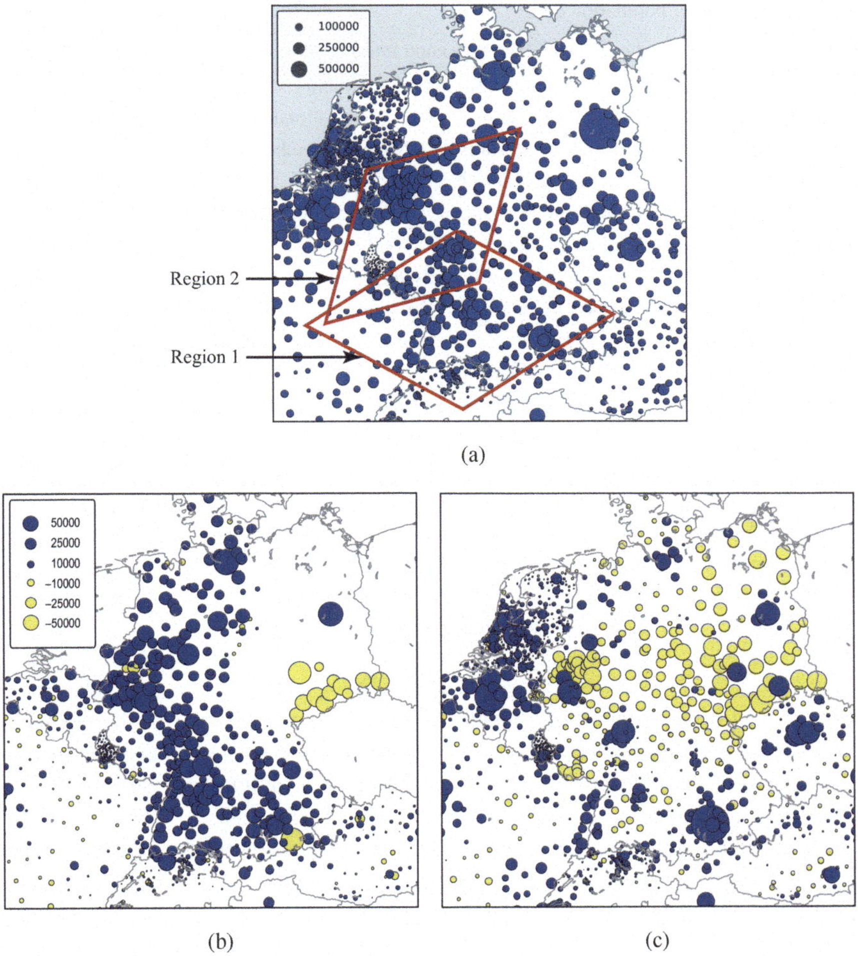


**Fig. 6.8** The population data map of Germany and neighboring countries (in (**a**), a blue circle denotes the size of the population; in (**b**) and (**c**), a blue circle denotes a positive component; a yellow circle indicates a negative one; the area of a circle expresses the magnitude of the incremental population). (**a**) Population in 2011. (**b**) Increment between 1987 to 2000. (**c**) Increment between 2000 to 2011

distinct peaks of the spectrum for the megalopolis pattern $\boldsymbol{q}^{(1,0)}$ and for the core–satellite pattern $\boldsymbol{q}^{(2,1)}$, like the benchmark spectrum for the core–satellite pattern with an industrial belt (Fig. 6.5d). The megalopolis pattern in Fig. 6.9d expresses a large agglomeration along an industrial belt between Frankfurt and Stuttgart. The core–satellite pattern in Fig. 6.9e, with seven blue circular or elliptic zones, captures agglomeration at four large cities: München, Frankfurt, Stuttgart, and Nürnberg with a rhombic shape. Saarbrücken is another place of agglomeration in this pattern. However, we encounter a problem in that Zürich is located in the middle of two

**Table 6.2** Sources of population data

| Country | Data bank (date, type) and internet address |
|---|---|
| Germany | Statistisches Bundesamt Deutschland |
| | (1987/5/25, Census; 2001/12/31, Estimate; 2011/5/9, Census) |
| | https://www.destatis.de/EN/Homepage.html |
| Austria | Statistik Austria (1991/5/15, 2001/5/15, 2011/10/31, Census) |
| | http://www.statistik.at/web_de/statistiken/index.html |
| Belgium | Statistics Belgium (2010/1/1, Estimate) |
| | http://statbel.fgov.be/en/statistics/figures/ |
| France | Institut National de la Statistique et des Études Économiques |
| | (1990/3/5, 1999/3/8, Census; 2012/1/1, Estimate) |
| | http://www.insee.fr/fr/ |
| Netherlands | Centraal Bureau voor de Statistiek (2011/1/1, Estimate) |
| | http://statline.cbs.nl/StatWeb/publication/?VW |
| | =T&DM=SLNL&PA=70233ned&LA=NL |
| Switzerland | Swiss Statistics (1990/12/4, 2000/12/5, 2010/12/31, Census) |
| | http://www.bfs.admin.ch/bfs/portal/en/index.html |
| Luxembourg | Le Portail des Statistiques du Luxembourg |
| | (1991/3/1, 2001/2/15, 2011/2/1, Census) |
| | http://www.statistiques.public.lu/en/index.html |

elliptic zones of agglomeration, which express a chain of cities: Konstanz, Zürich, and Mulhouse. Thus, the core–periphery pattern alone is not sufficient to capture the real population distribution in the south, although it is an authentic self-organizing pattern bifurcating from the uniform state.

In a further search of agglomeration patterns, we resort to the principal vector $\boldsymbol{q}_4^* = \sum_{\mu \in R^*} \boldsymbol{q}^\mu$ in (6.6) with $m = 4$ terms to characterize agglomeration patterns based on the analysis of the benchmark spectra (Sect. 6.3). The spatial patterns $\boldsymbol{q}^\mu$ for $\mu = (1,0), (1,1), (2,0)$, and $(2,1)$ for these four terms are presented in Fig. 6.9d–g, while those for $\mu = (3,0)$ and $(2,2)$ are shown in Fig. 6.9h for reference.

To testify to the sufficiency of the use of $m = 4$ terms, the vector $\boldsymbol{q}_m^*$ in the present case is observed for various values of $m$, as depicted in Fig. 6.10; the pattern for $\boldsymbol{q}_4^*$ looks quite close to that for $\boldsymbol{q}_5^*$, thereby ensuring that $m = 4$ is already large enough to grasp agglomeration characteristics. Note that the number $m = 4$ is small in comparison with the total number of $18 \times 18$ $(= 324)$ terms of the Fourier analysis. The distribution becomes more scattered for large $m$ $(= 15)$.

The principal vector $\boldsymbol{q}_4^*$ (Fig. 6.11a), which is the superposition of the four components, displays several blue zones of agglomeration. This vector expresses a spatial pattern for which Stuttgart is surrounded by Frankfurt, Nürnberg, München, Zürich. This pattern is close to the pattern envisaged by Christaller (Fig. 6.11b), in which five places surround Stuttgart. Thus, the principal vector proposed in this chapter is capable of capturing spatial agglomeration patterns and is more flexible

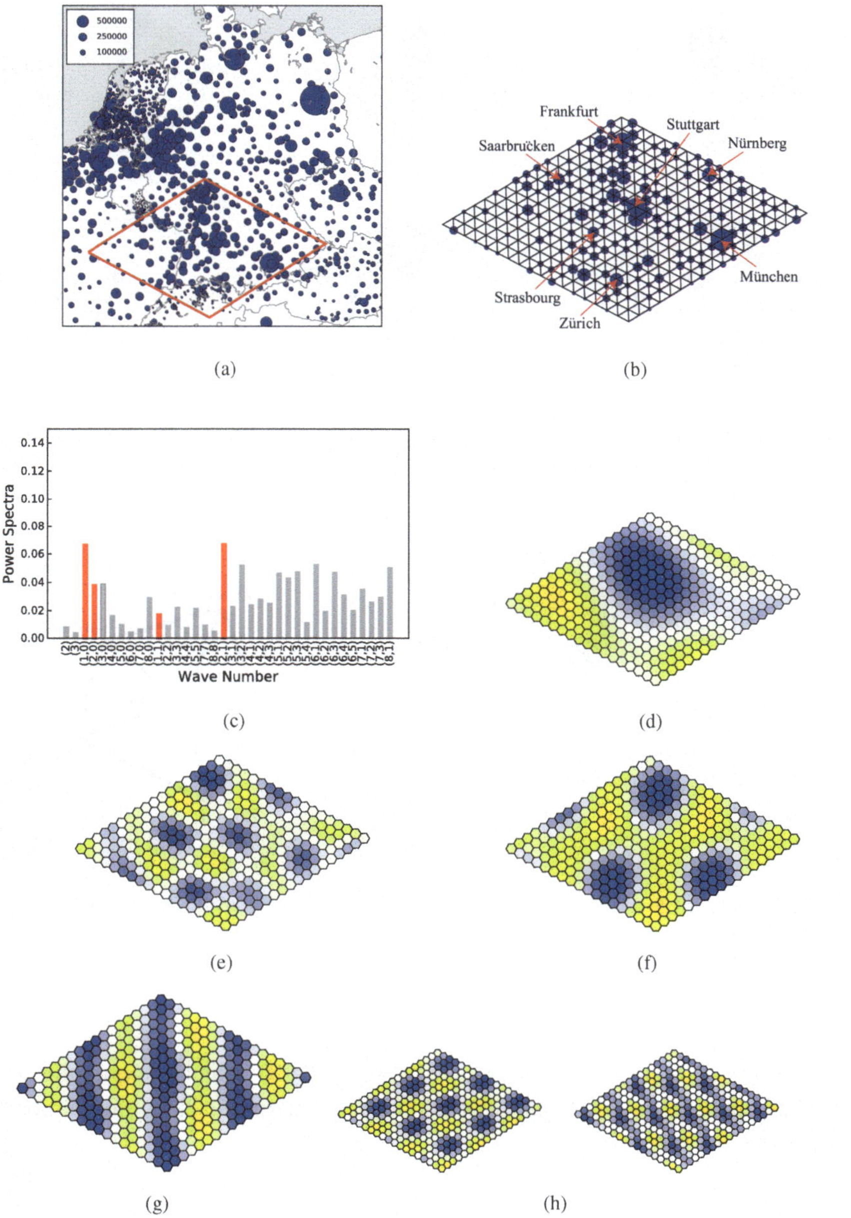


**Fig. 6.9** Spectrum analysis of Region 1 in Southern Germany in 2011 (Ikeda et al., 2022) (in (**c**), the red bars correspond to the spectra with the first to the fourth largest eigenvalues $e^{\mu}$ and the brown bars to others; in (**d**)–(**h**), a blue hexagonal area denotes a positive component; a yellow one indicates a negative one; the magnitude of the component increases as the color becomes darker). (**a**) Population map and Region 1 (red rhombus). (**b**) Discretized population. (**c**) Spectrum. (**d**) $\boldsymbol{q}^{(1,0)}$ (Megalopolis, 1st largest $e^{\mu}$). (**e**) $\boldsymbol{q}^{(2,1)}$ (Core–satellite, 4th largest $e^{\mu}$). (**f**) $\boldsymbol{q}^{(1,1)}$ (2nd largest $e^{\mu}$). (**g**) $\boldsymbol{q}^{(2,0)}$ (3rd largest $e^{\mu}$). (**h**) $\boldsymbol{q}^{(3,0)}$ and $\boldsymbol{q}^{(2,2)}$ (5th and 6th largest $e^{\mu}$, respectively)

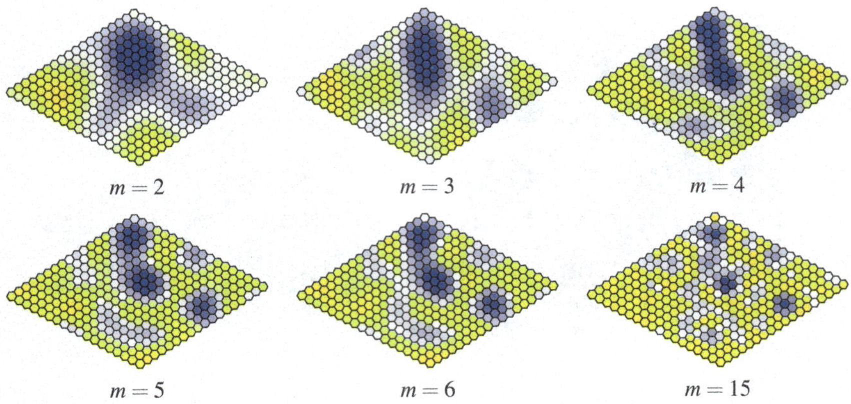


**Fig. 6.10** Vectors $\boldsymbol{q}_m^*$ for principal components for Region 1 (a blue hexagonal area denotes a positive component; a yellow one indicates a negative one; the magnitude of the component increases as the color becomes darker)

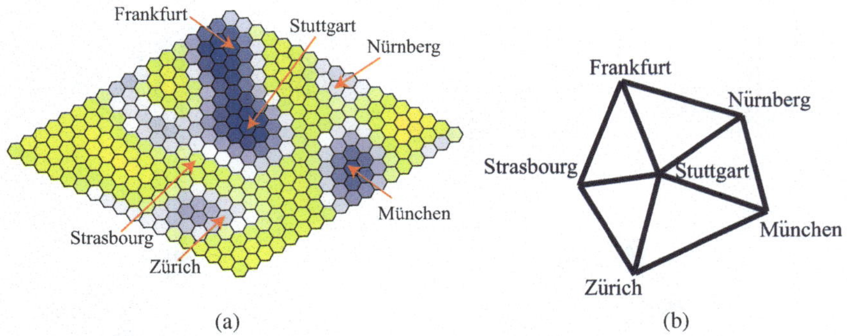


**Fig. 6.11** Comparison of principal vector $\boldsymbol{q}_4^*$ of Region 1 with Christaller's distribution of cities (Christaller, 1933, p.224–225) (in (**a**), a blue hexagonal area denotes a positive component; a yellow one indicates a negative one; the magnitude of the component increases as the color becomes darker). (**a**) Principal vector $\boldsymbol{q}_4^*$. (**b**) Christaller's distribution

than the use of a single term $\boldsymbol{q}_1^*$ in Ikeda et al. (2018), which chooses the core–periphery pattern for $\boldsymbol{q}^{(2,1)}$ as the predominant pattern and fixes the number of satellite places to be six.

### Time Evolution of Population During 1987 to 2011

We observe the evolution of the spectra over time for the incremental population in Southern Germany (Region 1) during the periods 1987–2000 and 2000–2011. In the period 1987–2000, which contains an epoch-making event of the German reunification in 1990, an increase in the population is spread over the region

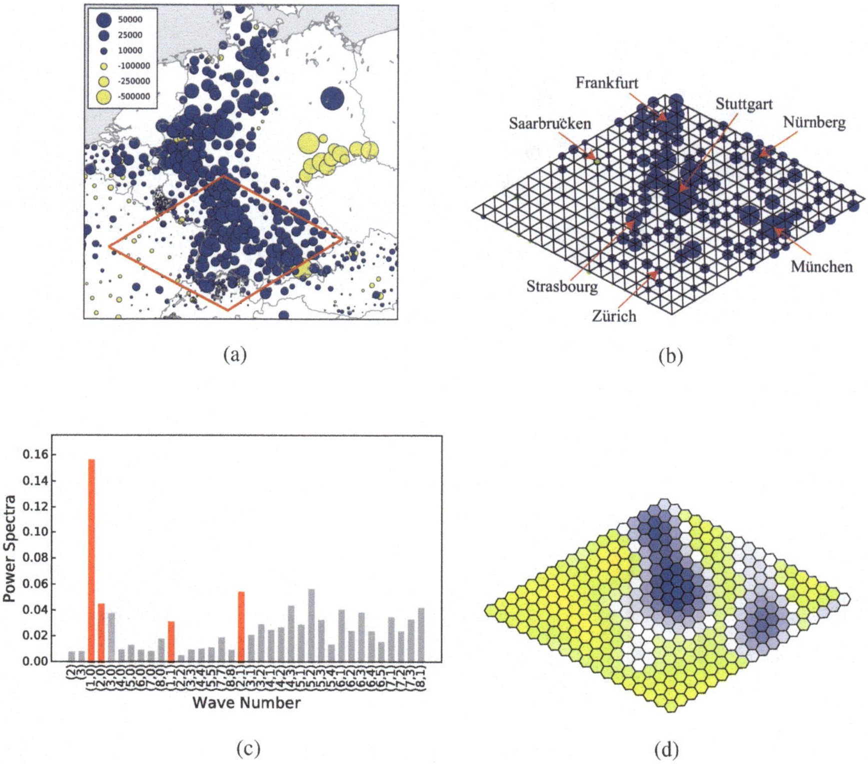


**Fig. 6.12** Time evolution of spectra in Region 1 (1987–2000) (Ikeda et al., 2022) (in (**c**), the red bars correspond to the spectra with the first to the fourth largest eigenvalues $e^{\mu}$ and the brown bars to others; in (**d**), a blue hexagonal area denotes a positive component; a yellow one indicates a negative one; the magnitude of the component increases as the color becomes darker). (**a**) Map of incremental population. (**b**) Discretized incremental population. (**c**) Spectrum. (**d**) Principal component $\boldsymbol{q}_4^*$

(Fig. 6.12b). As shown in Fig. 6.12c, the megalopolis pattern $\boldsymbol{q}^{(1,0)}$ gives the peak spectrum, whereas other spectra are similar to those for 2011 in Fig. 6.9c.

Such similarity also exists in the principal vector $\boldsymbol{q}_4^*$ in the present case in Fig. 6.12d and that in 2011 in Fig. 6.11a. There is a large increase in the population near Stuttgart. The four rhombic cities, München, Frankfurt, Stuttgart, and Nürnberg, display an increase in population proportional to the population in 2011. On the other hand, Saarbrücken, Strasbourg, and Zürich display a slight increase in population.

A phase shift in population increase pattern can be observed in the period 2000–2011. The core–satellite pattern $\boldsymbol{q}^{(2,1)}$ becomes the peak of spectrum (Fig. 6.13c) in this period, and an agglomeration to München is observed from $\boldsymbol{q}_4^*$ in Fig. 6.13d. By contrast, the megalopolis pattern $\boldsymbol{q}^{(1,0)}$ is predominant from 1987 to 2000. The

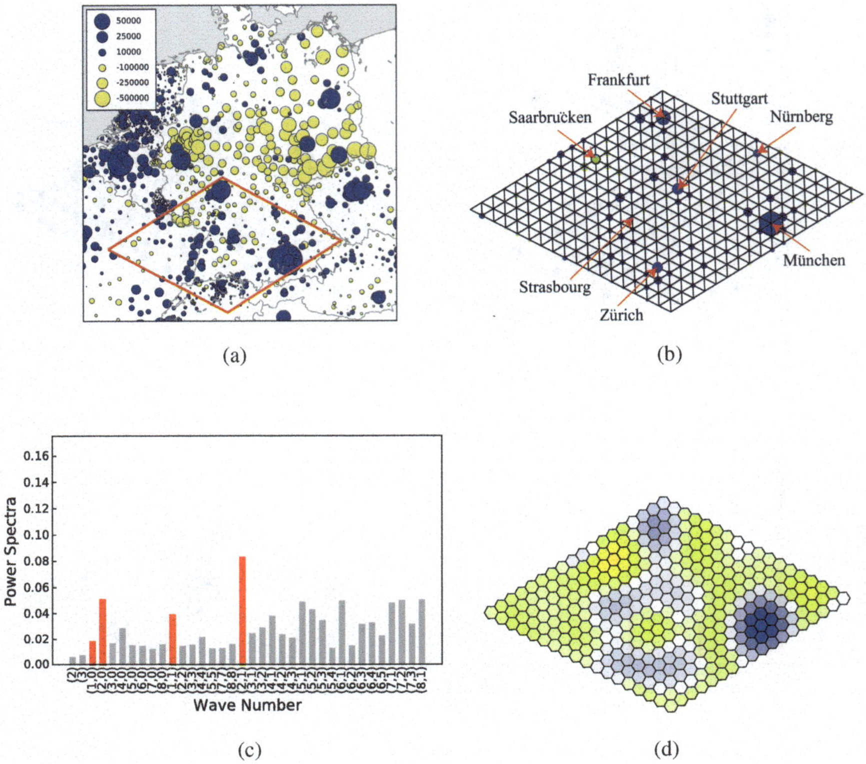


**Fig. 6.13** Time evolution of spectra in Region 1 (2000–2011) (Ikeda et al., 2022) (in (**c**), the red bars correspond to the spectra with the first to the fourth largest eigenvalues $e^{\mu}$ and the brown bars to others; in (**d**), a blue hexagonal area denotes a positive component; a yellow one indicates a negative one; the magnitude of the component increases as the color becomes darker). (**a**) Map of incremental population. (**b**) Discretized incremental population. (**c**) Spectrum. (**d**) Principal component $\boldsymbol{q}_4^*$

core of agglomeration shifts from Stuttgart during 1987–2000 to München during 2000–2011.

Thus, we have successfully analyzed the time evolution of agglomerating places. Whereas central place theory is static, the present spectrum analysis presents a quasi-dynamic view.

### *6.4.3 Middle Germany*

Figure 6.14c depicts the spectrum for the population data in 2011 in Region 2 of Middle Germany in Fig. 6.14b. The largest peak for the spectrum corresponds to the megalopolis pattern $\boldsymbol{q}^{(1,0)}$ in (e), accompanied by other peaks for the rhombic

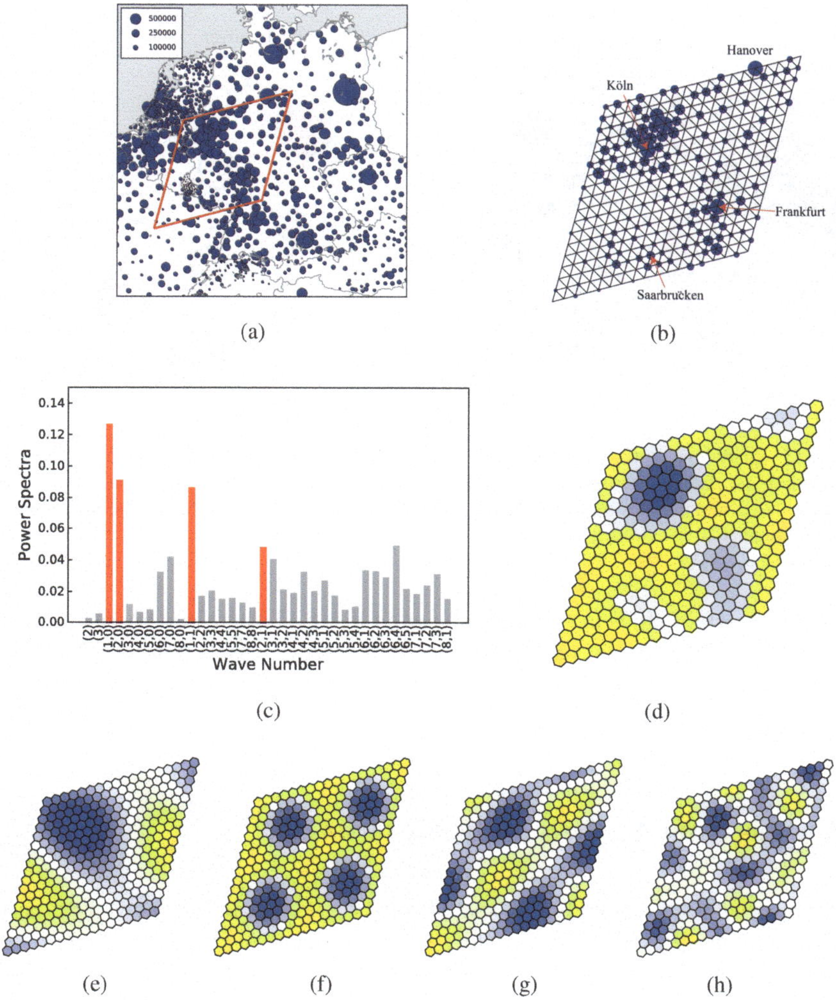


**Fig. 6.14** Spectrum analysis of Region 2 in Middle Germany in 2011 (Ikeda et al., 2022) (in (**b**), a blue circle denotes the size of the population; in (**c**), the red bars correspond to the spectra with the first to the fourth largest eigenvalues $e^{\mu}$ and the brown bars to others; in (**d**)–(**h**), a blue hexagonal area denotes a positive component; a yellow one indicates a negative one; the magnitude of the component increases as the color becomes darker). (**a**) Population map and Region 2 (red rhombus). (**b**) Discretized population. (**c**) Spectrum. (**d**) Principal component $\boldsymbol{q}_4^*$. (**e**) Megalopolis. (**f**) Rhombic. (**g**) Skewed rhombic. (**h**) Core–satellite pattern

pattern $\boldsymbol{q}^{(2,0)}$ in (f), and the skewed rhombic-like pattern $\boldsymbol{q}^{(1,1)}$ in (g). The principal vector $\boldsymbol{q}_4^*$ in (d) displays four blue zones of agglomeration forming a distorted rhombic shape comprising four cities: Köln, Saarbrücken, Frankfurt, and Hanover.

Combining this rhombic pattern of Region 2 with the core–satellite pattern for Region 1 expressed by the principal vector (Fig. 6.11a), we can construct the

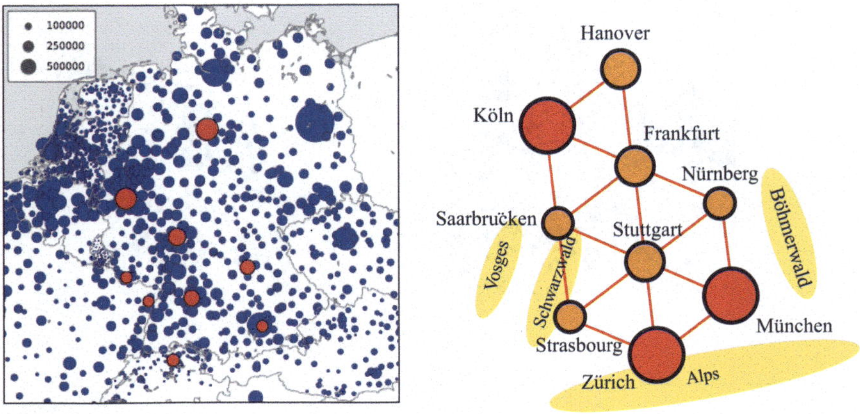


**Fig. 6.15** Spatial network of cities in Germany (on the left, circles indicate the size of the population and red circles express the location of major cities that appear in the network on the right; a blue circle denotes a positive component; the area of a circle expresses the magnitude of the component)

spatial network of cities in Germany shown in Fig. 6.15, thereby demonstrating the usefulness of the principal vector. This network is skewed toward the southwest due to the geographical borders of the Alps in the south and Rhine River, Schwarzwald, and Vosges in the west.

## 6.5 Spectrum Analysis of the Eastern USA

After the successful search for the spatial network of cities in Germany, we proceed to deal with the Eastern USA, in which a large number of cities are spread over a large expanse of flat area, as a summary of Ikeda et al. (2018).

We utilize the population data of the Eastern USA in 2014 taken from the City Population website (http://www.citypopulation.de/). We classify cities from A-center to F-center in decreasing order of population size.

Group-theoretic spectrum analysis of the population data is conducted to detect the 18 rhombic domains accommodating core–satellite patterns of cities and hexagonal ones. These domains are selected based on the two criteria: (i) the existence of a predominant spectrum and (ii) the compatibility of the spatial pattern for this spectrum with the real population distribution. For simplicity, we set $m = 1$ in (6.5) in choosing the principal vector $\boldsymbol{q}_m^*$.

A gigantic city network emerges there as an assemblage of a series of hexagonal and core–satellite patterns, like a giant jigsaw puzzle with many pieces (Fig. 6.16). There is a hierarchy of subnetworks with various geographical scales that look like the overlapping hexagons of Lösch (1940). Thus, this vast expanse of flat area

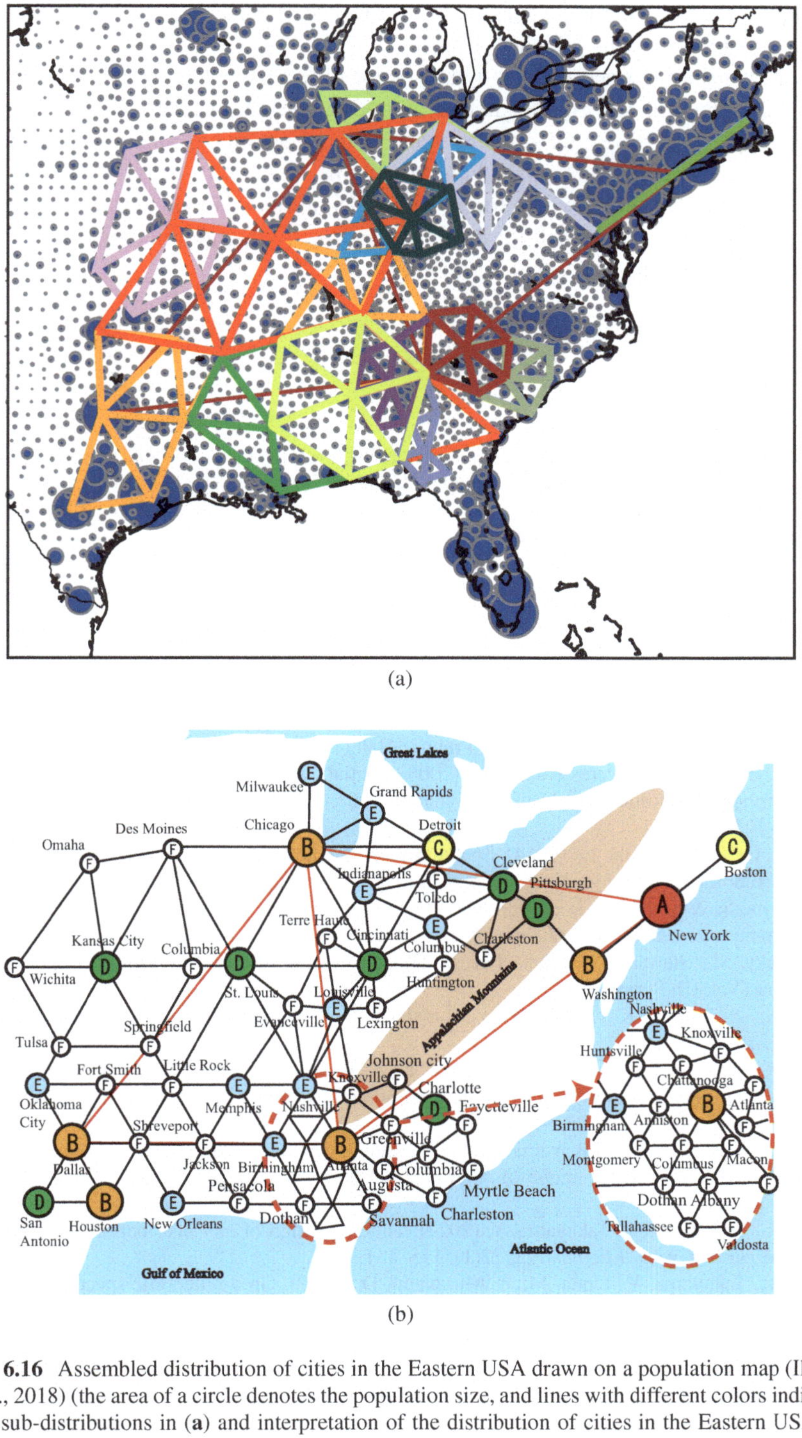


**Fig. 6.16** Assembled distribution of cities in the Eastern USA drawn on a population map (Ikeda et al., 2018) (the area of a circle denotes the population size, and lines with different colors indicate city sub-distributions in (**a**) and interpretation of the distribution of cities in the Eastern USA in (**b**))

accommodates hierarchical hexagonal patterns, just as hexagonal patterns in physics emerge from uniformity (Bénard, 1900).

The distribution encompasses many major cities, except for some parts of the Atlantic seaboard, the Appalachian Mountains, and the Florida Peninsula. The Mississippi River does not influence the patterns so much, and the distribution becomes sparser near the Appalachian Mountains.

Among the 18 city networks found herein, 15 of them are core–satellite patterns, and there is a single network for each of the three, four, and nine hexagons. Thus, the variety of dominant spectra is limited.

## References


Barbero, J., & Zofío, J. M. (2016). The multiregional core-periphery model: The role of the spatial topology. *Networks and Spatial Economics, 16*, 469–496.

Bénard, H. (1900). Les tourbillons cellulaires dans une nappe liquide. *Revue Générale des Sciences Pures et Appliquées, 11*, 1261–1271, 1309–1328.

Bosker, M., Brakman, S., Garretsen, H., & Schramm, M. (2008). A century of shocks: The evolution of the German city size distribution 1925–1999. *Regional Science and Urban Economics, 38*(4), 330–347.

Bosker, M., Brakman, S., Garretsen, H., & Schramm, M. (2010). Adding geography to the new economic geography: Bridging the gap between theory and empirics. *Journal of Economic Geography, 10*(6), 793–823.

Christaller, W. (1933). *Die zentralen Orte in Süddeutschland*. Gustav Fischer. English translation: Central places in Southern Germany. Prentice Hall (1966).

Dray, S., Legendre, P., & Peres-Neto, P. R. (2006). Spatial modelling: A comprehensive framework for principal coordinate analysis of neighbour matrices (PCNM). *Ecological Modelling, 196*, 483–493.

Ducruet, C., & Beauguitte, L. (2014). Spatial science and network science: Review and outcomes of a complex relationship. *Networks and Spatial Economics, 14*, 297–316.

Findeisena, S., & Südekum, J. (2008). Industry churning and the evolution of cities: Evidence for Germany. *Journal of Urban Economics, 64*(2), 326–339.

Golubitsky, M., Stewart, I., & Schaeffer, D. G. (1988). *Singularities and groups in bifurcation theory* (Vol. II). Springer.

Ikeda, K., & Murota, K. (2014). *Bifurcation theory for hexagonal agglomeration in economic geography*. Springer.

Ikeda, K., Murota, K., Akamatsu, T., Kono, T., & Takayama, Y. (2014). Self-organization of hexagonal agglomeration patterns in new economic geography models. *Journal of Economic Behavior and Organization, 99*, 32–52.

Ikeda, K., Murota, K., & Takayama, Y. (2017). Stable economic agglomeration patterns in two dimensions: Beyond the scope of central place theory. *Journal of Regional Science, 57*(1), 132–172.

Ikeda, K., Osawa, M., & Takayama, Y. (2022). Time evolution of city distributions in Germany. *Networks and Spatial Economics, 22*(1), 125–151.

Ikeda, K., Takayama, Y., Onda, M., & Murakami, D. (2018). Group-theoretic spectrum analysis of population distribution in Southern Germany and Eastern USA. *International Journal of Bifurcation and Chaos, 28*(14), 18300458.

Lösch, A. (1940). *Die räumliche Ordnung der Wirtschaft*. Gustav Fischer. English translation: *The Economics of Location*. Yale University Press (1954).

Mulligan, G. F., Partridge, M. D., & Carruthers, J. I. (2012). Central place theory and its reemergence in regional science. *The Annals of Regional Science, 48*(2), 405–431.

Murakami, D., & Griffith, D. A. (2015). Random effects specifications in eigenvector spatial filtering: A simulation study. *Journal of Geographical Systems, 17*, 311–331.

Redding, S. J., & Sturm, D. M. (2008). The costs of remoteness: Evidence from German division and reunification. *The American Economic Review, 98*(5), 1766–1797.

Stelder, D. (2005). Where do cities form? A geographical agglomeration model for Europe. *Journal of Regional Science, 45*, 657–679.

Tiefelsdorf, M., & Griffith, D. A. (2007). Semiparametric filtering of spatial autocorrelation: The eigenvector approach. *Environment Planning, A 39*, 1193–1221.

# Chapter 7
# Locational Heterogeneity and Catastrophe in Two-Place Economy

**Abstract** As an independent topic on direct bifurcation, this chapter presents an imperfect pitchfork bifurcation in a two-strategy population game (two-place economy) and its application to an economic geography model by Pflüger (Reg Sci Urban Econ 34(5):565–573, 2004). We derive two types of approximating games: a cusp catastrophe form and a general catastrophe form, considering universal unfolding. The effects of an arbitrary number of regional asymmetries are expressed using only three asymmetry parameters, thereby allowing for theoretical analysis. We propose behavior with hysteresis for the general form and demonstrate its usefulness in the economic geography model with exogenous locational asymmetries. A numerical recipe to construct a cusp catastrophe form, applicable to general economic geography models, is presented.

**Keywords** Analytically solvable model · Cusp catastrophe · Exogenous asymmetry · Perturbed pitchfork bifurcation · Two-place economy · Two-strategy population game · Universal unfolding

## 7.1 Introduction

In economic geography and spatial economics, one commonly considers a *two-place economy*: a game with two locations that are initially symmetric in all economic aspects (Fig. 1.5 in Sect. 1.6.1). Its bifurcation engenders a *core–periphery pattern*.[1] However, any real system is subject to locational asymmetries that unbalance these locations. Related studies have reported that various asymmetries blur the bifurcation and make one location more advantageous than the other (Sect. 7.2). This chapter elucidates the effects of asymmetries that are yet to be integrated quantitatively and scientifically in economic geography and spatial economics.

[1] See, e.g., Krugman (1991), Fujita et al. (1999), and Baldwin et al. (2003).

 
K. Ikeda. Y. Takayama, *Bifurcation Theory of Pattern Formation in Economic Geography*, Interdisciplinary Applied Mathematics 61,
https://doi.org/10.1007/978-3-031-71686-7_7

The importance of *catastrophe theory* in economics caught the attention of Rosser (2007), who concluded that "the baby of catastrophe theory was largely thrown out with the bathwater of its inappropriate applications." This chapter revives the usefulness of the cusp catastrophe in economic problems, summarizing Ikeda et al. (2022). We demonstrate how to apply catastrophe theory to economic geography models.

As the first contribution of this chapter, we propose a procedure to integrate the effects of an arbitrary number of locational asymmetries on the *perturbed pitchfork bifurcation* (both subcritical and supercritical) in a two-strategy population game (two-place economy). For this purpose, we refer to the *cusp catastrophe* form:[2]

$$G(x, y, \epsilon) = -x^3 - yx + \epsilon = 0. \tag{7.1}$$

Here, $x$ is an independent variable; $y$ is a normalized transportation cost (bifurcation) parameter; and $\epsilon$ is a parameter expressing an asymmetry. We analytically formulate the influence of model parameters of an economic system using the asymmetry parameter $\epsilon$, based on imperfect bifurcation theory (Ikeda & Murota, 2019).

The second contribution of this chapter is to address behavior that has yet to be acknowledged in economic geography and spatial economics. First, we observe the well-known qualitative behavior that can be captured by the cusp catastrophe form in (7.1). The left panel of Fig. 7.1a depicts the solution curves (dashed curves) of the symmetric system ($\epsilon = 0$) with the *supercritical pitchfork* bifurcation point "a" in the $(x, y)$-space. This bifurcation is perturbed into two separate (solid) curves for an asymmetric system ($\epsilon \neq 0$). Stable perturbed curves, shown as bold red curves, are of great economic interest. Next, we observe the behavior for multiple asymmetry parameters known as the *universal unfolding* of the pitchfork (Golubitsky & Schaeffer, 1985). When pertinent asymmetries are introduced, a curve with hysteresis emerges. It has a pair of fold points (B and C on the left panel of Fig. 7.1b) and a discontinuous jump from the fold point (limit point) B when $y$ decreases across the fold point.

As the third contribution of this chapter, we obtain the general form for an analytically solvable economic geography model (Pflüger, 2004) by explicitly expressing the variables $x$, $y$, $\epsilon$, and so on in terms of economic variables. We describe the qualitative behavior of this model using this form. Section 7.5 proposes a numerical recipe to construct the cusp catastrophe form in order to make this form applicable to general economic geography models.

This chapter is organized as follows: Sect. 7.2 introduces related studies. A two-strategy population game and its approximating games are presented in Sect. 7.3. Section 7.4 advances explicit forms of the approximating games in an economic geography model. In Sect. 7.5, a numerical recipe to construct the cusp catastrophe form is proposed.

[2] See, e.g., Baldwin et al. (2003) and Pflüger and Südekum (2008, 2011).

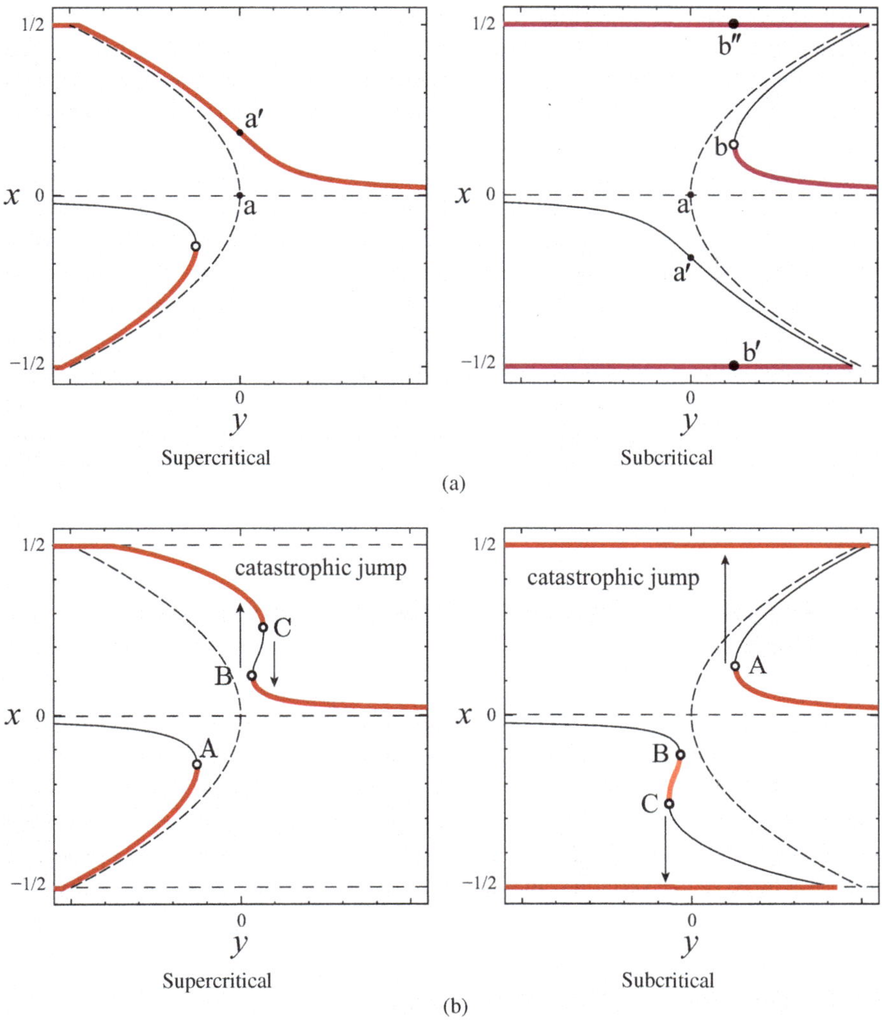


**Fig. 7.1** Illustration of perturbed supercritical and subcritical pitchfork bifurcations (dashed curve: symmetric system; black solid curve: asymmetric system (unstable); red bold solid curve: asymmetric system (stable); (∘): fold point). (**a**) Cusp catastrophe. (**b**) Behavior with hysteresis

## 7.2 Related Studies

The cusp catastrophe for a single asymmetry parameter has been studied extensively.[3] The perturbed bifurcation is explained by *catastrophe (singularity) theory* (Thom, 1968 and Poston & Stewart, 1978).

[3] See, e.g., Poston and Stewart (1978), Thompson (1982), and Guckenheimer and Holmes (1983).

An early effort at modeling the unstable behavior of stock exchanges was undertaken by Zeeman (1974). After that, several studies related to catastrophe theory have been conducted in economics as reviewed in books (Wilson, 1981 and Rosser, 2000). Berliant and Kung (2009) use the cusp catastrophe to explain the mechanism of the influence of exogenous differences in economic geography.

The Pflüger model (2004) encounters the *supercritical pitchfork*, whereas the subcritical pitchfork is observed in the Krugman model (1991), the Forslid and Ottaviano (2003) model, and other similar models. The pitchfork bifurcation of a two-place economy is vulnerable to various exogenous locational asymmetries: exogenous size asymmetry of the population of unskilled workers, together with asymmetries in transportation costs (Forslid & Ottaviano, 2003), asymmetries in regional amenities (Tabuchi & Thisse, 2002), a class of forward-looking agents subject to sufficiently small frictions in migration decisions (Oyama, 2009a,b), and location advantages (Picard & Zeng, 2010). Such asymmetries induce discontinuous transitions of industry between different locations (e.g., Forslid & Ottaviano, 2003 and Picard & Zeng, 2010).

## 7.3 Two-Strategy Population Game and Approximating Games

We present the formulation of a two-strategy population game (two-place economy) with an arbitrary number of exogenous locational asymmetry parameters (Sect. 7.3.1). We derive two forms of approximating games (Sect. 7.3.2): cusp catastrophe form and general catastrophe form in light of *universal unfolding* (Golubitsky & Schaeffer, 1985). Nonlinear effects of many parameters are projected onto only three scalar variables, making it possible to handle them analytically.

### *7.3.1 Two-Strategy Population Game*

We consider a game consisting of two actions, "1" and "2" with a continuum of homogeneous players for this game, as described by Weibull (1995) and Sandholm (2010). A population game is defined by utility (payoff) functions $(v_1, v_2)$, in which homogeneous players choose between two actions, 1 and 2. The utility functions $(v_1, v_2)$ are determined solely by the fraction $x_1$ of players choosing action 1 and $x_2$ choosing action 2. A given state in a one-dimensional simplex $(x_1, x_2) \in \{(x_1, x_2) \mid x_1 + x_2 = 1,\ x_1, x_2 \in [0, 1]\}$ is expressed as a single scalar variable:[4]

[4] See, e.g., Fujita et al. (1999, §5.5) and Baldwin et al. (2003, §2.2.1).

$$x = \frac{x_1 - x_2}{2} \quad (-1/2 \leq x \leq 1/2).$$

We employ the transportation cost parameter $Z$ (see (7.11)) as the bifurcation parameter and consider a set of $k$ asymmetry parameters $\boldsymbol{\alpha} = (\alpha_1, \ldots, \alpha_k)$, expressing locational asymmetries.

We express the indirect utility in each place using these variables as $v_i = v_i(x, Z, \boldsymbol{\alpha})$ $(i = 1, 2)$, assumed to be a sufficiently smooth function. We introduce the following *indirect utility difference function*:

$$v(x, Z, \boldsymbol{\alpha}) = v_1(x, Z, \boldsymbol{\alpha}) - v_2(x, Z, \boldsymbol{\alpha}). \tag{7.2}$$

An *equilibrium* (cf. (1.8)) is expressed by a state $x \in [-1/2, 1/2]$ that satisfies

$$\begin{cases} v(x, Z, \boldsymbol{\alpha}) \geq 0 \ \text{ if } \ x = 1/2 & (x_1 = 1, \ x_2 = 0), \\ v(x, Z, \boldsymbol{\alpha}) = 0 \ \text{ if } \ x \in (-1/2, 1/2) & (0 < x_1 < 1, \ 0 < x_2 < 1), \\ v(x, Z, \boldsymbol{\alpha}) \leq 0 \ \text{ if } \ x = -1/2 & (x_1 = 0, \ x_2 = 1). \end{cases}$$

For an interior equilibrium $x \in (-1/2, 1/2)$, we have utility equalization $v(x, Z, \boldsymbol{\alpha}) = v_1 - v_2 = 0$. For corner solutions, the inequality conditions in this equation ensure that no incentive exists for players to migrate toward the empty location.

### *7.3.2 Catastrophe Forms*

We derive a general catastrophe form at a pitchfork bifurcation point $Z = Z_c$ in the homogeneous state $x = 0$ $(x_1 = x_2)$ as follows:

- Define an incremental variable

$$y = Z - Z_c.$$

- Expand the utility difference function $v$ in (7.2) into a power series at this bifurcation point.
- Truncate higher order terms in this power series.

Then, we arrive at a general catastrophe form:

$$F(x, y, \epsilon, \eta, \xi) = Axy + Bx^3 + \epsilon + y\eta + x^2\xi = 0 \tag{7.3}$$

(e.g., Golubitsky & Schaeffer, 1985). Here, $A$ and $B$ are functions in economic parameters; and $\epsilon$, $\eta$, and $\xi$ are generalized asymmetry parameters. The form in (7.3) holds accurately for $|x| \ll 1$, $|y| \ll 1$, $|\epsilon| \ll 1$, $|\eta| \ll 1$, and $|\xi| \ll 1$ in an asymptotic sense. The effects of an arbitrary number of asymmetry parameters can be projected onto only three scalar variables $\epsilon$, $\eta$, and $\xi$ based on *imperfect bifurcation theory* (Ikeda & Murota, 2019). Even for many asymmetry parameters, the general form in (7.3) can encompass all behaviors that can arise. It embodies the usefulness of the general form.

When the condition

$$|\epsilon| \gg |y\eta + x^2\xi| \tag{7.4}$$

holds, it is possible to truncate the term $y\eta + x^2\xi$ in the general form in (7.3). In turn, one can arrive at the *cusp catastrophe* form:[5]

$$G(x, y, \epsilon) = Axy + Bx^3 + \epsilon = 0. \tag{7.5}$$

The solution curves of (7.5) are depicted in Fig. 7.1a for a homogeneous state with $\epsilon = 0$ by the dashed curves and a nonhomogeneous state with $\epsilon \neq 0$ by the solid curves. For the homogeneous state, a pitchfork bifurcation point exists at the intersection point $(x, y) = (0, 0)$ of the straight line $x = 0$ (symmetric solution) and the bifurcating curve $Ay + Bx^2 = 0$. To be consistent with the economic geography model in Sect. 7.4, we assume that

$$A < 0, \quad B \neq 0$$

(cf. Remark 7.1 below). That is, the symmetric solution $(x, y) = (0, y)$ encounters a *pitchfork bifurcation* at $y = 0$ and is stable for $y > 0$ and unstable for $y < 0$ (cf. $\partial G/\partial x \mid_{x=0} = Ay < 0$ for $y > 0$). Then, the pitchfork bifurcation can be classified into

$$\begin{cases} B < 0 & \text{supercritical,} \\ B > 0 & \text{subcritical.} \end{cases} \tag{7.6}$$

**Remark 7.1** Higher order catastrophes emerge in exceptional cases of $A = 0$ and/or $B = 0$. However, in the economic geography model in Sect. 7.4, these catastrophes do not exist because $A \neq 0$ and $B \neq 0$ hold for any values of model parameters. □

When pertinent asymmetries are introduced into the general form (7.3) as guided by Proposition 7.1, there appear three fold points (the left panel of Fig. 7.1b). As $y$ decreases, such hysteresis entails a catastrophic jump at fold point B. We investigate the interactive influence of $\epsilon$, $\eta$, and $\xi$ on the number of critical points, which are either a fold point (limit point of $y$), a stationary point, or a bifurcation point. The location $(x, y)$ of a critical point is given by the simultaneous solutions to Eq. (7.3) and the criticality condition:

[5] See, e.g., Baldwin et al. (2003) and Pflüger and Südekum (2008, 2011).

$$\frac{\partial F}{\partial x}(x, y, \epsilon, \eta, \xi) = Ay + 3Bx^2 + 2\xi x = 0. \tag{7.7}$$

(Higher order terms, such as $x\varepsilon$ and $xy\eta$, are not influential on the results and are suppressed in (7.3)). The elimination of $y$ from (7.3) and (7.7) gives a cubic polynomial equation in $x$ as

$$-2Bx^3 - (\xi - 3e\eta)x^2 - \left(\frac{2}{A}\eta\xi\right)x + \epsilon = 0. \tag{7.8}$$

Here $e = -B/A$. The number of critical points, which is equal to the number of real roots of (7.8), is given by the following lemma referring to the discriminant $D$ of (7.8):

$$D = a\epsilon^2 + b\epsilon + c,$$

being defined using a series of variables:

$$a = -108B^2 < 0, \quad b = b(\eta, \xi) = 4\,(\xi - 3e\eta)\,(\xi^2 + 12e\eta\xi + 9e^2\eta^2),$$
$$c = c(\eta, \xi) = \frac{4}{A^2}\eta^2\xi^2\,(\xi + e\eta)\,(\xi + 9e\eta)\,.$$

**Lemma 7.1** *The number $n_{\text{critical}}$ of critical points is 1, 2, or 3 according to the value of D as*

$$n_{\text{critical}} = \begin{cases} 1 & D < 0, \\ 2 & D = 0, \\ 3 & D > 0. \end{cases}$$

This lemma leads to the following proposition.

**Proposition 7.1** *Hysteresis with three fold points appears when $D > 0$ holds.*

The emergence of the hysteresis ($n_{\text{critical}} = 3$) in Proposition 7.1 includes the case of $\epsilon = 0$.

## 7.4 Cusp Catastrophe and Hysteresis in Economic Geography

The *cusp catastrophe* form and the general catastrophe form are applicable to an economic geography model with locational asymmetries in trade costs, market size, and locational amenities, and so on. As a concrete example, we consider an analytically solvable variant of the seminal model (Krugman, 1991) by Pflüger

(2004) (called the Pf model hereafter). Appendix gives details of this model, which features a *supercritical* pitchfork bifurcation. A companion paper gives a study on subcritical one (Ikeda et al., 2022).

### 7.4.1 Economic Geography Model with Locational Asymmetries

There are $x_1$ mobile (skilled) workers in place 1 and $x_2$ in place 2 ($x_1 + x_2 = 1$). In addition, a mass $L$ of (unskilled) workers is immobile inter-locationally. We use $\ell_i > 0$ to designate the exogenously given mass of immobile workers at location $i \in \{1, 2\}$ so that $L = \ell_1 + \ell_2$. These mobile and immobile workers conduct economic activities, such as production, consumption, and interplace trading of goods. There are two types of consumption goods: manufactural and agricultural goods. Preferences for manufactural and agricultural sector goods are identical among individuals.

We define the *trade freeness* of place $i$ (cf. $m(i, j) = 1$ in (2.12)):

$$\phi_i = \tau_i^{1-\sigma} \in (0, 1) \quad (\sigma > 1).$$

Here, place $j$ must export $\tau_i > 1$ units for one unit of goods to reach place $i$ ($\neq j$). The trade freeness $\phi_i$ is inversely proportional to $\tau_i$.

All individuals located in place $i$ have identical preference:

$$U_i(C_i^{\mathrm{M}}, C_i^{\mathrm{A}}) = \mu \log \left( \frac{C_i^{\mathrm{M}}}{\mu} \right) + C_i^{\mathrm{A}} + \gamma_i, \tag{7.9}$$

where $\mu > 0$ is a constant parameter; $\gamma_i \in \mathbb{R}$ signifies the utility of locational amenity; $C_i^{\mathrm{A}}$ stands for the consumption of the agricultural goods; and $C_i^{\mathrm{M}}$ denotes the consumption of the CES composite of manufactural varieties (see (2.2)).

The indirect utility $v_i$ ($i = 1, 2$) is defined as

$$v_i = w_i - \mu \log \rho_i + \bar{A} - \mu + \gamma_i.$$

Here, $\rho_i$ denotes the regional price index; and $\bar{A}(> \mu)$ signifies the initial endowment of the numéraire; and $w_i$ stands for the market equilibrium wage paid to mobile workers in (7.26) (cf. Appendix):

$$w_i = \frac{\mu}{\sigma} \left\{ \frac{1}{\Delta_i}(x_i + \ell_i) + \frac{\phi_j}{\Delta_j}(x_j + \ell_j) \right\}.$$

We introduce antisymmetric variable $x$, antisymmetric parameter vector $\boldsymbol{\alpha}$, and symmetric parameter vector $\bar{\boldsymbol{\alpha}}$, being defined respectively as

$$x = \frac{x_1 - x_2}{2}, \quad \boldsymbol{\alpha} = \begin{pmatrix} \phi \\ \ell \\ \gamma \end{pmatrix} \equiv \frac{1}{2} \begin{pmatrix} \phi_1 - \phi_2 \\ \ell_1 - \ell_2 \\ \gamma_1 - \gamma_2 \end{pmatrix}, \quad \bar{\boldsymbol{\alpha}} = \begin{pmatrix} \bar{\phi} \\ \bar{\ell} \\ \bar{\gamma} \end{pmatrix} \equiv \frac{1}{2} \begin{pmatrix} \phi_1 + \phi_2 \\ \ell_1 + \ell_2 \\ \gamma_1 + \gamma_2 \end{pmatrix}. \tag{7.10}$$

We have $\phi \in (-1/2, 1/2)$, $\bar{\phi} \in (0, 1)$, $\ell \in (-L/2, L/2)$, and $\bar{\ell} = L/2$. For example, parameter $\ell$ expresses an exogenous regional asymmetry in economic geography.

As the bifurcation parameter, we use the trade cost index (Fujita et al., 1999):

$$Z \equiv \frac{1 - \bar{\phi}}{1 + \bar{\phi}} \in (0, 1). \tag{7.11}$$

The index $Z$ is an increasing function of $\tau$.

With the use of antisymmetric variables and parameters in (7.10), the *utility difference function* $v(x, Z, \boldsymbol{\alpha}) = v_1 - v_2$ in (7.2) is expressed as

$$v(x, Z, \boldsymbol{\alpha}) = w_1 - w_2 + \frac{\mu}{\sigma - 1} \log \frac{\Delta_1}{\Delta_2} + 2\gamma, \tag{7.12}$$

where $\Delta_1 = x_1 + \phi_1 x_2$ and $\Delta_2 = x_2 + \phi_2 x_1$. In (7.12), the variables $x$, $Z$, and $\boldsymbol{\alpha}$ are contained implicitly in $w_i$ and $\Delta_i$ $(i = 1, 2)$.

## 7.4.2 General Catastrophe Form

We obtain the explicit forms of the approximating game for the two-place Pf model (Sect. 7.4.1) and simulate *perturbed pitchfork* bifurcations of this model. For this purpose, we recall the general form in (7.3):

$$F(x, y, \epsilon, \eta, \xi) = Axy + Bx^3 + \epsilon + y\eta + x^2\xi = 0 \tag{7.13}$$

and express the abstract variables $(x,\ y,\ A,\ B,\ \epsilon,\ \eta,\ \xi)$ as functions of economic parameters for the two-place Pf model.

In the homogeneous state with no local asymmetry ($\boldsymbol{\alpha} = \mathbf{0}$), the utility difference function $v(x, Z, \boldsymbol{\alpha})$ in (7.12) becomes

$$v(x, Z, \mathbf{0}) = 4x \left\{(\tilde{\mu} + \mu') - \tilde{\mu}(1 + L)Z\right\} Z; \quad \tilde{\mu} = \frac{\mu}{\sigma} > 0, \quad \mu' = \frac{\mu}{\sigma - 1} > 0.$$

The critical value (break point) of $Z$ at the pitchfork bifurcation point is obtained from $v(x, Z, \mathbf{0})/x = 0$ as

$$Z_{\rm c} = \frac{\tilde{\mu} + \mu'}{(1 + L)\tilde{\mu}} = \frac{2\sigma - 1}{(1 + L)(\sigma - 1)} > 0. \tag{7.14}$$

We define the incremental variable $y$ by setting $y = Z - Z_{\rm c}$.

The coefficients of the general form: $F(x, y, \epsilon, \eta, \xi) = 0$ in (7.13) are given as

$$A = -4(\tilde{\mu} + \mu') < 0, \qquad B = -\frac{32}{3}\mu' Z_{\mathrm{c}}^3 < 0.$$

The present bifurcation with $B < 0$ is a supercritical pitchfork by (7.6). Because $A \neq 0$ and $B \neq 0$ hold for any parameter value, we do not have higher order catastrophes, as explained in Remark 7.1 in Sect. 7.3.2.

The effects of locational asymmetry parameters $\boldsymbol{\alpha} = (\phi, \ell, \gamma)$ in (7.10) can be projected onto integrated locational asymmetry parameters $\epsilon$, $\eta$, and $\xi$ through the formulas:

$$\epsilon = C_\phi \phi + C_\ell \ell + C_\gamma \gamma, \quad \eta = D_\phi \phi + D_\ell \ell + D_\gamma \gamma, \quad \xi = E_\phi \phi + E_\ell \ell + E_\gamma \gamma;$$

$$\begin{aligned}
C_\phi &= -\tilde{\mu}(2+L)(1+Z_{\mathrm{c}}) < 0, & C_\ell &= 4\tilde{\mu} Z_{\mathrm{c}} > 0, & C_\gamma &= 2 > 0;\\
D_\phi &= -2(2+L)\tilde{\mu} - \mu' < 0, & D_\ell &= 4\tilde{\mu} > 0, & D_\gamma &= 0;\\
E_\phi &= -4\left(\tilde{\mu} + 2\mu'\right) Z_{\mathrm{c}}(1+Z_{\mathrm{c}})^2 < 0, & E_\ell &= 16\tilde{\mu} Z_{\mathrm{c}}^3 > 0, & E_\gamma &= 0.
\end{aligned}$$

The value of $\epsilon$, for example, is reduced by an increase of $\phi$ with a negative coefficient $C_\phi$ ($< 0$) and is increased by a rise in $\ell$ and/or $\gamma$ with positive coefficients $C_\ell$ and $C_\gamma$.

We investigate the agglomeration behavior expressed by the general form in (7.13). For $(\sigma, \mu, L) = (3.0, 0.8, 5.0)$ ($Z_{\mathrm{c}} = 0.4167$ from (7.14)), this form becomes

$$\begin{aligned}
\hat{F}(x, y, \phi, \ell, \gamma) = &-2.667xy - 0.3086x^3 + (-2.644\,\phi + 0.4444\,\ell + 2.000\,\gamma)\\
&+ y(-4.133\,\phi + 1.067\,\ell) + x^2(-3.568\,\phi + 0.3086\,\ell) = 0.
\end{aligned} \tag{7.15}$$

We provide an example of hysteresis with threefold points denoted by (∘) in Fig. 7.2 for $(\phi, \ell, \gamma) = (-0.11, -0.30, -0.040)$ and $(\sigma, \mu, L) = (4.0, 0.8, 4.0)$, which give $(\epsilon, \eta, \xi) \approx (0.0016, 0.053, 0.23)$. The condition $|\epsilon| \gg |y\eta + x^2\xi|$ in (7.4) for reducibility to the cusp catastrophe is violated because $|\epsilon| \ll |\eta|$ and $|\epsilon| \ll |\xi|$. Accordingly, it is necessary to resort to the general form in (7.13), i.e., (7.15). Because the set of asymmetric parameters $(\epsilon, \eta, \xi)$ satisfies $D \approx 3.0 \times 10^{-5} > 0$, three critical points are ensured to exist by Proposition 7.1.

### 7.4.3 Cusp Catastrophe Form

We deal with a simple but commonly encountered case of changing the difference $\ell$ in the exogenously given mass of immobile workers (Forslid & Ottaviano, 2003).

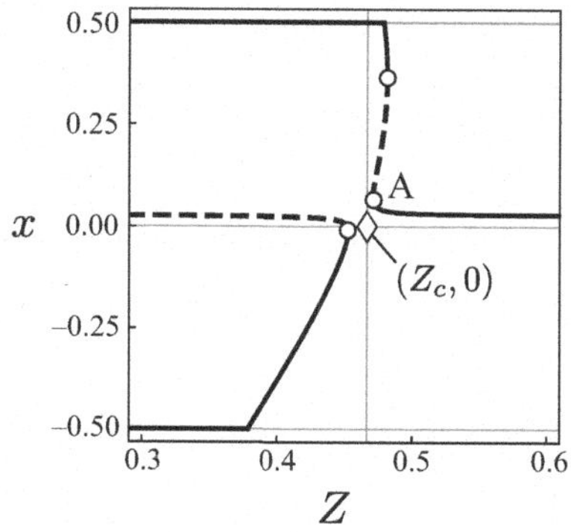


**Fig. 7.2** Hysteresis in a solution curve for the Pf model with parameter values of $(\sigma, \mu, L) = (4.0, 0.8, 4.0)$ and $\boldsymbol{\alpha} = (\phi, \ell, \gamma) = (-0.11, -0.30, -0.040)$ (Ikeda et al., 2022) (solid curve: stable; dashed curve: unstable; (∘): fold point; (⋄): pitchfork bifurcation point)

We set $\ell > 0$ and $\phi = \gamma = 0$ to express an exogenous advantage for place 1. Then, under conditions $\ell \ll 1$, $|x| \ll 1$, and $|y| \ll 1$, the general form in (7.15) reduces to a cusp catastrophe form[6]

$$\hat{G}(x, y, \ell) = -2.667xy - 0.3086x^3 + 0.4444\,\ell = 0$$

(cf. (7.5)). Figure 7.3 presents the curves expressed by this form for $\ell = 0.0025$ and 0.025. The curves of the original game (black curves) and of this form (red curves) are in good agreement. Such agreement demonstrates the usefulness of the cusp catastrophe form. Each aloof curve in $x < 0$ has only onefold point at $(x_{\text{fold}}, Z_{\text{fold}})$ shown by (∘), whereas each main curve in $x > 0$ has no fold point. A unique interior stable equilibrium exists for $Z > Z_{\text{fold}}$.

This perturbed behavior presents the following economic implications. If trade costs are initially very high ($Z > Z_{\text{fold}}$), place 1 has more industry than place 2 ($x > 0$). When $Z$ falls below $Z_{\text{fold}}$, a further agglomeration to place 1 is a smooth and gradual process until eventually place 1 gathers all mobile population (see the upper stable solid curve with a kink in Fig. 7.3).

## 7.5 Numerical Recipe to Obtain the Cusp Catastrophe Form

In a quantitative application of the cusp catastrophe form in (7.5), we must evaluate the values of coefficients $A$ and $B$. In the analytically solvable model presented in Sect. 7.4.1, we can analytically evaluate these coefficients. However, such analytical evaluation is not always possible in general economic geography models. To alleviate this shortcoming, we propose a numerical method to evaluate the values of these coefficients.

[6] We have $(\epsilon, y\eta, x^2\xi) = (0.4444\,\ell, 1.067\,y\ell, 0.3086x^2\ell)$; accordingly, the condition $|\epsilon| \gg |y\eta + x^2\xi|$ in (7.4) for the applicability of the cusp catastrophe form is satisfied for $|x| \approx 0$ and $|y| \approx 0$.

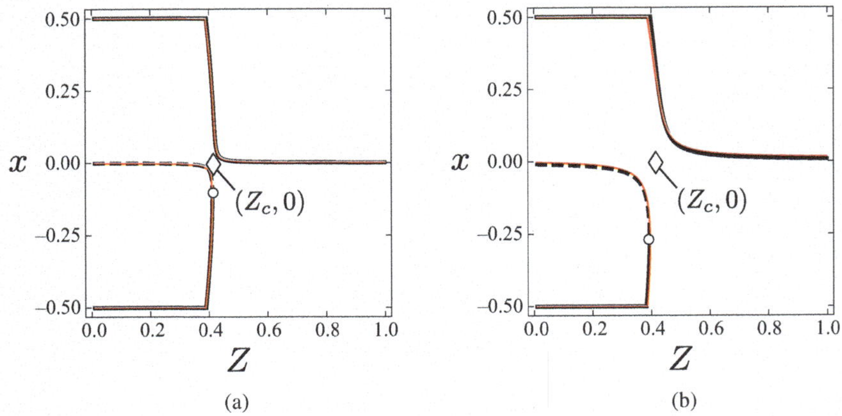


**Fig. 7.3** Solution curves of original and approximating games for the Pf model for $(\sigma, \mu, L) = (3.0, 0.8, 5.0)$ and $(\phi, \gamma) = (0, 0)$ (Ikeda et al., 2022) (solid curve: stable; dashed curve: unstable; black curve: original game; red thin curve: approximating game; (○): fold point; (◇): pitchfork bifurcation point). (**a**) $\ell = 0.0025$. (**b**) $\ell = 0.025$

In general economic geography models with locational asymmetries, we can obtain solution curves through numerical analysis, as conducted in the studies of locational asymmetries (see Sect. 7.2). Then, we can numerically construct the cusp catastrophe form in such models as described below.

**Single Asymmetry Parameter**

Recall that a solution curve in the neighborhood of the pitchfork bifurcation point has a fold point (a local maximum or minimum point of the transportation cost parameter $Z$ or $y$) for a given generalized asymmetry parameter $\epsilon = \hat{\epsilon}$, as depicted in Fig. 7.1a. Then, as explained in the following proposition, it is possible to evaluate the values of coefficients $A$ and $B$ of the cusp catastrophe form.

**Lemma 7.2** *Using the location $(x^{\text{fold}}, y^{\text{fold}})$ of the fold point for a given $\epsilon = \hat{\epsilon}$ $(\sim 0)$, we can approximate the coefficients as*

$$A = \frac{-3\hat{\epsilon}}{2x^{\text{fold}}\, y^{\text{fold}}}, \qquad B = \frac{\hat{\epsilon}}{2(x^{\text{fold}})^3}. \tag{7.16}$$

***Proof*** The location of the fold point is obtained by simultaneously solving $G(x, y, \epsilon) = 0$ and $\partial G/\partial x(x, y, \epsilon) = 0$ as (Koiter, 1945 and Ikeda & Murota, 2019, §3.2.3)

$$(x^{\text{fold}}, y^{\text{fold}}) = \left(\left(\frac{\epsilon}{2B}\right)^{1/3}, -\frac{3B}{A}\left(\frac{\epsilon}{2B}\right)^{2/3}\right),$$

which relates this location with the generalized asymmetry parameter $\epsilon \approx 0$. This relation is solved for $A$ and $B$ for $\epsilon = \hat{\epsilon}$ to obtain (7.16). □

For a single asymmetry parameter, say, $\epsilon = C_j\alpha_j$ for some $j$ with unknown $C_j$, we can numerically obtain location $(x_j^{\text{fold}}, y_j^{\text{fold}})$ for a given $\alpha_j$. Then, by Lemma 7.2, we can construct a "scaled" cusp catastrophe form, which differs from the original form by a scalar scaling constant $C_j$ as

$$A_j xy + B_j x^3 + \alpha_j = 0 \tag{7.17}$$

with numerically computed scaled coefficients, obtained using (7.16) and $\epsilon = C_j\alpha_j$, as

$$A_j = \frac{A}{C_j} = \frac{-3\alpha_j}{2x_j^{\text{fold}}\, y_j^{\text{fold}}}, \qquad B_j = \frac{B}{C_j} = \frac{\alpha_j}{2(x_j^{\text{fold}})^3}. \tag{7.18}$$

Such scaling is not influential on the solution curves parameterized by $\alpha_j$.

**Multiple Asymmetry Parameters**

We extend this scaled form for a single asymmetry parameter to multiple asymmetry parameters $\boldsymbol{\alpha} = (\alpha_1, \alpha_2, \ldots, \alpha_k)$ with the generalized asymmetry parameter:

$$\epsilon = \sum_{j=1}^{k} C_j\, \alpha_j .$$

For each given $\alpha_j$, we can compute $A_j$ and $B_j$ using (7.18).

**Proposition 7.2** *For the multiple asymmetry parameters $\epsilon = \sum_{j=1}^{k} C_j\, \alpha_j$, we have a scaled cusp catastrophe form:*

$$A_1 xy + B_1 x^3 + \alpha_1 + B_1 \sum_{j=2}^{k} \frac{\alpha_j}{B_j} = 0. \tag{7.19}$$

***Proof*** From (7.18), we have the relation for the scaling $B = C_1 B_1 = C_2 B_2 = \cdots$, which gives $C_j/C_1 = B_1/B_j$. Then, a scaled asymmetry parameter is expressed as

$$\frac{\epsilon}{C_1} = \frac{1}{C_1}\sum_{j=1}^{k} C_j\alpha_j = \sum_{j=1}^{k} \frac{C_j}{C_1}\alpha_j = \alpha_1 + B_1 \sum_{j=2}^{k} \frac{\alpha_j}{B_j}.$$

The use of this equation in (7.17) for $j = 1$, $A_1 xy + B_1 x^3 + \frac{\varepsilon}{C_1} = 0$, gives (7.19).

□

**Example 7.1** We numerically compute $(x^{\text{fold}}, y^{\text{fold}})$ in two different cases: (1) $\alpha_1 = \ell = 0.0025 \neq 0$ and $\phi = \gamma = 0$ in Fig. 7.3a with $(x^{\text{fold}}, y^{\text{fold}}) = (-0.1225, 0.4116 - 0.4167)$. (2) $\alpha_2 = \gamma = 0.00111$ ($\ell = \phi = 0$) with $(x^{\text{fold}}, y^{\text{fold}}) = (-0.1570, -0.00809)$. Using (7.18), we can evaluate $A_1 = -6.00$, $B_1 = -0.680$, and $B_2 = -0.1434$. Then, the scaled cusp catastrophe in (7.19) in the present case is constructed as

$$-6.00xy - 0.680x^3 + \ell + 4.74\gamma = 0.$$

This form produces almost the same curve as in Fig. 7.3a for $\alpha_1 = \ell = 0.0025 \neq 0$ and $\phi = \gamma = 0$ and is quite accurate. □

## Appendix: Details of Analytically Solvable Model

In the Pf model, the derivation takes a similar but different course than the FO model in Sect. 2.2. The budget constraint for a worker is given as an extension of (2.3) by adding $\bar{A}$ that signifies the initial endowment of the numéraire as

$$p_i^{\mathrm{A}} C_i^{\mathrm{A}} + \sum_{j=1}^{2} \int_0^{n_j} p_{ji}(k) q_{ji}(k) dk = Y_i + \bar{A}. \tag{7.20}$$

Here, $(\cdot)_i$ denotes a variable in place $i$; $p_i^{\mathrm{A}}$ stands for the price of agricultural sector goods; $p_{ji}(k)$ denotes the price of a variety $k$ in place $i$ produced in place $j$; $Y_i$ expresses the income of an individual; and we assume that $\bar{A} > \mu$ to ensure the consumption of both types of goods.

An individual in place $i$ maximizes the utility in (7.9) subject to (7.20), which yields the following demand functions:

$$C_i^{\mathrm{A}} = \frac{Y_i + \bar{A}}{p_i^{\mathrm{A}}} - \mu, \quad C_i^{\mathrm{M}} = \mu \frac{p_i^{\mathrm{A}}}{\rho_i}, \quad q_{ji}(k) = \mu p_i^{\mathrm{A}} \frac{p_{ji}(k)^{-\sigma}}{\rho_i^{1-\sigma}}. \tag{7.21}$$

Therein, $\rho_i$ denotes the price index of the differentiated product, which is given as

$$\rho_i = \left( \sum_{j=1}^{2} \int_0^{n_j} p_{ji}(k)^{1-\sigma} dk \right)^{1/(1-\sigma)}. \tag{7.22}$$

Firms in the agricultural sector produce the agricultural goods under perfect competition and constant returns to scale, involving a marginal input of one unit of immobile labor. The agricultural goods are traded freely among locations so that the nominal labor wage is equalized in the two locations and is chosen as the numéraire. Then, in equilibrium, we have $p_i^{\mathrm{A}} = 1$ $(i = 1, 2)$. The wage of unskilled workers is unity.

A monopolistically competitive sector produces the manufactural goods. Production of various manufactural goods in place $i$ is assumed to involve a fixed input of one mobile worker and a marginal input of $\beta$ units of unskilled labor. Therefore, the total cost of production of $m_i$ units is given as $w_i + \beta m_i$, where $w_i$ is the reward (market equilibrium wage) for a skilled worker. Labor market clearing implies that $n_i = x_i$ ($i = 1, 2$) in the market equilibrium.[7]

The profit of a manufactural sector firm is equal to

$$\Pi_i(k) = \sum_{j=1}^{2} \left(p_{ij}(k) - \tau_{ij}\beta\right) q_{ij}(k) \left(x_j + \ell_j\right) - w_i.$$

Here, $\tau_{ij} = \tau_j \in (0, +\infty)$ for $i \neq j$ and $\tau_{ij} = 1$ for $i = j$ (Sect. 7.4.1). Because we have a continuum of firms, each firm is negligible in the sense that its action does not affect the market. Therefore, the first-order condition for profit maximization yields the following optimal prices:

$$p_{ii} = \frac{\sigma\beta}{\sigma - 1}, \qquad p_{ij} = \frac{\sigma\beta}{\sigma - 1}\tau_j. \tag{7.23}$$

The substitution of (7.23) into (7.22), together with $n_j = x_j$, yields the following price index:

$$\rho_i = \frac{\sigma\beta}{\sigma - 1}\left(x_i + \phi_i x_j\right)^{\frac{1}{1-\sigma}} = \frac{\sigma\beta}{\sigma - 1}\Delta_i^{\frac{1}{1-\sigma}}. \tag{7.24}$$

Here, $\Delta_i = x_i + \phi_i x_j$.

The spatial distribution of mobile workers $(x_1, x_2)$ is given in the market equilibrium. Given free entry and exit in the manufactural sector, firms earn zero profits at equilibrium, which means that operating profits are absorbed completely by the nominal wage bill as

$$w_i = \sum_{j=1}^{2} \left(p_{ij}(k) - \tau_{ij}\beta\right) q_{ij}(k) \left(x_j + \ell_j\right). \tag{7.25}$$

Using the demands in $q_{ij}$ from (7.21), together with (7.23), (7.24), and (7.25), we reach the market equilibrium wage:

$$w_i = \frac{\mu}{\sigma}\left\{\frac{1}{\Delta_i}(x_i + \ell_i) + \frac{\phi_j}{\Delta_j}(x_j + \ell_j)\right\}. \tag{7.26}$$

---

[7] Wage equalization holds as long as the agricultural sector goods are produced in all locations. We assume that this condition is satisfied and that $m_i n_i < l_i$ ($i = 1, 2$) holds.

## References


Baldwin, R., Forslid, R., Martin, P., Ottaviano, G., & Robert-Nicoud, F. (2003). *Economic geography and public policy*. Princeton University Press.

Berliant, M., & Kung, F.-C. (2009). Bifurcations in regional migration dynamics. *Regional Science and Urban Economics, 39,* 714–720.

Forslid, R., & Ottaviano, G. I. P. (2003). An analytically solvable core-periphery model. *Journal of Economic Geography, 3,* 229–240.

Fujita, M., Krugman, P., & Venables, A. J. (1999). *The spatial economy: Cities, regions, and international trade*. MIT Press .

Golubitsky, M., & Schaeffer, D. G. (1985). *Singularities and groups in bifurcation theory* (Vol. I). Springer.

Guckenheimer, J., & Holmes, P. (1983). *Nonlinear oscillations, dynamical systems, and bifurcations of vector fields*. Springer.

Ikeda, K., & Murota, K. (2019). *Imperfect bifurcation in structures and materials* (3rd ed.). Springer.

Ikeda, K., Takayama, Y., Gaspar, J. M., & Osawa, M. (2022). Perturbed cusp catastrophe in a population game: Spatial economics with locational asymmetries. *Journal of Regional Science, 62*(4), 961–980.

Koiter, W. T. (1945). *On the stability of elastic equilibrium.* Dissertation, Delft. English translation: NASA Technical Translation F10: 833 (1967).

Krugman, P. (1991). Increasing returns and economic geography. *Journal of Political Economy, 99,* 483–499.

Oyama, D. (2009a). History versus expectations in economic geography reconsidered. *Journal of Economic Dynamics and Control, 33,* 394–408.

Oyama, D. (2009b). Agglomeration under forward-looking expectations: Potentials and global stability. *Regional Science and Urban Economics, 39,* 696–713.

Picard, P. M., & Zeng, D.-Z. (2010). A harmonization of first and second natures. *Journal of Regional Science, 50*(5), 973–994.

Pflüger, M. (2004). A simple, analytically solvable, Chamberlinian agglomeration model. *Regional Science and Urban Economics, 34*(5), 565–573.

Pflüger, M., & Südekum, J. (2008). A synthesis of footloose-entrepreneur new economic geography models: when is agglomeration smooth and easily reversible? *Journal of Economic Geography, 8*(1), 39–54.

Pflüger, M., & Südekum, J. (2011). On pitchforks and tomahawks. *Journal of Regional Science, 51*(2), 292–298.

Poston, T., & Stewart, I. (1978). *Catastrophe theory and its applications*. Pitman.

Rosser Jr., J. B. (2000). *From catastrophe to chaos: A general theory of economic discontinuities* (2nd ed.). Kluwer.

Rosser Jr., J. B. (2007). The rise and fall of catastrophe theory applications in economics: Was the baby thrown out with the bathwater? *Journal of Economic Dynamics and Control, 31,* 3255–3280.

Sandholm, W. H. (2010). *Population games and evolutionary dynamics*. MIT Press.

Tabuchi, T., & Thisse, J.-F. (2002). Taste heterogeneity, labor mobility and economic geography. *Journal of Development Economics, 69*(1), 155–177.

Thom, R. (1968). Topological methods in biology. *Biology, 8,* 313–335.

Thompson, J. M. T. (1982). *Instabilities and catastrophes in science and engineering*. Wiley.

Weibull, J. W. (1995). *Evolutionary game theory*. MIT Press.

Wilson, A. G. (1981). *Catastrophe theory and bifurcation: Applications to Urban and regional systems*. Croom Helm and University of California Press.

Zeeman, E. C. (1974). On the unstable behaviour of stock exchanges. *Journal of Mathematical Economics, 1*(1), 39–49.

# Chapter 8
# Analysis of Break Points

**Abstract** As another independent topic on direct bifurcation, this chapter deals with the break point, which denotes the value of the transportation cost at the onset of breaking uniformity through bifurcation. It occurs in symmetric spatial platforms, such as the racetrack economy and square and hexagonal lattices. Analytical formulas for this point are derived using replicator and logit dynamics and an analytically solvable economic geography model by Forslid and Ottaviano (J Econ Geogr 3:229–240, 2003). The formulas for the racetrack economy and the square lattice are synthesized. The formulas for a hexagonal lattice are advanced independently. We employ these formulas in describing numerical agglomeration behaviors for the lattices. Chapters 3, 4, and 5 are foundations for the bifurcation of the spatial platforms. This chapter covers only direct bifurcation, whereas Chap. 12 will study the break points of secondary and further bifurcations.

**Keywords** Analytically solvable model · Bifurcation · Break point · Economic geography model · Hexagonal lattice · Logit dynamics · Racetrack economy · Replicator dynamics · Square lattice

## 8.1 Introduction

The value of the transportation cost at the onset of a breaking bifurcation in the two-place economy is called the *break point* (Fujita et al., 1999). This point indexes the functioning of the investment in transportation infrastructure. Formulas to determine the break point for a class of footloose-entrepreneur models are given in Pflüger and Südekum (2008). Tabuchi and Thisse (2011) show the occurrence of spatial period doubling in racetrack economy by investigating the break point.

In the analytical study of the break point, among various models, we employ the Forslid and Ottaviano (2003) for its analytical tractability. Through the numerical



K. Ikeda, Y. Takayama, *Bifurcation Theory of Pattern Formation in Economic Geography*, Interdisciplinary Applied Mathematics 61,
https://doi.org/10.1007/978-3-031-71686-7_8

analysis of this model, we ensure the usefulness of the formulas for the break point. The present methodology is extendable to other models.[1]

As spatial platforms, we consider the racetrack economy and square and hexagonal lattices, both with periodic boundary conditions. These platforms have a uniform state that undergoes direct bifurcations resulting in various bifurcating patterns (Chaps. 3, 4, and 5).

We present synthetic formulas for break points of the racetrack economy and the square lattice, both in replicator dynamics. This alludes to the racetrack analogy in describing the agglomeration behavior of the square lattice (Chap. 13). Using the logit dynamics, we derive the formulas for break points for bifurcations of a $6 \times 6$ hexagonal lattice and employ them to determine the predominant bifurcation mode.

This chapter is organized as follows: Sect. 8.2 presents economic modeling. Jacobian matrices and their eigenvectors are studied in Sect. 8.3. Formulas for break points for the racetrack economy and the square lattice are advanced in Sect. 8.4 and for the hexagonal lattice in Sect. 8.5.

## 8.2 Economy Modeling

We use replicator and logit dynamics and employ the Forslid and Ottaviano model (2003) for its analytical tractability.

**Two Types of Dynamics**

The static governing equation is

$$\boldsymbol{F}(\boldsymbol{\lambda}, \tau) = \boldsymbol{0}, \tag{8.1}$$

where $\boldsymbol{F} = (F_i) \in \mathbb{R}^N$ is a sufficiently smooth function in $\boldsymbol{\lambda} = (\lambda_i) \in \mathbb{R}^N$ and the transportation cost $\tau \in \mathbb{R}$.

In defining the governing equation, we use two types of dynamics: The *replicator dynamics* in (1.7):

$$F_i(\boldsymbol{\lambda}, \tau) = (v_i(\boldsymbol{\lambda}, \tau) - \bar{v}(\boldsymbol{\lambda}, \tau))\lambda_i \quad (i = 1, \ldots, N) \tag{8.2}$$

($\bar{v} = \sum_{i=1}^{N} \lambda_i v_i$) and the *logit dynamics* in (1.9):

$$\boldsymbol{F}(\boldsymbol{\lambda}, \tau) = \boldsymbol{P}(\boldsymbol{v}(\boldsymbol{\lambda}, \tau)) - \boldsymbol{\lambda}. \tag{8.3}$$

Here, $\boldsymbol{P}(\boldsymbol{v}) = (P_1, \ldots, P_N)^\top$ is defined by the *logit choice function* in (1.10):

[1] See, e.g., the application to the Pflüger (2004) by Akamatsu et al. (2012), and to the Murata and Thisse model (2005) by Takayama et al. (2020).

$$P_i(\boldsymbol{v}) = \frac{\exp(\theta v_i)}{\sum_{j=1}^{N} \exp(\theta v_j)}.$$

**Analytically Solvable Model**

We briefly introduce an analytically solvable model by Forslid and Ottaviano (2003) (see Sect. 2.2 for more details).

We recall the *trade freeness* in (2.11):

$$\phi = \exp[-\tau(\sigma - 1)\tilde{L}] \quad (0 < \phi < 1), \tag{8.4}$$

which is inversely proportional to the transportation cost $\tau > 0$. Here, $\tilde{L}$ is the distance unit and $\sigma > 1$ is the constant elasticity of substitution between any two varieties. We recall the *spatial discount factor* in (2.12):

$$d_{ij} \equiv \tau_{ij}^{1-\sigma} = \phi^{m(i,j)}, \tag{8.5}$$

where $m(i, j)$ is an integer road distance measure between places $i$ and $j$. This factor represents friction between places $i$ and $j$ that decays in proportion to the transportation distance. The *spatial discount matrix* $D = (d_{ij})$ plays a key role in the derivation of formulas for break points.

We recall the indirect utility $v_i$ in (2.17):

$$v_i = \frac{\mu}{\sigma - 1} \log \Delta_i + \log w_i. \tag{8.6}$$

Here, $\mu$ is a constant parameter expressing the expenditure share of manufacturing sector goods; $\Delta_i = \sum_{j=1}^{N} d_{ji}\lambda_j$ denotes the market size of the M-sector in place $i$; and $w_i$ is the *market equilibrium wage* that can be expressed by (2.16):[2]

$$w_i = \frac{\mu}{\sigma} \sum_{j=1}^{N} \frac{d_{ij}}{\Delta_j}(w_j\lambda_j + 1). \tag{8.7}$$

We can write this equation as $\boldsymbol{w} = \frac{\mu}{\sigma} D\Delta^{-1}(\Lambda \boldsymbol{w} + \mathbf{1})$ and solve it for $\boldsymbol{w}$ as

$$\boldsymbol{w} = \frac{\mu}{\sigma}\left(I - \frac{\mu}{\sigma} D\Delta^{-1}\Lambda\right)^{-1} D\Delta^{-1}\mathbf{1} \tag{8.8}$$

with the notations:

[2] Unskilled workers are immobile and equally distributed across all places with the unit density (i.e., $L = 1 \times N$). This choice of the total population $L$ of unskilled workers is not influential on the results as the payoff is linear in $L$ (see Gaspar et al., 2020, p. 9 for details).

$$\boldsymbol{w} = (w_i), \quad \Delta = \mathrm{diag}(\Delta_1, \ldots, \Delta_N), \quad D = (d_{ij}),$$

$$\Lambda = \mathrm{diag}(\lambda_1, \ldots, \lambda_N), \quad \mathbf{1} = (1, \ldots, 1)^\top, \quad I = \mathrm{diag}(1, \ldots, 1).$$

The use of (8.8) in (8.6) gives the indirect utility function vector $\boldsymbol{v} = \boldsymbol{v}(\boldsymbol{\lambda}, \tau) = (v_1(\boldsymbol{\lambda}, \tau), \ldots, v_N(\boldsymbol{\lambda}, \tau))^\top$.

To avoid the nonexistence of a break point, we assume the *no-black-hole condition* (Forslid & Ottaviano, 2003):

$$\mu < \sigma - 1. \tag{8.9}$$

**Remark 8.1** Anderson and van Wincoop (2004), for instance, find that $\sigma$ is likely to range between 5 and 10, thereby satisfying the no-black-hole condition (8.9). □

## 8.3 Jacobian Matrices and Their Eigenvectors

We are interested in the break point at the uniform state:

$$\boldsymbol{\lambda} = \boldsymbol{\lambda}^* = \frac{1}{N}\mathbf{1} = \frac{1}{N}(1, \ldots, 1)^\top$$

and consider the *Jacobian matrix*: $J(\boldsymbol{\lambda}^*, \tau) = \frac{\partial \boldsymbol{F}}{\partial \boldsymbol{\lambda}}(\boldsymbol{\lambda}^*, \tau)$ of the governing equation at this state. This matrix is related to the Jacobian matrix $V(\boldsymbol{\lambda}^*) = \partial \boldsymbol{v}/\partial \boldsymbol{\lambda}(\boldsymbol{\lambda}^*, \tau)$ of the indirect utility as (see Appendix)

$$J(\boldsymbol{\lambda}^*, \tau) = \begin{cases} \left(\frac{1}{N}I - \frac{1}{N^2}\mathbf{1}\mathbf{1}^\top\right) V(\boldsymbol{\lambda}^*) - \frac{\bar{v}}{N}\mathbf{1}\mathbf{1}^\top, & \text{replicator dynamics,} \\ -\frac{\theta}{N^2}\mathbf{1}\mathbf{1}^\top V(\boldsymbol{\lambda}^*) + \frac{\theta}{N} V(\boldsymbol{\lambda}^*) - I, & \text{logit dynamics.} \end{cases} \tag{8.10}$$

The other Jacobian matrix $V(\boldsymbol{\lambda}^*)$ enjoys an explicit form:

$$V(\boldsymbol{\lambda}^*) = N\left[\mu' \frac{D}{d} + \left(I - \tilde{\mu}\frac{D}{d}\right)^{-1} \cdot \frac{D}{d}\left(\tilde{\mu} I - \frac{D}{d}\right)\right]; \tag{8.11}$$

$$d = d(\phi) = \sum_{j=1}^{N} d_{1j}, \quad \tilde{\mu} = \frac{\mu}{\sigma} \quad (0 < \tilde{\mu} < 1), \quad \mu' = \frac{\mu}{\sigma - 1} > 0. \tag{8.12}$$

Thus, (8.10) expresses the difference in dynamics. The difference in spatial platforms leads to different $D$ and is reflected on $V(\boldsymbol{\lambda}^*)$ through (8.11).

Using the symmetry group $G$ and its $N \times N$ representation matrices $T(g)$ ($g \in G$),[3] we can express the symmetries of the functions $\boldsymbol{F}$ and $\boldsymbol{v}$ by the *equivariance* conditions:

$$T(g)\boldsymbol{F}(\boldsymbol{\lambda}, \tau) = \boldsymbol{F}(T(g)\boldsymbol{\lambda}, \tau) \quad (g \in G),$$
$$T(g)\boldsymbol{v}(\boldsymbol{\lambda}, \tau) = \boldsymbol{v}(T(g)\boldsymbol{\lambda}, \tau) \quad (g \in G).$$

Then, the Jacobian matrices $J(\boldsymbol{\lambda}^*, \tau)$ and $V(\boldsymbol{\lambda}^*, \tau)$ at the uniform state $\boldsymbol{\lambda} = \boldsymbol{\lambda}^*(T(g)\boldsymbol{\lambda}^* = \boldsymbol{\lambda}^*)$ satisfy the commutativity with $T(g)$ as follows:

$$T(g)J(\boldsymbol{\lambda}^*, \tau) = J(\boldsymbol{\lambda}^*, \tau)T(g) \quad (g \in G),$$
$$T(g)V(\boldsymbol{\lambda}^*, \tau) = V(\boldsymbol{\lambda}^*, \tau)T(g) \quad (g \in G).$$

The spatial discount matrix $D$ also has the commutativity:

$$T(g)D = DT(g) \quad (g \in G).$$

These commutativities lead to the following lemma.

**Lemma 8.1** *$J(\boldsymbol{\lambda}^*)$, $V(\boldsymbol{\lambda}^*)$, and $D$ have the same set of eigenvectors $\boldsymbol{\eta}_i$ ($i = 1, \ldots, N$).*

***Proof*** Ikeda and Murota (2014, §13.1) assess the existence of the same set of eigenvectors for any matrix $Q$ satisfying commutativity $T(g)Q = QT(g)$ ($g \in G$). □

## 8.4 Racetrack Economy and Square Lattice with Replicator Dynamics

We develop formulas for the *break points* for the racetrack economy and the square lattice with the replicator dynamics. The method in Akamatsu et al. (2012) for the racetrack economy is adapted to the present case.

The total degree of freedom $N$ is given by

$$N = \begin{cases} n, & \text{racetrack economy}, \\ n^2, & \text{square lattice}. \end{cases}$$

Here, $n$ characterizes the size of the economy. We choose $n = 4m$ in the derivation of the formulas and $n = 4m - 2$ for comparison ($m = 1, 2, \ldots$).

[3] Equation (3.2) gives the concrete form of $T(g)$ for the racetrack economy.

**Assumption 8.1** $n = 4m$ $(m = 1, 2, \ldots)$.

The total length $\mathscr{L}$ of the road in the racetrack economy, as well as the side length of the square lattice, is set to unity ($\mathscr{L} = 1$). Neighboring places are connected by an interplace road of length $\tilde{L} = 1/n$.

### 8.4.1 Spatial Discount Matrix

The *spatial discount matrix* $D_{\mathrm{R}}$ for the racetrack economy is related to the spatial discount matrix $D_{\mathrm{L}}$ for the square lattice. The matrix $D_{\mathrm{R}}$ enjoys a simple formula:

$$D_{\mathrm{R}} = (\phi^{m(i,j)})$$

with the trade freeness $\phi$ in (8.4) and the integer road distance measure $m(i, j)$ between places $i$ and $j$, being defined as

$$m(i, j) = \min(|i - j|, n - |i - j|) \quad (i, j = 1, \ldots, n).$$

The concrete forms of $D_{\mathrm{R}}$ for $n = 2, 4$, and 8 are given, respectively, as

$$D_{\mathrm{R}} = \begin{bmatrix} 1 & \phi \\ \phi & 1 \end{bmatrix}, \quad \begin{bmatrix} 1 & \phi & \phi^2 & \phi \\ \phi & 1 & \phi & \phi^2 \\ \phi^2 & \phi & 1 & \phi \\ \phi & \phi^2 & \phi & 1 \end{bmatrix}, \quad \begin{bmatrix} 1 & \phi & \phi^2 & \phi^3 & \phi^4 & \phi^3 & \phi^2 & \phi \\ \phi & 1 & \phi & \phi^2 & \phi^3 & \phi^4 & \phi^3 & \phi^2 \\ \phi^2 & \phi & 1 & \phi & \phi^2 & \phi^3 & \phi^4 & \phi^3 \\ \phi^3 & \phi^2 & \phi & 1 & \phi & \phi^2 & \phi^3 & \phi^4 \\ \phi^4 & \phi^3 & \phi^2 & \phi & 1 & \phi & \phi^2 & \phi^3 \\ \phi^3 & \phi^4 & \phi^3 & \phi^2 & \phi & 1 & \phi & \phi^2 \\ \phi^2 & \phi^3 & \phi^4 & \phi^3 & \phi^2 & \phi & 1 & \phi \\ \phi & \phi^2 & \phi^3 & \phi^4 & \phi^3 & \phi^2 & \phi & 1 \end{bmatrix}.$$

Owing to the symmetry, $D_{\mathrm{L}}$ is given in terms of $D_{\mathrm{R}}$ for the same size $n$ as

$$D_{\mathrm{L}} = D_{\mathrm{R}} \otimes D_{\mathrm{R}}, \tag{8.13}$$

where $\otimes$ is the tensor product. For example, $D_{\mathrm{L}}$ for the square lattice with the size $N = n^2 = 4$ is

$$D_{\mathrm{L}} = D_{\mathrm{R}} \otimes D_{\mathrm{R}} = \begin{bmatrix} 1 & \phi \\ \phi & 1 \end{bmatrix} \otimes \begin{bmatrix} 1 & \phi \\ \phi & 1 \end{bmatrix} = \begin{bmatrix} 1 & \phi & \phi & \phi^2 \\ \phi & 1 & \phi^2 & \phi \\ \phi & \phi^2 & 1 & \phi \\ \phi^2 & \phi & \phi & 1 \end{bmatrix}. \tag{8.14}$$

### 8.4.2 Formulas for Break Point

*Break point* at the uniform state $\boldsymbol{\lambda}^*$ is given by a zero eigenvalue of the Jacobian matrix $J(\boldsymbol{\lambda}^*)$ in (8.10):

$$J(\boldsymbol{\lambda}^*) = \left(\frac{1}{N}I - \frac{1}{N^2}\mathbf{1}\mathbf{1}^\top\right) V(\boldsymbol{\lambda}^*) - \frac{\bar{v}}{N}\mathbf{1}\mathbf{1}^\top. \tag{8.15}$$

Equation (8.11), which gives $V(\boldsymbol{\lambda}^*)$ in this equation, is rewritten using $\hat{D} = D/d$ as

$$V(\boldsymbol{\lambda}^*) = N\left[\mu'\hat{D} + \left(I - \tilde{\mu}\hat{D}\right)^{-1} \cdot \hat{D}\left(\tilde{\mu}I - \hat{D}\right)\right]. \tag{8.16}$$

By Lemma 8.1, the matrices $J(\boldsymbol{\lambda}^*)$, $V(\boldsymbol{\lambda}^*)$, and $D = d\hat{D}$ have the common eigenvector $\boldsymbol{\eta}$. It depends on spatial platforms as follows:

$$\boldsymbol{\eta} = \begin{cases} \boldsymbol{\eta}_{\mathrm{R1}} & \text{(racetrack economy)}, \\ \boldsymbol{\eta}_{\mathrm{H1}} & \text{(half doubling of square lattice)}, \\ \boldsymbol{\eta}_{\mathrm{F1}} & \text{(full doubling of square lattice)}. \end{cases} \tag{8.17}$$

Here, $\boldsymbol{\eta}_{\mathrm{R1}}$, $\boldsymbol{\eta}_{\mathrm{H1}}$, and $\boldsymbol{\eta}_{\mathrm{F1}}$ are given, respectively, by (13.6), (13.10), and (13.13):

$$\boldsymbol{\eta}_{\mathrm{R1}} = (1, -1, \ldots, 1, -1) \in \mathbb{R}^N,$$

$$\boldsymbol{\eta}_{\mathrm{H1}} = \underbrace{(\boldsymbol{h}_1, \ldots, \boldsymbol{h}_1)}_{n/2 \text{ times}} \quad \text{with } \boldsymbol{h}_1 = (\boldsymbol{\eta}_{\mathrm{R1}}, -\boldsymbol{\eta}_{R1}),$$

$$\boldsymbol{\eta}_{\mathrm{F1}} = \underbrace{(\boldsymbol{f}_1, \ldots, \boldsymbol{f}_1)}_{n/2 \text{ times}} \quad \text{with } \boldsymbol{f}_1 = (1, 0, \ldots, 1, 0;\ 0, -1, \ldots, 0, -1).$$

**Lemma 8.2** *The eigenvalues $\beta$, $\gamma$, and $\epsilon$ of the matrices $J(\boldsymbol{\lambda}^*)$, $V(\boldsymbol{\lambda}^*)$, and $\hat{D}$, respectively, with the common eigenvector $\boldsymbol{\eta}$ in* (8.17) *are related as*

$$\beta = \Psi(\epsilon) = \frac{\epsilon\{\tilde{\mu} + \mu' - (\tilde{\mu}\mu' + 1)\epsilon\}}{1 - \tilde{\mu}\epsilon}, \tag{8.18}$$

$$\gamma = N\Psi(\epsilon) = \frac{N\epsilon\{\tilde{\mu} + \mu' - (\tilde{\mu}\mu' + 1)\epsilon\}}{1 - \tilde{\mu}\epsilon}. \tag{8.19}$$

***Proof*** Multiplying (8.15) by $\boldsymbol{\eta}$ from the right and using $\mathbf{1}^\top\boldsymbol{\eta} = 0$ and $\mathbf{1}^\top V(\boldsymbol{\lambda}^*)\cdot\boldsymbol{\eta} = \gamma\mathbf{1}^\top\boldsymbol{\eta} = 0$, we obtain $J(\boldsymbol{\lambda}^*)\cdot\boldsymbol{\eta} = \frac{\gamma}{N}\boldsymbol{\eta}$, which gives $\beta = \gamma/N$. Multiplying $V(\boldsymbol{\lambda}^*)$ in (8.16) by $\boldsymbol{\eta}$ from the right and using $\hat{D}\boldsymbol{\eta} = \epsilon\boldsymbol{\eta}$, we obtain $V(\boldsymbol{\lambda}^*)\cdot\boldsymbol{\eta} = \gamma\boldsymbol{\eta}$ with $\gamma = N[\mu'\epsilon + (1-\tilde{\mu}\epsilon)^{-1}\cdot\epsilon(\tilde{\mu}-\epsilon)]$, giving (8.19). Then, $\beta = \gamma/N$ gives (8.18). □

We derive the formulas for break point $\tau^*$ for $n = 4m$ (Assumption 8.1) by using the following procedure:

- Obtain $\epsilon = \epsilon^*$ for the break point in Lemma 8.3.
- Express the trade freeness $\phi^*$ for the break point as a function of $\epsilon^*$ in Lemma 8.4.
- Derive the formula for break point $\tau^*$ in Proposition 8.1, using $\phi^*$ in (8.4) with $\tilde{L} = 1/n$.

The dependence of the variables is

$$\beta \overset{(8.18)}{\longrightarrow} \epsilon^* \overset{(8.23)}{\longrightarrow} \phi^* \overset{(8.4)}{\longrightarrow} \tau^*.$$

First, $\epsilon = \epsilon^*$ for the break point satisfying $\beta = \Psi(\epsilon^*) = 0$ is given in the following lemma.

**Lemma 8.3** $\epsilon = \epsilon^*$ *for the break point is parameterized by* $\sigma$ *and* $\mu$ *as*

$$\epsilon^* = \epsilon^*(\sigma, \mu) = \frac{\mu(2\sigma - 1)}{\sigma(\sigma - 1) + \mu^2} < 1. \tag{8.20}$$

***Proof*** $\epsilon^* = (\tilde{\mu} + \mu')/(\tilde{\mu}\mu' + 1)$, which nullifies $\beta$ in (8.18), is rewritten using $\tilde{\mu} = \frac{\mu}{\sigma}$ and $\mu' = \frac{\mu}{\sigma - 1}$ to arrive at (8.20). Although $\beta = 0$ is satisfied also by $\epsilon = 0$ from (8.19), this case is excluded since $\epsilon > 0$ holds by (8.23) for $n = 4m$. □

Since $\epsilon < 1$ by (8.23), the break point exists when the parameters $\sigma$ and $\mu$ satisfy the inequality condition in (8.20).

Next, we advance the relation between $\epsilon$ and $\phi$, referring to the results in Appendix:

- In the racetrack economy, the eigenvalue of $\hat{D}$ for the eigenvector $\eta = \eta_{\mathrm{R1}}$ is given by (8.53):

$$\epsilon = \begin{cases} \left(\dfrac{1-\phi}{1+\phi}\right)^2 & (\text{if } n = 4m), \\ \left(\dfrac{1-\phi}{1+\phi}\right)^2 \dfrac{1+\phi^{n/2}}{1-\phi^{n/2}} & (\text{if } n = 4m - 2), \end{cases} \tag{8.21}$$

  where $m$ is a positive integer. Since $0 < \phi < 1$, we have $0 < \epsilon < 1$.
- For the half doubling in the square lattice economy in the case of $n = 4m$, by the tensor product property in (8.13), the eigenvalue is the square of the eigenvalue in (8.21), i.e., $\epsilon = \{(1 - \phi)/(1 + \phi)\}^4$, and $0 < \epsilon < 1$ holds.
- For the full doubling in the square lattice economy in the case of $n = 4m$, we have $\epsilon = \left(\frac{1-\phi}{1+\phi}\right)^2$, which is the same form as the racetrack economy with $n = 4m$ in (8.21).

To sum up, we advance the relation between $\epsilon$ and $\phi$ using a variable $p$, expressing the squared tensor product linkage, as

$$p = \begin{cases} 1 & \text{for the racetrack and the square lattice (full doubling),} \\ 2 & \text{for the square lattice (half doubling).} \end{cases} \tag{8.22}$$

**Lemma 8.4** *For $n = 4m$ $(m = 1, 2, \ldots)$, we have*

$$\epsilon = \left(\frac{1-\phi}{1+\phi}\right)^{2p} < 1, \quad \textit{i.e.,} \quad \phi = \Phi(\epsilon) = \frac{1-\epsilon^{\frac{1}{2p}}}{1+\epsilon^{\frac{1}{2p}}}. \tag{8.23}$$

Lastly, we substitute (8.23) for $\phi = \phi^*$ into (8.4) with $\tilde{L} = 1/n$ to arrive at the following proposition for the break point $\tau^*$.

**Proposition 8.1** *The break points for the racetrack economy and for the square lattice with $n = 4m$ $(m = 1, 2, \ldots)$ are formulated synthetically as*

$$\frac{\tau^*}{n} = \frac{\tau^*(n, \sigma, \mu)}{n} = \frac{1}{\sigma - 1} \log \left( \frac{1 + (\epsilon^*)^{\frac{1}{2p}}}{1 - (\epsilon^*)^{\frac{1}{2p}}} \right) \quad (p = 1, 2) \tag{8.24}$$

*with $\epsilon^* = \epsilon^*(\sigma, \mu)$ in (8.20) and $p$ in (8.22).*

**Proposition 8.2** *As $\tau$ decreases from a large value for the square lattice, the onset of half doubling precedes that of full doubling, i.e., $\tau^*_{\mathrm{H1}} > \tau^*_{\mathrm{F1}}$.*

***Proof*** For a given $\epsilon^*$, (8.24) gives a larger $\tau^* = \tau^*_{\mathrm{H1}}$ for $p = 2$ than $\tau^* = \tau^*_{\mathrm{F1}}$ for $p = 1$, which shows $\tau^*_{\mathrm{H1}} > \tau^*_{\mathrm{F1}}$. □

In a customary situation of decreasing transportation cost $\tau$ (Sect. 8.4.3), half doubling bifurcation precedes the full doubling bifurcation and is more important.

The formula in (8.24) contains the influence $\sigma$ and $\mu$ implicitly through $\epsilon^* = \epsilon^*(\sigma, \mu)$ in (8.20). We propose the following approximate formulas that can explicitly express the influence of these parameters on the break point $\tau^*$.

**Proposition 8.3** *Under the assumption $\sigma \gg 1$, the break point $\tau^*$ for $n = 4m$ $(m = 1, 2, \ldots)$ is approximated by*

$$\frac{\tau^*_{\mathrm{R1}}}{n} = \frac{\tau^*_{\mathrm{F1}}}{n} \approx \frac{2^{3/2}\mu^{1/2}}{(\sigma - 1)^{3/2}}, \quad \frac{\tau^*_{\mathrm{H1}}}{n} \approx \frac{2^{5/4}\mu^{1/4}}{(\sigma - 1)^{5/4}}. \tag{8.25}$$

***Proof*** Under the assumption $\sigma \gg 1$, we have $\sigma(\sigma - 1) + \mu^2 \approx \sigma(\sigma - 1)$ and $(2\sigma - 1)/\sigma \approx 2$. Then, (8.20) is approximated by $\epsilon^* \approx \frac{2\mu}{\sigma - 1} \ll 1$. The logarithmic term in (8.24) is approximated by $2(\epsilon^*)^{\frac{1}{2p}} \approx 2\left(\frac{2\mu}{\sigma-1}\right)^{\frac{1}{2p}}$, and (8.25) holds for $p = 1, 2$ in (8.22). □

Formula (8.25) for $n = 4m$ indicates that the onset of agglomeration is hastened by a smaller $\sigma$ and/or a larger $\mu$. We observe the same tendency for the two-place economy as explained below.

**Proposition 8.4** *The break point of the two-place economy is*

$$\frac{\tau^*}{2} = \frac{1}{\sigma - 1} \log\left(\frac{1+\epsilon^*}{1-\epsilon^*}\right). \tag{8.26}$$

*Under the assumption $\sigma \gg 1$, we can approximate $\tau^*$ by*

$$\frac{\tau^*}{2} \approx \frac{4\mu}{(\sigma-1)^2}. \tag{8.27}$$

***Proof*** From (8.53) in Appendix, we get the eigenvalue $\epsilon = (1-\phi)/(1+\phi)$ and the eigenvector $\eta = (1, -1)^\top$. We rewrite this relation for the eigenvalue in terms of $\tau^*$ using (8.4) to arrive at (8.26). For $\sigma \gg 1$, the use of $\epsilon^* \approx \frac{2\mu}{\sigma-1} \ll 1$ in (8.26) gives (8.27). □

### 8.4.3 Break Points Observed in Numerical Analysis

We conducted a numerical analysis of the break point for the racetrack economy and the square lattice. We set their size as $n = 2^m$ $(m = 1, 2, 3, 4)$, the length of an interplace road as $\tilde{L} = 1/n$, and economic parameter values as $(\sigma, \mu) = (10.0, 0.4)$, satisfying the no-black-hole condition in (8.9).

Figures 8.1 and 8.2 show the solution curves for the racetrack economy and the square lattice, respectively. A solid line and a dashed curve show stable and unstable solution curves. The horizontal lines A to I represent stable ones. There is a stable curve for the uniform state in $\tau > \tau^*$ (state A), and the break bifurcation point is located at $\tau = \tau^*$ (see, e.g., point a in Fig 8.1b).

Normalized break points $\tau^*/n$ of the uniform state A are listed in Table 8.1. Their numerically computed values are in complete agreement with the theoretical ones by (8.24) or (8.26) and are in good agreement with the approximate ones by (8.25) or (8.27). The values of $\tau^*/n$ for $n = 2$ (Proposition 8.4) are apparently different from those for $n = 4m$ $(m = 1, 2, 4)$.

## 8.5 Hexagonal Lattice with Logit Dynamics

We present formulas for *break points* for a $6 \times 6$ *hexagonal lattice* with the *logit dynamics*. The methodology advanced herein for a specific value of $n = 6$ is extendable to the hexagonal lattice of any size $n$. We show that the hexagon for

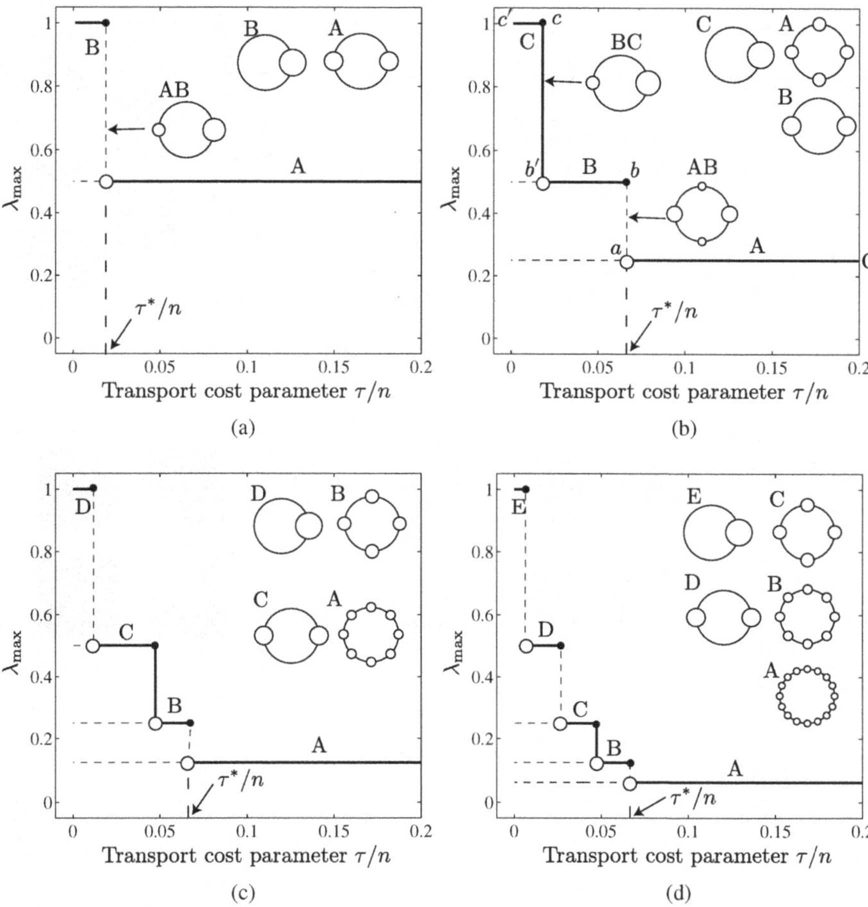


**Fig. 8.1** Solution curves for the racetrack economy with $n = 2, 4, 8,$ and 16 (Ikeda et al., 2018) (solid line: stable; dashed curve: unstable; (∘): simple bifurcation point; (•): sustain point; $\lambda_{max} = \lambda_1$). **(a)** $n = 2$. **(b)** $n = 4$. **(c)** $n = 8$. **(d)** $n = 16$

Christaller's $k = 3$ system (1933) emerges first (Fig. 2.1a) and the megalopolis emerges last under decreasing transportation cost $\tau$.

### 8.5.1 Eigenanalysis of Spatial Discount Matrix

We present the *spatial discount matrix* $D = (d_{ij}) = (\phi^{m(i,j)})$ in (8.5) of the $6 \times 6$ hexagonal lattice as a function of the trade freeness $\phi = \exp[-\tau(\sigma - 1)\tilde{L}]$ in (8.4). The distance unit $\tilde{L}$ is chosen as $\tilde{L} = 1/n = 1/6$.

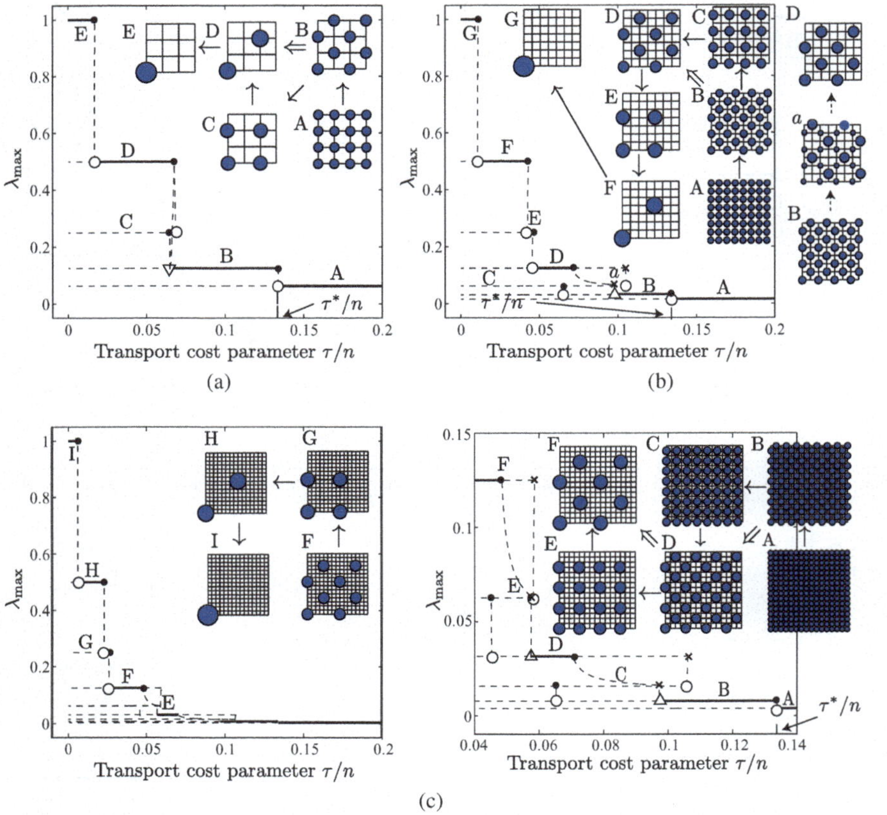


**Fig. 8.2** Solution curves for the square lattice with $n = 4$, 8, and 16 (Ikeda et al., 2018) (solid line: stable; dashed curve: unstable; (∘): simple bifurcation point; (•): sustain point; (△): double bifurcation point; (▽): triple bifurcation point; ×: sustaining bifurcation point; $\lambda_{\max} = \lambda_1$). (**a**) $n = 4$. (**b**) $n = 8$. (**c**) $n = 16$

**Table 8.1** Comparison of numerical, theoretical, and approximate break points

| (a) Racetrack economy | | | | | |
|---|---|---|---|---|---|
| | Number $n$ of places | 2 | 4 | 8 | 16 |
| $\tau^*/n$ | Numerically computed | 0.019 | 0.066 | 0.066 | 0.066 |
| | Theoretical formula in (8.24) or (8.26) | 0.019 | 0.066 | 0.066 | 0.066 |
| | Approximate formula in (8.25) or (8.27) | 0.020 | 0.066 | 0.066 | 0.066 |
| (b) Square lattice | | | | | |
| | Number $n$ of places | 2 | 4 | 8 | 16 |
| $\tau^*/n$ | Numerically computed | 0.066 | 0.134 | 0.134 | 0.134 |
| | Theoretical formula in (8.24) or (8.26) | 0.066 | 0.134 | 0.134 | 0.134 |
| | Approximate formula in (8.25) or (8.27) | 0.066 | 0.121 | 0.121 | 0.121 |

This matrix takes the form

$$D = \begin{bmatrix} D_0 & D_1 & D_2 & D_3 & D_4 & D_5 \\ D_5 & D_0 & D_1 & D_2 & D_3 & D_4 \\ D_4 & D_5 & D_0 & D_1 & D_2 & D_3 \\ D_3 & D_4 & D_5 & D_0 & D_1 & D_2 \\ D_2 & D_3 & D_4 & D_5 & D_0 & D_1 \\ D_1 & D_2 & D_3 & D_4 & D_5 & D_0 \end{bmatrix}$$

and is a block-circulant matrix made up of circulant matrices:

$$D_0 = \begin{bmatrix} 1 & \phi & \phi^2 & \phi^3 & \phi^2 & \phi \\ \phi & 1 & \phi & \phi^2 & \phi^3 & \phi^2 \\ \phi^2 & \phi & 1 & \phi & \phi^2 & \phi^3 \\ \phi^3 & \phi^2 & \phi & 1 & \phi & \phi^2 \\ \phi^2 & \phi^3 & \phi^2 & \phi & 1 & \phi \\ \phi & \phi^2 & \phi^3 & \phi^2 & \phi & 1 \end{bmatrix}, \quad D_1 = D_5{}^\top = \begin{bmatrix} \phi & \phi & \phi^2 & \phi^3 & \phi^3 & \phi^2 \\ \phi^2 & \phi & \phi & \phi^2 & \phi^3 & \phi^3 \\ \phi^3 & \phi^2 & \phi & \phi & \phi^2 & \phi^3 \\ \phi^3 & \phi^3 & \phi^2 & \phi & \phi & \phi^2 \\ \phi^2 & \phi^3 & \phi^3 & \phi^2 & \phi & \phi \\ \phi & \phi^2 & \phi^3 & \phi^3 & \phi^2 & \phi \end{bmatrix},$$

$$D_2 = D_4{}^\top = \begin{bmatrix} \phi^2 & \phi^2 & \phi^2 & \phi^3 & \phi^4 & \phi^3 \\ \phi^3 & \phi^2 & \phi^2 & \phi^2 & \phi^3 & \phi^4 \\ \phi^4 & \phi^3 & \phi^2 & \phi^2 & \phi^2 & \phi^3 \\ \phi^3 & \phi^4 & \phi^3 & \phi^2 & \phi^2 & \phi^2 \\ \phi^2 & \phi^3 & \phi^4 & \phi^3 & \phi^2 & \phi^2 \\ \phi^2 & \phi^2 & \phi^3 & \phi^4 & \phi^3 & \phi^2 \end{bmatrix}, \quad D_3 = \phi^3 \begin{bmatrix} 1 & 1 & 1 & 1 & 1 & 1 \\ 1 & 1 & 1 & 1 & 1 & 1 \\ 1 & 1 & 1 & 1 & 1 & 1 \\ 1 & 1 & 1 & 1 & 1 & 1 \\ 1 & 1 & 1 & 1 & 1 & 1 \\ 1 & 1 & 1 & 1 & 1 & 1 \end{bmatrix}.$$

The direct bifurcation from the uniform state $\boldsymbol{\lambda}^* = \frac{1}{N}\mathbf{1}$ ($N = n^2 = 6^2$) is associated with the *irreducible representations* $\mu$ of the hexagonal lattice (Sect. 5.3). The set of $\mu$ for $n = 6$ is given by (5.7):

$$R_6 = \{(2), (3), (6; 1, 0), (6; 1, 1), (6; 2, 0), (12; 2, 1)\}. \tag{8.28}$$

The direction $\boldsymbol{\eta}$ of a bifurcating hexagonal pattern is associated with the eigenvector of the Jacobian matrix $J(\boldsymbol{\lambda}^*)$ for each $\mu$ (cf. (5.4)), as depicted in Fig. 8.3.

By Lemma 8.1, the vector $\boldsymbol{\eta}$ is also an eigenvector of the spatial discount matrix $D$, i.e.,

$$D\boldsymbol{\eta} = \tilde{\epsilon}^{(\mu)}\boldsymbol{\eta} \quad (\mu \in R_6) \tag{8.29}$$

with the eigenvalue $\tilde{\epsilon}^{(\mu)}$ given by

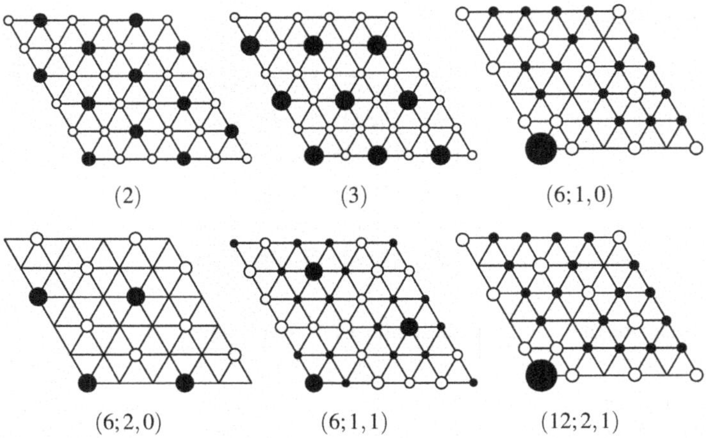


**Fig. 8.3** Eigenvectors for bifurcating hexagonal patterns on the 6 × 6 hexagonal lattice (a black circle denotes a positive component; a white circle indicates a negative component; the area of a circle expresses the magnitude of the component)

$$\tilde{\epsilon}^{(\mu)} = \tilde{\epsilon}^{(\mu)}(\phi) = \begin{cases} 1 - 3\phi + 3\phi^2 - 3\phi^3 + 2\phi^4 & \text{for } (2), \\ 1 - 2\phi + 4\phi^2 - 5\phi^3 + 2\phi^4 & \text{for } (3), \\ 1 + 4\phi + \phi^2 - 5\phi^3 - \phi^4 & \text{for } (6;\, 1,\, 0), \\ 1 + \phi - 5\phi^2 + \phi^3 + 2\phi^4 & \text{for } (6;\, 1,\, 1), \\ 1 - 3\phi^2 + 3\phi^3 - \phi^4 & \text{for } (6;\, 2,\, 0), \\ 1 - 2\phi + \phi^2 + \phi^3 - \phi^4 & \text{for } (12;\, 2,\, 1). \end{cases} \tag{8.30}$$

The sum $d$ of the entries of a column of $D$ becomes

$$d = d(\phi) = \sum_{i=1}^{N} \phi^{m(i,j)} = 1 + 6\phi + 12\phi^2 + 15\phi^3 + 2\phi^4. \tag{8.31}$$

Then, for the vector $\boldsymbol{\eta}$ in (8.29), we have a normalized eigenproblem

$$\hat{D}\boldsymbol{\eta} = \epsilon\boldsymbol{\eta}$$

($\hat{D} = D/d$) with a normalized eigenvalue:

$$\epsilon = \frac{\tilde{\epsilon}^{(\mu)}(\phi)}{d(\phi)} < 1. \tag{8.32}$$

Because $0 < \phi < 1$, we have $0 < \epsilon < 1$ by (8.30) and (8.31). By the implicit function theorem, the relation in (8.32) between $\epsilon$ and $\phi$ is solvable for $\phi$ as

$$\phi = \Phi^{(\mu)}(\epsilon) \tag{8.33}$$

with some function $\Phi^{(\mu)}$.

### 8.5.2 Formulas for Break Point

A bifurcation point at the uniform state $\boldsymbol{\lambda} = \boldsymbol{\lambda}^* = \frac{1}{36}(1, \ldots, 1)^\top$ is determined as the break point $\tau = \tau^*$ at which the Jacobian matrix $J(\boldsymbol{\lambda}^*) = \frac{\partial \boldsymbol{F}}{\partial \boldsymbol{\lambda}}(\boldsymbol{\lambda}^*)$ of the governing equation has one or more zero eigenvalues.

At the uniform state $\boldsymbol{\lambda}^*$, $J(\boldsymbol{\lambda}^*)$ in (8.10) in the logit dynamics is evaluated as

$$J(\boldsymbol{\lambda}^*) = -\frac{\theta}{N^2}\boldsymbol{1}\boldsymbol{1}^\top V(\boldsymbol{\lambda}^*) + \frac{\theta}{N}V(\boldsymbol{\lambda}^*) - I \tag{8.34}$$

with the expression in (8.16):

$$V(\boldsymbol{\lambda}^*) = N\left[\mu'\hat{D} + \left(I - \tilde{\mu}\hat{D}\right)^{-1} \cdot \hat{D}\left(\tilde{\mu}I - \hat{D}\right)\right]. \tag{8.35}$$

Multiplying (8.34) by the vector $\boldsymbol{\eta}$ from the right and using (8.29), (8.35), and $\boldsymbol{1}^\top V(\boldsymbol{\lambda}^*) \cdot \boldsymbol{\eta} = \gamma \boldsymbol{1}^\top \boldsymbol{\eta} = 0$, we obtain $J(\boldsymbol{\lambda}^*) \cdot \boldsymbol{\eta} = \beta \boldsymbol{\eta}$ with the eigenvalue

$$\beta = \Psi(\epsilon) = \theta\left(\mu'\epsilon + \frac{\epsilon(\tilde{\mu} - \epsilon)}{1 - \tilde{\mu}\epsilon} - \frac{1}{\theta}\right). \tag{8.36}$$

We determine break point $\tau^*$ from the condition that the eigenvalue $\beta$ vanishes. The dependence of variables is given by

$$\beta \overset{(8.36)}{\longrightarrow} \epsilon^* \overset{(8.32)}{\longrightarrow} \phi^* \overset{(8.4)}{\longrightarrow} \tau^*. \tag{8.37}$$

To begin with, from (8.36), we can obtain the value of $x = \epsilon^*$ satisfying $\Psi(\epsilon^*) = 0$ as a solution to the quadratic equation:

$$\theta(bx - ax^2) - 1 = 0. \tag{8.38}$$

Here,

$$a = \tilde{\mu}\mu' + 1 > 0, \quad b = \tilde{\mu} + \mu' + \theta^{-1}\tilde{\mu} > 0. \tag{8.39}$$

Equation (8.38) has two solutions

$$\epsilon_+^* = \frac{b + \sqrt{b^2 - 4a\theta^{-1}}}{2a}, \quad \epsilon_-^* = \frac{b - \sqrt{b^2 - 4a\theta^{-1}}}{2a}. \tag{8.40}$$

These solutions are related to break points when $0 < \epsilon_-^* < \epsilon_+^* < 1$ is satisfied (cf. $\epsilon < 1$ in (8.32)). The conditions for the existence is expressed, using $\tilde{\mu} = \frac{\mu}{\sigma} < 1$ and $\mu' = \frac{\mu}{\sigma - 1} > 0$ ($\sigma > 1$ and $0 < \mu < 1$) in (8.12).

**Lemma 8.5** *We have* $0 < \epsilon_-^* < \epsilon_+^* < 1$ *under the following conditions:*

$$\frac{\mu}{\sigma - 1} < 1 + \theta^{-1}, \tag{8.41}$$

$$2 - \frac{\mu}{\sigma(\sigma - 1)}\{2\sigma - 1 - 2\mu + (\sigma - 1)\theta^{-1}\} > 0, \tag{8.42}$$

$$\mu^2 \left[\frac{1}{\sigma}(1 + \theta^{-1}) + \frac{1}{\sigma - 1}\right]^2 - 4\theta^{-1}\left(\frac{\mu^2}{\sigma(\sigma - 1)} + 1\right) > 0. \tag{8.43}$$

***Proof*** The condition $b^2 - 4a\theta^{-1} > 0$ with (8.12) and (8.39) gives (8.43). $\epsilon_+^* < 1$ gives the conditions $2a - b > 0$ and $(2a - b)^2 > b^2 - 4a\theta^{-1}$. The former condition leads to and (8.42). The latter condition leads to (8.41) as follows:

$$\begin{aligned}(2a - b)^2 - (b^2 - 4a\theta^{-1}) &= 4a(a - b + \theta^{-1}) \\ &= 4a\{\tilde{\mu}\mu' + 1 - (\tilde{\mu} + \mu' + \theta^{-1}\tilde{\mu}) + \theta^{-1}\} \\ &= 4a(1 - \tilde{\mu})(1 - \mu' + \theta^{-1}) > 0.\end{aligned}$$

Since $a > 0$ and $\tilde{\mu} < 1$, we have $1 - \mu' + \theta^{-1} > 0$, i.e., $\frac{\mu}{\sigma - 1} < 1 + \theta^{-1}$. □

In the limit of $\theta \to +\infty$, corresponding to the case without heterogeneity, the condition (8.43) is satisfied, and (8.41) reduces to the no-black-hole condition $\mu/(\sigma - 1) < 1$ in (8.9).

Under the conditions in Lemma 8.5, $\epsilon_+^*$ and $\epsilon_+^*$ can be associated with two break points $\tau_+$ and $\tau_-$ ($\tau_+ > \tau_- > 0$). When the transportation cost parameter $\tau$ is reduced continuously from $+\infty$ to 0, two break points $\tau_+$ and $\tau_-$ are encountered for each irreducible representation $\mu$.

All irreducible representations $\mu \in R_6$ in (8.28) follow the same conditions for the existence of break points. Accordingly, no bifurcation occurs if these conditions are not satisfied. Then, we encounter the worst scenario for downtown development through social investment in that no agglomeration emerges at whatever investment.

Next, by the dependence of variables in (8.37), we obtain the break point $\tau_\pm^{(\mu)}$ in the following proposition.

**Proposition 8.5** *Under the conditions in Lemma 8.5, break points* $\tau_+^{(\mu)}$ *and* $\tau_-^{(\mu)}$ *for each* $\mu \in R_6$ *in* (8.28) *are given as*

$$\tau_+^{(\mu)} = -\frac{6}{\sigma - 1}\log(\Phi^{(\mu)}(\epsilon_+^*)), \quad \tau_-^{(\mu)} = -\frac{6}{\sigma - 1}\log(\Phi^{(\mu)}(\epsilon_-^*)). \tag{8.44}$$

***Proof*** The value of $\phi = \phi_\pm^{(\mu)}$ corresponding to $\epsilon = \epsilon_\pm^{(\mu)}$ is given as $\phi_\pm^{(\mu)} = \Phi^{(\mu)}(\epsilon_\pm^*)$ by (8.33). Then, from $\phi = \exp[-\tau(\sigma - 1)\tilde{L}]$ in (8.4) with $\tilde{L} = 1/n = 1/6$, we obtain $\tau_\pm^{(\mu)}$ in (8.44). □

The following approximate formula for $\tau_+^{(2)}$ helps study the *parameter dependence* of the break point for $\tau_+^{(2)}$ associated with Christaller's $k = 3$ system.

**Proposition 8.6** *Under the condition $\theta \gg \frac{\sigma}{\mu} \gg 1$, the larger break point $\tau_+^{(2)}$ is given approximately as*

$$\tau_+^{(2)} = 18 \cdot 2^{1/3} \frac{\mu^{1/3}}{(\sigma - 1)^{4/3}}. \tag{8.45}$$

***Proof*** See Appendix for the proof. □

The formula in (8.45) indicates that the onset of agglomeration is hastened by a smaller $\sigma$ and/or a larger $\mu$, like the racetrack economy and the square lattice (Proposition 8.3 in Sect. 8.4.2). This fact concurs with economic intuition and the numerical examples in Sect. 8.5.4.

### 8.5.3 Stability and Order of Emerging Hexagons

When $\phi$ (or $\tau$) changes continuously, the first and the last bifurcations are the most important regarding the stability of the system. We explain how to predetermine these bifurcations.

**Lemma 8.6** *Under the conditions* (8.41) *and* (8.43), *the uniform state is stable for $\tau = +0$ and $+\infty$.*

***Proof*** We show that the eigenvalue $\beta$ in (8.36) is negative for any irreducible representation $\mu$ in (8.28) in two cases $\tau = +0$ and $+\infty$. The first case of $\tau = +0$ entails $\phi = 1$ by (8.4) and $\epsilon \to 0$ by (8.32) with (8.30) and, in turn, $\beta = -1 < 0$ from (8.36). The other case $\tau = +\infty$ entails $\phi = +0$ by (8.4) and $\epsilon \to 1$ by (8.32) with (8.30) and (8.31); the eigenvalue $\beta$ in (8.36) becomes negative because $\beta = \theta(\mu' - 1 - \theta^{-1}) = \theta(\frac{\mu}{\sigma - 1} - 1 - \theta^{-1}) < 0$ under condition (8.41). □

The functions $\epsilon(\phi) = \tilde{\epsilon}^{(\mu)}(\phi)/d (0 < \phi < 1)$ for irreducible representations $\mu \in R_6$ are plotted in Fig. 8.4. For a given value of $\epsilon = \epsilon_\pm^*$, the curve for $\mu = (2)$ gives the smallest value of $\phi$ and that for $\mu = (6; 1, 0)$ gives the largest value among $\mu$'s. Then, we have

$$\phi_+^{(2)} < \phi_+^{(\mu)} < \phi_+^{(6;1,0)}, \quad \phi_-^{(2)} < \phi_-^{(\mu)} < \phi_-^{(6;1,0)};$$
$$\mu \in \{(3), (6; 1, 1), (6; 2, 0), (12; 2, 1)\}.$$

Since $\phi$ is inversely proportional to $\tau$ by (8.4), this relation is rewritten as

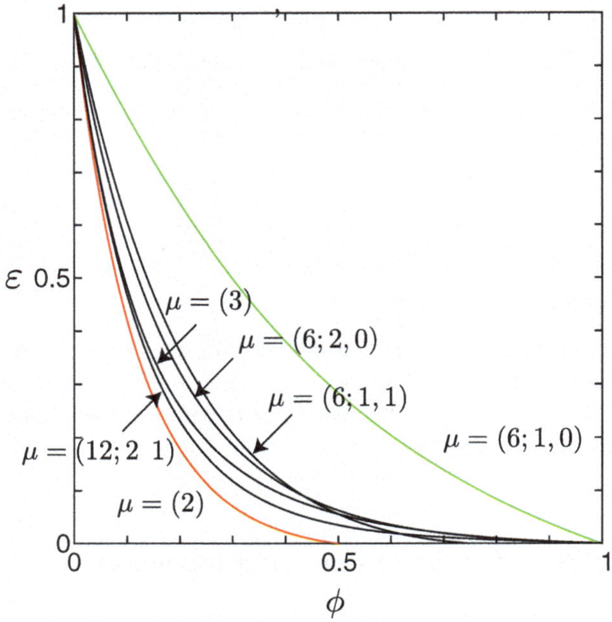


**Fig. 8.4** Curves of $\epsilon$ plotted against $\phi$ for irreducible representations $\mu \in R_6$ $(0 < \epsilon < 1,\ 0 < \phi < 1)$

$$\tau_+^{(2)} > \tau_+^{(\mu)} > \tau_+^{(6;1,0)}, \quad \tau_-^{(2)} > \tau_-^{(\mu)} > \tau_-^{(6;1,0)};$$
$$\mu \in \{(3), (6;1,1), (6;2,0), (12;2,1)\}.$$

Therefore, when $\tau$ decreases from a large value, the first bifurcation is associated with $\tau_+^{(2)}$ $(> \tau_-^{(2)})$ for Christaller's $k = 3$ system, and the last one with $\tau_-^{(6;1,0)}$ $(< \tau_+^{(6;1,0)})$ for megalopolis pattern (see Fig. 8.3).

**Proposition 8.7** *When $\tau$ is reduced continuously from $+\infty$ to 0, bifurcations, when they exist, occur in the following order:*

*(i) Bifurcation producing Christaller's $k = 3$ occurs first at $\tau = \tau_+^{(2)}$.*
*(ii) Bifurcation producing a megalopolis occurs last at $\tau = \tau_-^{(6;1,0)}$.*

The smallest hexagon in Christaller's $k = 3$ system is the most important. Another hexagon for $\mu = (6; 1, 0)$, not considered in central place theory, is also important because this hexagon expresses centralization for small $\tau$.

We combine Lemma 8.6 and Proposition 8.7 to arrive at the following proposition for stability of the uniform state for a large $\tau$ $(> \tau_+^{(2)})$ and a small $\tau$.

**Proposition 8.8** *Under the conditions* (8.41) *and* (8.43), *the uniform state is stable for $\tau_+^{(2)} < \tau < \infty$ and $0 < \tau < \tau_-^{(6;1,0)}$ and is unstable for $\tau_-^{(6;1,0)} < \tau < \tau_+^{(2)}$.*

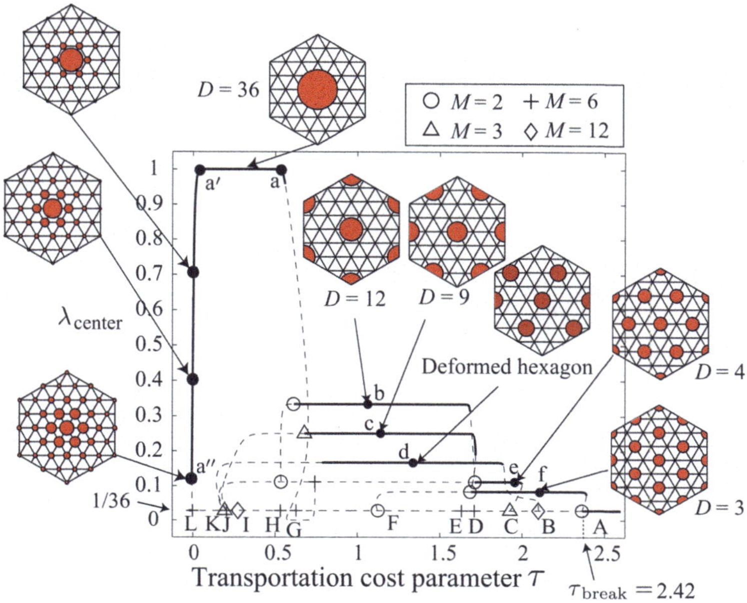


**Fig. 8.5** Solution curves for hexagonal patterns of the infinite hexagonal lattice (Ikeda et al., 2017) (solid curve: stable; dashed one: unstable; $M$ is the multiplicity of the critical point)

### *8.5.4 Break Points Observed in Numerical Analysis*

In Fig. 8.5, we revisit the numerical analysis of the infinite hexagonal lattice (Fig. 2.7 in Sect. 2.3.2 for $(\sigma, \mu) = (5.0, 0.4)$ and $\theta = 1000$). In agreement with the theoretical prediction in Sect. 8.5.3, a bifurcation producing Christaller's $k = 3$ system occurs first at point A, and the hexagon producing the megalopolis occurs last at point L, when $\tau$ decreases from a large value (i.e., from right to left in this figure). The uniform state is unstable during AL and is stable for $\tau > \tau_{\text{break}} = \tau_+^{(2)}$, in agreement with Proposition 8.8. There are stable bifurcated hexagons in the zone $\tau_-^{(6;1,0)} < \tau < \tau_+^{(2)}$.

We specifically examine the break point $\tau = \tau_+^{(2)}$ for the first bifurcation producing Christaller's $k = 3$ system (Proposition 8.7(i)). The dependence of $\tau_+^{(2)}$ on the parameters $\sigma$ and $\mu$ expressed by the theoretical law in (8.44) is depicted by the dashed curve in Fig. 8.6. As $\sigma$ increases, the economic balance shifts from in favor of agglomeration to dispersion and $\tau_+^{(2)}$ decreases. As $\mu$ increases, it shifts in favor of agglomeration and $\tau_+^{(2)}$ increases. The solid curve of the approximate law in (8.45) is in fair agreement with the dashed curve of the theoretical law. Such agreement ensures the validity of the approximate law.

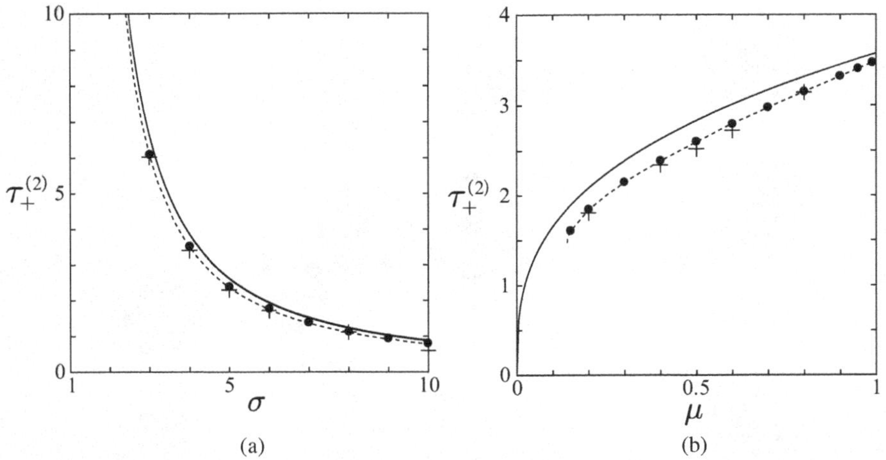


**Fig. 8.6** Dependence of break point $\tau_+^{(2)}$ on the parameters $\sigma$ and $\mu$ ($\theta = 1000$) (Ikeda et al., 2017) ((•) denotes the break point obtained through the numerical analysis; (+) denotes that for the finite lattice in Sect. 2.4; the dashed curve means that for the theoretical law in (8.44); the solid curve means that for the approximate law in (8.45)). (**a**) Dependence on $\sigma$ ($\mu = 0.4$). (**b**) Dependence on $\mu$ ($\sigma = 5.0$)

The values of $\tau_+^{(2)}$ for the infinite lattice shown by (•) in Fig. 8.6 are in good agreement with the theoretical curve. Those values for the finite lattice read from Fig. 2.10 in Sect. 2.4 (shown by (+)) are in good agreement with the curve.

Lemma 8.5 indicates that the existence of the break point depends on the values of parameters. For $\sigma = 5.0$ and $\theta = 1000$ in Fig. 8.6b, (8.43) provides a condition $\mu > 0.141$ for the existence of the break point. Although this condition is violated only in exceptional cases, due regard should be given to such cases, in which investment in transportation is wasted without realizing centralization.

## Appendix: Theoretical Details

### Proof of (8.10)

In the replicator dynamics, the differentiation of the governing equation $\boldsymbol{F}$ in (8.1) with (8.2) gives

$$\frac{\partial F_i}{\partial \lambda_j} = \left( v_i - \sum_{k=1}^{N} \lambda_k v_k \right) \delta_{ij} + \lambda_i \left( \frac{\partial v_i}{\partial \lambda_j} - v_j - \sum_{k=1}^{N} \lambda_k \frac{\partial v_k}{\partial \lambda_j} \right),$$

where $\delta_{ij}$ is the Kronecker delta. Therefore, the Jacobian matrices $J(\boldsymbol{\lambda}) = \partial \boldsymbol{F}/\partial \boldsymbol{\lambda}$ and $V(\boldsymbol{\lambda}) = \partial \boldsymbol{v}/\partial \boldsymbol{\lambda}$ are related as follows:

$$J(\boldsymbol{\lambda}) = \mathrm{diag}(v_1 - \bar{v}, \ldots, v_N - \bar{v}) + (\Lambda - \boldsymbol{\lambda}\boldsymbol{\lambda}^\top) V(\boldsymbol{\lambda}) - \boldsymbol{\lambda}\boldsymbol{v}^\top,$$

where $\Lambda = \mathrm{diag}(\lambda_1, \ldots, \lambda_N)$. At the uniform state $\boldsymbol{\lambda} = \boldsymbol{\lambda}^* = \frac{1}{N}\mathbf{1}$ with $v_1 = \cdots = v_N = \bar{v}$, this equation gives the formula for the replicator dynamics in (8.10).

In the logit dynamics, the differentiation of the governing equation $\boldsymbol{F}$ in (8.3) leads to

$$J(\boldsymbol{\lambda}) = -\theta \begin{bmatrix} P_1 \\ \vdots \\ P_N \end{bmatrix} \begin{bmatrix} P_1 \cdots P_N \end{bmatrix} V(\boldsymbol{\lambda}) + \theta \begin{bmatrix} P_1 & & \\ & \ddots & \\ & & P_N \end{bmatrix} V(\boldsymbol{\lambda}) - I. \tag{8.46}$$

At the uniform state, (8.46) yields the formula for the logit dynamics in (8.10).

**Proof of (8.11)**

Differentiation of (8.6) and of (8.7) with respect to $\lambda_j$ respectively yields

$$\frac{\partial v_i}{\partial \lambda_j} = \mu' \frac{d_{ji}}{\Delta_i} + \frac{1}{w_i}\frac{\partial w_i}{\partial \lambda_j}, \tag{8.47}$$

$$\frac{\partial w_i}{\partial \lambda_j} = \tilde{\mu} \sum_{k=1}^{N} \frac{d_{ik}}{{\Delta_k}^2} \left[ \left( \frac{\partial w_k}{\partial \lambda_j} \lambda_k + w_k \delta_{kj} \right) \Delta_k - (w_k \lambda_k + 1) d_{jk} \right], \tag{8.48}$$

where $\tilde{\mu} = \frac{\mu}{\sigma}$ and $\mu' = \frac{\mu}{\sigma - 1}$. We have $0 < \tilde{\mu} < 1$ and $\mu' > 0$ because $\sigma > 1$ and $0 < \mu < 1$.

Matrix $V(\boldsymbol{\lambda}^*)$ in (8.10) can be evaluated as below. At $\boldsymbol{\lambda} = \boldsymbol{\lambda}^*$, we have

$$\Delta_i = \Delta_i(\boldsymbol{\lambda}^*, \tau) = \sum_{j=1}^{N} d_{ji} \lambda_j = \frac{d}{N}. \tag{8.49}$$

Here, $d = d(\phi) = \sum_{j=1}^{N} d_{1j}$. Because $w_j$ is independent of $j$, we write $w_j = w$. Then, (8.7) becomes

$$w = \tilde{\mu} \sum_{j=1}^{N} \frac{N}{d} d_{ij} \left( \frac{w}{N} + 1 \right) = \tilde{\mu}\,(w + N). \tag{8.50}$$

At $\boldsymbol{\lambda} = \boldsymbol{\lambda}^*$, (8.48) becomes

$$\frac{\partial w_i}{\partial \lambda_j} = \tilde{\mu} \sum_{k=1}^{N} \frac{N^2}{d^2} d_{ik} \left[ \left( \frac{1}{N} \frac{\partial w_k}{\partial \lambda_j} + w \delta_{kj} \right) \frac{d}{N} - \left( \frac{w}{N} + 1 \right) d_{jk} \right],$$

which in matrix form reads

$$W = \tilde{\mu}\frac{N^2}{d^2} D \left[ \frac{d}{N} \left( \frac{1}{N} W + wI \right) - \frac{w+N}{N} D \right]$$
$$= \tilde{\mu}\frac{D}{d} W + Nw \frac{D}{d} \left( \tilde{\mu} I - \frac{\tilde{\mu}(w+N)}{w} \frac{D}{d} \right)$$

with $W = (\partial w_i / \partial \lambda_j)$. Using $\tilde{\mu}(w+N) = w$ in (8.50), we can rewrite this equation as

$$\left( I - \tilde{\mu}\frac{D}{d} \right) W = Nw \frac{D}{d} \left( \tilde{\mu} I - \frac{D}{d} \right),$$

and the second term $\frac{1}{w_i}\frac{\partial w_i}{\partial \lambda_j}$ in (8.47) can be obtained in a matrix form as

$$\frac{W}{w} = N \left( I - \tilde{\mu}\frac{D}{d} \right)^{-1} \cdot \frac{D}{d} \left( \tilde{\mu} I - \frac{D}{d} \right).$$

Then, with the use of this equation and (8.49), (8.47) can be evaluated in the matrix form in (8.11).

**Proof of Lemma 8.4**

First, for $\eta = \eta_{\mathrm{R1}}$ for the racetrack economy with $p = 1$, the odd-numbered components of $D\eta_{\mathrm{R1}}$ are equal to

$$\begin{aligned} \tilde{\epsilon} &= \sum_{j=1}^{n} (-1)^{j-1} \phi^{m(i,j)} \\ &= 1 + 2\left( -\phi + \phi^2 - \cdots - (-1)^{n/2} \phi^{n/2-1} \right) + (-1)^{n/2} \phi^{n/2} \\ &= \frac{1-\phi}{1+\phi} \left( 1 - (-\phi)^{n/2} \right) \end{aligned} \tag{8.51}$$

and even-numbered components are $-\tilde{\epsilon}$. Hence, $\tilde{\epsilon}$ is the eigenvalue of $D$ for the eigenvector $\eta = \eta_{\mathrm{R1}}$.

We consider a normalized spatial discount matrix $\hat{D} = D/d$. Here, $d$ denotes the sum of the entries of a column of $D$ and is given for $n$ even as

$$\begin{aligned} d &= \sum_{i=1}^{n} \phi^{m(i,j)} = 1 + 2(\phi + \phi^2 + \cdots + \phi^{n/2-1}) + \phi^{n/2} \\ &= 2(1 + \phi + \phi^2 + \cdots + \phi^{n/2-1}) + \phi^{n/2} - 1 = \frac{1+\phi}{1-\phi} \left( 1 - \phi^{n/2} \right). \end{aligned} \tag{8.52}$$

The eigenvalue of $\hat{D}$ for the eigenvector $\boldsymbol{\eta} = \boldsymbol{\eta}_{\mathrm{R1}}$ is given by

$$\epsilon = \frac{\tilde{\epsilon}}{d} = \left(\frac{1-\phi}{1+\phi}\right)^2 \frac{1-(-1)^{n/2}\phi^{n/2}}{1-\phi^{n/2}}$$
$$= \begin{cases} \left(\dfrac{1-\phi}{1+\phi}\right)^2 & (\text{if } n = 4m), \\ \left(\dfrac{1-\phi}{1+\phi}\right)^2 \dfrac{1+\phi^{n/2}}{1-\phi^{n/2}} & (\text{if } n = 4m-2), \end{cases} \tag{8.53}$$

where (8.51) and (8.52) are used and $m$ is a positive integer. Since $0 < \phi < 1$, we have $0 < \epsilon < 1$.

Next, we consider $\boldsymbol{\eta} = \boldsymbol{\eta}_{\mathrm{H1}}$ for the half doubling in the square lattice economy with $p = 2$ in the case of $n = 4m$. By the tensor product property in (8.13), the eigenvalue in this case is the square of the eigenvalue in (8.53), i.e., $\epsilon = \{(1-\phi)/(1+\phi)\}^4$ for $p = 2$ and $0 < \epsilon < 1$ holds.

Last, we consider $\boldsymbol{\eta} = \boldsymbol{\eta}_{\mathrm{F1}}$ for the full doubling in the square lattice economy with $p = 1$ in the case of $n = 4m$. The eigenvalue is equal to the first component of $D\boldsymbol{\eta}_{\mathrm{F1}}$ and is evaluated by its positive and negative components separately as $(q = n/2)$

$$\begin{aligned} \tilde{\epsilon} &= (1 + 2\phi^2 + \cdots + 2\phi^{q-1} + \phi^q)^2 - 4\phi^2(1 + \phi^2 + \cdots + \phi^{q-1})^2 \\ &= [(1 + \phi^2 + \cdots + \phi^{q-1})(1+\phi^2)]^2 - 4\phi^2(1 + \phi^2 + \cdots + \phi^{q-1})^2 \\ &= [(1 + \phi^2 + \cdots + \phi^{q-1})(1-\phi^2)]^2 = (1-\phi^{n/2})^2. \end{aligned}$$

By (8.52) and the tensor product property in (8.13), we obtain

$$d = \left\{\frac{1+\phi}{1-\phi}\left(1-\phi^{n/2}\right)\right\}^2.$$

The eigenvalue of $\hat{D}$ for the eigenvector $\boldsymbol{\eta} = \boldsymbol{\eta}_{\mathrm{F1}}$ is given by

$$\epsilon = \frac{\tilde{\epsilon}}{d} = \left(\frac{1-\phi}{1+\phi}\right)^2.$$

**Proof of Proposition 8.6**

We search for an approximate formula of $\tau_+^{(2)}$ in (8.45) under the following condition:

$$\theta \gg \frac{\sigma}{\mu} \gg 1. \tag{8.54}$$

Then, (8.39) and (8.40) give

$$a \approx 1, \quad b \approx \tilde{\mu} + \mu' \approx \frac{2\mu}{\sigma - 1}, \quad \epsilon_+^* \approx \frac{b}{a} \approx \frac{2\mu}{\sigma - 1} \ll 1. \tag{8.55}$$

Because $\epsilon_+^* > 0$ and the numerator $\tilde{\epsilon}^{(2)}$ of (8.32) is equal to $(1 - \phi)(1 - 2\phi)(1 + \phi^2)$, we have $0 < \phi_+^{(2)} < \frac{1}{2}$, and it is possible to introduce a fairly accurate assumption

$$(\phi_+^{(2)})^3 \ll 1. \tag{8.56}$$

From (8.32) for $\mu = (2)$ with (8.30) and (8.31), we have

$$\epsilon = \frac{(1 - 3\phi + 3\phi^2 - \phi^3) - 2\phi^3 + 2\phi^4}{(1 + 6\phi + 12\phi^2 + 8\phi^3) + 7\phi^3 + 2\phi^4} = \frac{(1 - \phi)^3 - 2\phi^3(1 - \phi)}{(1 + 2\phi)^3 + \phi^3(7 + 2\phi)}.$$

Then, for $\phi = \phi_+^{(2)}$ satisfying (8.56), we get

$$\epsilon_+^* \approx \left( \frac{1 - \phi_+^{(2)}}{1 + 2\phi_+^{(2)}} \right)^3,$$

which yields

$$\phi_+^{(2)} = \Phi^{(2)}(\epsilon_+^*) \approx \frac{1 - (\epsilon_+^*)^{1/3}}{1 + 2(\epsilon_+^*)^{1/3}}.$$

Then, from (8.44) with (8.55), we obtain (8.45):

$$\tau_+^{(2)} \approx -\frac{n}{\tilde{L}(\sigma - 1)} \log \left( \frac{1 - (\epsilon_+^*)^{1/3}}{1 + 2(\epsilon_+^*)^{1/3}} \right) \approx 18 \cdot 2^{1/3} \frac{\mu^{1/3}}{(\sigma - 1)^{4/3}}.$$

## References

Akamatsu, T., Takayama, Y., & Ikeda, K. (2012). Spatial discounting, Fourier, and racetrack economy: A recipe for the analysis of spatial agglomeration models. *Journal of Economic Dynamics and Control, 36*(11), 1729–1759.

Anderson, J. E., & van Wincoop, E. (2004). Trade costs. *Journal of Economic Literature, 42*(3), 691–751.

Christaller, W. (1933). *Die zentralen Orte in Süddeutschland*. Gustav Fischer. English translation: *Central places in Southern Germany*. Prentice Hall (1966).

Forslid, R., & Ottaviano, G. I. P. (2003). An analytically solvable core-periphery model. *Journal of Economic Geography, 3*, 229–240.

Fujita, M., Krugman, P., & Venables, A. J. (1999). *The spatial economy: Cities, regions, and international trade*. MIT Press.

Gaspar, J. M., Castro, S. B. S. D., & Correia-da-Silva, J. (2020). The footloose entrepreneur model with a finite number of equidistant regions. *International Journal of Economic Theory, 16*(4), 420–446.

Ikeda, K., & Murota, K. (2014). *Bifurcation theory for hexagonal agglomeration in economic geography*. Springer.

Ikeda, K., Murota, K., & Takayama, Y. (2017). Stable economic agglomeration patterns in two dimensions: Beyond the scope of central place theory. *Journal of Regional Science, 57*(1), 132–172.

Ikeda, K., Onda, M., & Takayama, Y. (2018). Spatial period doubling, invariant pattern, and break point in economic agglomeration in two dimensions. *Journal of Economic Dynamics and Control, 92,* 129–152.

Murata, Y., & Thisse, J.-F. (2005). A simple model of economic geography á la Helpman-Tabuchi. *Journal of Urban Economics, 58*(1), 137–155.

Pflüger, M. (2004). A simple, analytically solvable, Chamberlinian agglomeration model. *Regional Science and Urban Economics, 34*(5), 565–573.

Pflüger, M., & Südekum, J. (2008). A synthesis of footloose-entrepreneur new economic geography models: When is agglomeration smooth and easily reversible? *Journal of Economics Geography, 8,* 39–54.

Tabuchi, T., & Thisse, J.-F. (2011). A new economic geography model of central places. *Journal of Urban Economics, 69,* 240–252.

Takayama, Y., Ikeda, K., & Thisse, J.-F. (2020). Stability and sustainability of urban systems under commuting and transportation costs. *Regional Science and Urban Economics, 84,* 103553.

# Part II
# Hierarchical Bifurcations and Invariant Patterns

Part II of this book is devoted to a study of hierarchical (repeated) bifurcations and invariant patterns of symmetric spatial platforms with many locations. Symmetric platforms potentially undergo hierarchical bifurcations until they lose symmetry completely. As symmetric platforms, we consider racetrack, equidistant, square lattice, and hexagonal lattice. We elucidate, through the group-theoretic analysis, the mechanisms of hierarchical bifurcations in these platforms.

There are unique spatial patterns that are solutions to the governing equation for any value of the bifurcation (transportation cost) parameter. Such patterns are called invariant patterns. Symmetric spatial platforms with replicator dynamics have many invariant patterns. Solution curves of these patterns are connected by those of non-invariant ones to form a complex mesh-like structure, akin to warp and weft threads. Invariant patterns are studied in symmetric spatial platforms, including:

- A racetrack economy in Chap. 9
- Square and hexagonal lattices with periodic boundary conditions in Chap. 10

Chapter 11 develops a global–local spatial platform for economic agglomerations, representing a hierarchical structure of cities, towns, villages, and so on. A global system models the geographical distribution of a system of cities. An individual square lattice network models each city's population size and local geography. We employ an air network of cities with the seven largest hub airports in the USA as a realistic example.

The mechanism of an economic agglomeration of spatial platforms is investigated through a comparative study with a racetrack economy.

- Chapter 12 deals with a long-narrow economy. Formulas for break points for the direct bifurcation (Chap. 8) in the racetrack economy are upgraded to encompass the secondary and further bifurcations. These upgraded formulas can predict the occurrence of spatial period doubling in the long-narrow economy.
- Chapter 13 investigates a square lattice based on an analogy to the racetrack economy. A doubling of the spatial period is advanced as a common mechanism to engender fewer larger agglomerations.

Chapter 14 deals with an equidistant economy. We study the mechanism of bifurcations from invariant patterns. The stability of bifurcating branches is analyzed to demonstrate that most branches are asymptotically unstable. This economy is employed as a global network in Chap. 11.

# Chapter 9
# Racetrack Economy: Hierarchical Bifurcation and Invariant Patterns

**Abstract** The mechanism of the direct bifurcation from the uniform state in the racetrack economy was presented in Chap. 3. This chapter extends this mechanism to a broader context. First, we obtain invariant patterns that are solutions to the governing equation for any value of the bifurcation parameter. Among the plethora of theoretically possible invariant patterns, we identify those that become stable within economic geography models. Next, we introduce sustaining bifurcation, which is a unique bifurcation in the replicator dynamics. Last, we describe the hierarchical bifurcation from the uniform state. A likely course predicted for $2^k$ places is a gradual and successive process, called a spatial period-doubling cascade. This chapter is a prerequisite for sustaining bifurcations in Chaps. 11 and 14. This chapter provides the theoretical foundation of racetrack economy analogy for a long-narrow economy in Chap. 12 and for a square lattice in Chap. 13.

**Keywords** Bifurcation · Economic geography model · Hierarchical bifurcation · Invariant pattern · Racetrack economy · Replicator dynamics · Spatial period doubling · Sustaining bifurcation

## 9.1 Introduction

Chapter 3 investigated the onset of the breaking of uniformity in the racetrack economy. This chapter studies how spatial economic agglomerations proceed after that, based on Ikeda et al. (2012, 2019).

First, attention is devoted to invariant patterns that are solutions to the governing equation for any value of the bifurcation parameter. We propose a theory of invariant patterns and present a complete list of invariant patterns in the racetrack economy with an arbitrary number $n$ of places.

Next, we introduce sustaining bifurcation, which is a unique bifurcation in the replicator dynamics. Through this bifurcation, a place with no population gains



K. Ikeda, Y. Takayama, *Bifurcation Theory of Pattern Formation in Economic Geography*, Interdisciplinary Applied Mathematics 61,
https://doi.org/10.1007/978-3-031-71686-7_9

population. Although, in economic geography, this bifurcation is not given much attention, it will play an important role in subsequent chapters: the sustaining bifurcation of full agglomeration in Chap. 11 and the hierarchical bifurcation of $S_N$-symmetric system in Chap. 14.

Last, we revisit the direct bifurcating patterns from the uniform state in Chap. 3. The structure of this group depends on the number $n$ of places, and, in turn, bifurcating patterns vary with this number. When $n$ is a prime number, bifurcation triggers a spontaneous loss of spatial periodicity. When $n$ is a composite number, diverse bifurcations take place, leading to spatial period doubling, tripling, and so on, depending on the factorization of the number. We observe a sequence of recurrent bifurcations, called spatial period-doubling cascade.

We present the results of a numerical bifurcation analysis to observe stable equilibria of economic geography models: the Krugman model (e.g., Fujita et al., 1999) and the Forslid and Ottaviano (2003) model. Knowledge of invariant patterns is useful in this analysis. The spatial period-doubling cascade is highlighted as a key phenomenon.

This cascade is observed in various economic geography models.[1] The racetrack economy is studied comparatively with several spatial platforms: the star economy by Barbero and Zofío (2016), the line segment (long-narrow economy) in a continuous space by Mossay and Picard (2011), the long-narrow economy in a discrete space by Ikeda et al. (2017) (Chap. 12), and the square lattice by Ikeda et al. (2018) (Chap. 13).

This chapter is organized as follows: Sect. 9.2 advances invariant patterns in replicator dynamics. Sustaining bifurcation in the replicator dynamics is studied in Sect. 9.3. In Sect. 9.4, the mechanism of hierarchical symmetry-breaking bifurcation in the racetrack economy is advanced. Section 9.5 presents the numerical analysis in economic geography models.

## 9.2 Invariant Patterns in Replicator Dynamics

We propose a general theory of invariant patterns in replicator dynamics. This theory is applied to the racetrack economy to obtain its invariant patterns.

### 9.2.1 Replicator Dynamics and Corner Solutions

We consider the governing equation

$$\boldsymbol{F}(\boldsymbol{\lambda}, \tau) = \boldsymbol{0} \tag{9.1}$$

[1] See, e.g., Picard and Tabuchi (2010), Tabuchi and Thisse (2011), Akamatsu et al. (2012), Ikeda et al. (2012), and Osawa et al. (2017).

of the replicator dynamics with a sufficiently smooth function $\boldsymbol{F}(\boldsymbol{\lambda}, \tau) = (F_i(\boldsymbol{\lambda}, \tau) \mid i = 1, \ldots, n)$ defined as

$$F_i(\boldsymbol{\lambda}, \tau) = (v_i(\boldsymbol{\lambda}, \tau) - \bar{v}(\boldsymbol{\lambda}, \tau))\lambda_i.$$

Here, $\tau$ is the bifurcation (transportation cost) parameter; $\boldsymbol{\lambda}$ is the independent variable vector; and $\bar{v} = \sum_{i=1}^{n} \lambda_i v_i$.

In preparation for the description of the bifurcation mechanism, we classify stationary points $(\boldsymbol{\lambda}, \tau)$ in the replicator dynamics into *interior solutions* and *corner solutions*. In the former solutions, all places have positive populations, whereas in the latter solutions, one or more places have no population.

We classify stationary points into *invariant solutions* and *non-invariant solutions*. Stationary points that satisfy the governing equation (9.1) form solution curves $(\boldsymbol{\lambda}(\tau), \tau)$ parameterized by $\tau$. In general, the spatial pattern $\boldsymbol{\lambda}(\tau)$ varies with $\tau$ along a solution curve. By contrast, there can be a special solution curve $(\boldsymbol{\lambda}(\tau), \tau) = (\bar{\boldsymbol{\lambda}}, \tau)$ that has a constant spatial pattern $\boldsymbol{\lambda}(\tau) = \bar{\boldsymbol{\lambda}}$ along the curve. Such pattern $\bar{\boldsymbol{\lambda}}$ is called an *invariant pattern*. The solution curve of an invariant pattern exists for any value of $\tau > 0$. In contrast, a pattern $\boldsymbol{\lambda}(\tau)$ that varies with $\tau$ along a curve is called a *non-invariant pattern* and might not be a solution for a given $\tau$.

A pattern $\boldsymbol{\lambda}$ for a corner solution can be expressed without loss of generality by appropriately rearranging the order of its components as

$$\hat{\boldsymbol{\lambda}} = (\boldsymbol{\lambda}_m^+, \mathbf{0}_{n-m}) \qquad (1 \leq m < n) \tag{9.2}$$

with $\boldsymbol{\lambda}_m^+ = \{\lambda_i > 0,\ \sum_{1=1}^{m} \lambda_i = 1 \mid 1 \leq i \leq m\}$ and the $(n-m)$-dimensional zero vector $\mathbf{0}_{n-m}$. The static governing equation (9.1) can be rearranged accordingly as

$$\hat{\boldsymbol{F}}(\boldsymbol{\lambda}, \tau) = \{\boldsymbol{F}^+(\boldsymbol{\lambda}_m^+, \mathbf{0}_{n-m}, \tau), \boldsymbol{F}^0(\boldsymbol{\lambda}_m^+, \mathbf{0}_{n-m}, \tau)\} = \mathbf{0}. \tag{9.3}$$

We consider a special type of corner solution, the *core–periphery pattern*:

$$\boldsymbol{\lambda}_m^{\mathrm{CP}} = \left(\frac{1}{m}\mathbf{1}_m, \mathbf{0}_{n-m}\right) \qquad (1 \leq m < n) \tag{9.4}$$

$(\boldsymbol{\lambda}_m^+ = \frac{1}{m}\mathbf{1}_m)$ with a two-level hierarchy: $m$ core places with the same positive population and other peripheral places with no population. In particular, $\boldsymbol{\lambda}_1^{\mathrm{CP}} = (1, \mathbf{0}_{n-1})$ in the case where $m = 1$ is called *full agglomeration* to a single place. The core–periphery pattern can be an invariant pattern under some symmetry conditions, as explained in the next section. It is a unique feature in the replicator dynamics to accommodate many invariant patterns for symmetric spatial platforms.

### 9.2.2 General Theory of Invariant Patterns

We present a general theory of invariant patterns. In preparation for the discussion on these patterns, we introduce the *orbit decomposition* (Ikeda & Murota, 2019, §13.3.1). Let

$$P = \cup P_\ell \tag{9.5}$$

be the decomposition of a set $P$ of nodal points (places) into disjoint orbits $P_\ell$ regarding the permutation representation $T(g)$ ($g \in G$). That is, two points $i$ and $j$ belong to the same orbit $P_\ell$ if and only if $i$ is moved to $j$ by $T(g)$ for some $g \in G$. The following lemma demonstrates the use of the orbit in the discussion of invariant patterns.

**Lemma 9.1** *The core–periphery pattern* $\boldsymbol{\lambda}_m^{\mathrm{CP}} = (\frac{1}{m}\mathbf{1}_m, \mathbf{0}_{n-m})$ *is an invariant pattern if the set of points for* $\boldsymbol{\lambda}_m^{+} = \frac{1}{m}\mathbf{1}_m$ *belongs to an orbit of a subgroup* $G'$ *of* $G$.

***Proof*** Since $m$ points associated with $\frac{1}{m}\mathbf{1}_m$ belong to an orbit, we have $v_1 = \cdots = v_m$, as well as $\lambda_1 = \cdots = \lambda_m = 1/m$. Then, we have $\bar{v} = \sum_{i=1}^{m} \lambda_i v_i = v_1$ and, in turn, $v_i - \bar{v} = v_1 - \bar{v} = 0$ $(i = 1, \ldots, m)$, thereby satisfying $\boldsymbol{F}^{+}(\boldsymbol{\lambda}_m^{\mathrm{CP}}, \tau) = ((v_i - \bar{v})\lambda_i \mid i = 1, \ldots, m) = \mathbf{0}$ (see (9.3)). For $n - m$ places with no population, we have $\lambda_i = 0$ $(i = m + 1, \ldots, n)$, thereby satisfying $\boldsymbol{F}^{0}(\boldsymbol{\lambda}_m^{\mathrm{CP}}, \tau) = ((v_i - \bar{v})\lambda_i \mid i = m + 1, \ldots, n) = \mathbf{0}$. Thus, $(\boldsymbol{\lambda}_m^{\mathrm{CP}}, \tau)$ is a solution to the governing equation (9.1) for any $\tau$ and, accordingly, is an invariant pattern. □

Then, this lemma leads to the following proposition for the full agglomeration.

**Proposition 9.1** *A full agglomeration at any place is an invariant pattern for any spatial platform.*

***Proof*** A single point at place $i$ forms an orbit for a subgroup $G' = \{e\}$. Then, by Lemma 9.1, a full agglomeration in place $i$ is an invariant pattern. □

### 9.2.3 Invariant Patterns in Racetrack Economy

We consider a *racetrack economy* with $n$ places distributed evenly on the circumference of a circle (Fig. 9.1) and advance invariant patterns in the replicator dynamics in this economy. The symmetry of this economy plays a pivotal role, as explained in Lemma 9.1. The symmetry group of this economy is the *dihedral group* in (3.1):

$$\mathrm{D}_n = \langle r, s \rangle = \{r^i, sr^i \mid i = 0, 1, \ldots, n-1\}.$$

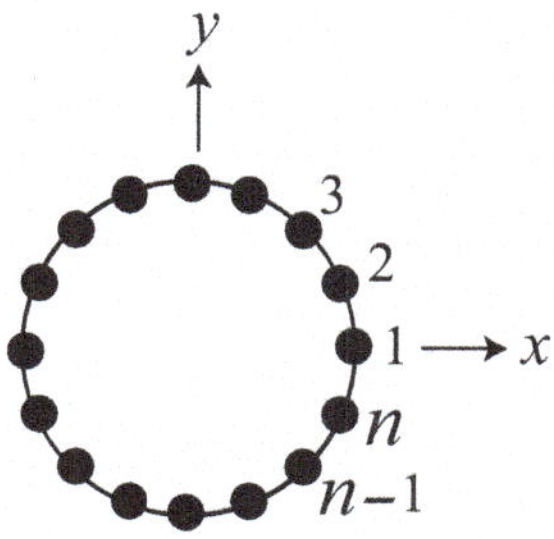


**Fig. 9.1** Racetrack economy with $n$ places

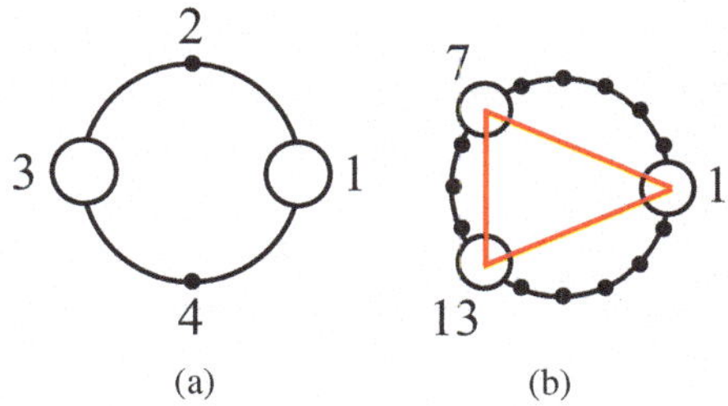


**Fig. 9.2** Examples of invariant pattern and non-invariant pattern (the area of a circle (○) denotes a population size; (•): a place with no population). (**a**) Invariant pattern. (**b**) Non-invariant pattern

Here, $\langle \cdots \rangle$ denotes a group generated by the geometrical transformations in the parentheses; $\{\cdot\}$ denotes a group consisting of the geometrical transformations therein; $r^i$ denotes a counterclockwise *rotation* about the center of the circle at an angle of $2\pi i/n$ $(i = 0, 1, \ldots, n-1)$; and $sr^i$ is the combined action of the rotation $r^i$ followed by the upside-down *reflection* $s$.

We begin with simple examples of the use of Lemma 9.1 in distinguishing invariant and non-invariant patterns.

**Example 9.1** We consider the racetrack economy with four places and its corner solution with equal complete agglomeration to places 1 and 3, i.e., $\boldsymbol{\lambda} = (\lambda_1, \lambda_2, \lambda_3, \lambda_4) = \frac{1}{2}(1, 0, 1, 0)$, as depicted in Fig. 9.2a. It reduces to a core–periphery pattern $\hat{\boldsymbol{\lambda}} = (\lambda_1, \lambda_3, \lambda_2, \lambda_4) = (\frac{1}{2}\mathbf{1}_2, \mathbf{0}_2) = \boldsymbol{\lambda}_2^{\mathrm{CP}}$ through a pertinent permutation of places. For $G' = \{e, r^2\}$ ($r^2$ is the $\pi$ rotation), we have permutation matrices:

$$T(e) = \begin{pmatrix} 1 & & & \\ & 1 & & \\ & & 1 & \\ & & & 1 \end{pmatrix}, \quad T(r^2) = \begin{pmatrix} & & 1 & \\ & & & 1 \\ 1 & & & \\ & 1 & & \end{pmatrix}.$$

Thus, $T(r^2)$ permutes the places 1 and 3, as well as 2 and 4. In (9.5), we have $P = \{1, 2, 3, 4\}$, $P_1 = \{1, 3\}$, and $P_2 = \{2, 4\}$. This core–periphery pattern is an invariant pattern by Lemma 9.1. □

**Example 9.2** The spatial pattern in Fig. 9.2b has the form $\hat{\boldsymbol{\lambda}} = \boldsymbol{\lambda}_3^{\rm CP} = (\frac{1}{3}\mathbf{1}_3, \mathbf{0}_{13})$ in (9.4) with $m = 3$ and $n = 16$. This pattern is invariant to the reflection $s$ for the horizontal axis, and the subgroup $\{e, s\}$ expresses its symmetry. This subgroup permutes places 7 and 13 but cannot permute places 1 and 7 (or 13). Hence, this subgroup cannot serve as the subgroup $G'$ in Lemma 9.1. Since it is not possible to find $G'$ in Lemma 9.1, this pattern is a non-invariant pattern. □

There are several important invariant patterns that exist for any $n$.

**Proposition 9.2** *A full agglomeration, the uniform state, and twin places are invariant patterns that exist for any number $n$ of places.*

***Proof*** The full agglomeration is an invariant pattern by Proposition 9.1. For the uniform state, we have $v_1 = \cdots = v_n = \bar{v}$, and the governing equation (9.1) is satisfied for any $\tau$; therefore, the uniform state is an invariant pattern for any $n$. For any twin places, we can find a pertinent reflection $s'$ that exchanges these places; Lemma 9.1 is satisfied by $G' = \langle s' \rangle$, thereby showing that the twin places are an invariant pattern for any $n$. □

We consider two types of $\mathrm{D}_m$-invariant patterns (for some $m$ divides $n$):

- *Regular m places*: A complete agglomeration to regular $m$-gonal places (Fig. 9.3a)
- *Skewed 2m places*: Agglomeration to a pair of places cyclically repeated $m$ times (Fig. 9.3b)

Two places of these types are called *regular twin places* and *skewed twin places*.

**Proposition 9.3** *The regular m places ($m$ divides $n$; $1 \leq m \leq n$) and the skewed 2m places ($m$ divides $n$; $1 \leq m \leq n/3$) are invariant patterns.*

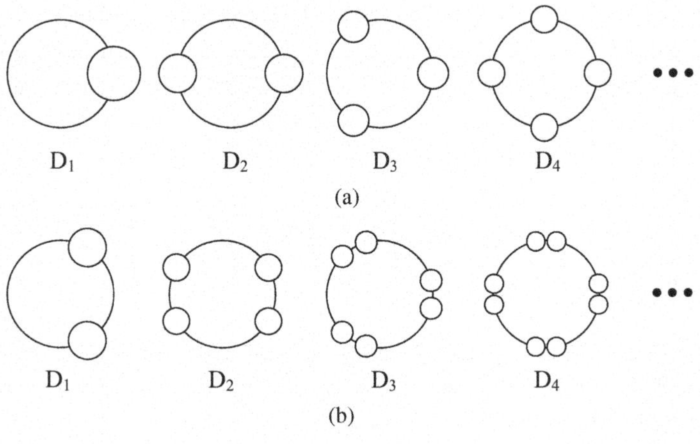


**Fig. 9.3** Invariant patterns with $\mathrm{D}_m$-invariance. (**a**) Regular $m$ places. (**b**) Skewed $2m$ places

**Table 9.1** Existence and non-existence of invariant patterns for several values of $n$ (○ indicates the existence and × denotes the non-existence)

(a) Regular $m$ places

| Number $n$ of places | $m$ = 12 | 6 | 5 | 4 | 3 | 2 | 1 |
|---|---|---|---|---|---|---|---|
| 2 | × | × | × | × | × | ○ | ○ |
| 3 | × | × | × | × | ○ | × | ○ |
| 4 | × | × | × | ○ | × | ○ | ○ |
| 5 | × | × | ○ | × | × | × | ○ |
| 6 | × | ○ | × | × | ○ | ○ | ○ |
| 12 | ○ | ○ | × | ○ | ○ | ○ | ○ |

(b) Skewed $2m$ places

| Number $n$ of places | $m$ = 12 | 6 | 5 | 4 | 3 | 2 | 1 |
|---|---|---|---|---|---|---|---|
| 2 | × | × | × | × | × | × | × |
| 3 | × | × | × | × | × | × | ○ |
| 4 | × | × | × | × | × | × | ○ |
| 5 | × | × | × | × | × | × | ○ |
| 6 | × | × | × | × | × | ○ | ○ |
| 12 | × | × | × | ○ | ○ | ○ | ○ |

***Proof*** Any two places of the regular $m$ places can be permuted by elements of $\mathrm{D}_m$. Any two places of the skewed $2m$ places are also permuted by $\mathrm{D}_m$, comprising a $2\pi/m$ rotation and a reflection that permutes two neighboring populated places. Hence, using $G' = \mathrm{D}_m$ in Lemma 9.1, we can ensure that they are invariant patterns. Skewed $3m$ places are $\mathrm{D}_m$-invariant, but no transformation permutes three neighboring agglomerated places; accordingly, they are non-invariant patterns. Likewise, regular $4m$ places and so on are non-invariant patterns. Since $\mathrm{D}_m$ is a subgroup of $\mathrm{D}_n$, $n$ is divisible by $m$. For the regular $m$ places, we have $1 \leq m \leq n$. For the skewed $2m$ places, $n/m = 1$ is infeasible since we have $n = m$ and violates $n \geq 2m$, and $n/m = 2$ is infeasible since it is nothing but the regular $n$ places as $2m = n$. Thus, $n/m \geq 3$ holds for the skewed $2m$ ones, thereby leading to $1 \leq m \leq n/3$. □

We can obtain invariant patterns of any number of places by these two propositions. Table 9.1 lists feasible values $m$ of invariant patterns for several numbers $n$ of places. For example, invariant patterns for $n = 12$ are depicted in Fig. 9.4.

## 9.3 Sustaining Bifurcation in Replicator Dynamics

We advance the mechanism of sustaining bifurcation at a corner solution in the replicator dynamics. This mechanism is general and is applicable to spatial platforms other than the racetrack economy.

Recall the appropriately rearranged vector $\hat{\boldsymbol{\lambda}} = (\boldsymbol{\lambda}_m^+, \mathbf{0}_{n-m})$ $(1 \leq m < n)$ in (9.2) and the associated rearranged governing equation $\hat{\boldsymbol{F}}(\boldsymbol{\lambda}, \tau) = \{\boldsymbol{F}^+, \boldsymbol{F}^0\} = \mathbf{0}$ in (9.3) for a corner solution. The Jacobian matrix of this rearranged equation becomes

$$\hat{J} = \frac{\partial \hat{\boldsymbol{F}}}{\partial \hat{\boldsymbol{\lambda}}} = \begin{bmatrix} J^+ & J^{+0} \\ O & J^0 \end{bmatrix}, \tag{9.6}$$

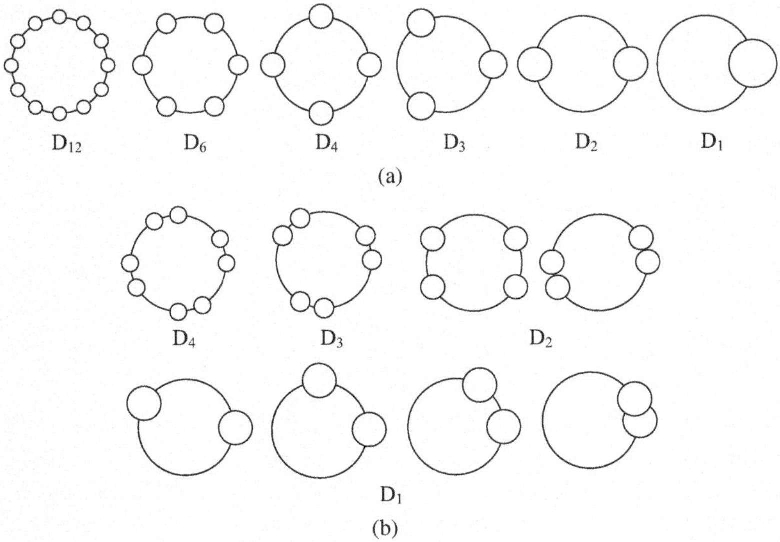


**Fig. 9.4** Invariant patterns for $n = 12$. (**a**) Regular $m$ places. (**b**) Skewed $2m$ places

where

$$J^{+} = \mathrm{diag}(\lambda_1, \ldots, \lambda_m)\Omega^{+}, \quad \Omega^{+} = \left\{ \frac{\partial(v_i - \bar{v})}{\partial \lambda_j} \mid i, j = 1, \ldots, m \right\}; \tag{9.7}$$

$$J^{+0} = \mathrm{diag}(\lambda_1, \ldots, \lambda_m)\Omega^{+0},$$

$$\Omega^{+0} = \left\{ \frac{\partial(v_i - \bar{v})}{\partial \lambda_j} \mid i = 1, \ldots, m;\ j = m+1, \ldots, n \right\};$$

$$J^{0} = \mathrm{diag}(v_{m+1} - \bar{v}, \ldots, v_n - \bar{v}).$$

**Lemma 9.2** *The eigenvalues of $J$ are $e_i$ $(i = 1, \ldots, m)$ of $J^{+}$ and $e_i = v_i - \bar{v}$ $(i = m+1, \ldots, n)$ of $J^{0}$.*

We can replace a corner solution's sustainability check with the Jacobian matrix's stability investigation. For a stable (corner) solution, all eigenvalues of $\hat{J}$ are negative; accordingly, $v_{m+1} - \bar{v}, \ldots, v_n - \bar{v}$ are all negative, thereby satisfying the sustainability condition $v_i - \bar{v} \leq 0$ $(i = 1, \ldots, n)$ in (1.8). Thus, we have:

- A stable solution is sustainable (Hofbauer & Sigmund, 1998).
- An unsustainable solution is unstable.

It leads to the following proposition.

**Proposition 9.4** *A stable corner solution with nonnegativity ($\lambda_i \geq 0$ for all $i$) is a stable spatial equilibrium.*

***Proof*** A solution to the replicator dynamics is a steady state. A stable solution is sustainable as stated above. This solution becomes a spatial equilibrium in (1.8), if the nonnegativity is satisfied additionally. □

By Lemma 9.2, we can classify the critical points of a system with the replicator dynamics as

$$\begin{cases} J^{+} \text{ has a zero eigenvalue}: & \textit{breaking bifurcation point} \text{ or } \textit{stationary point} \text{ of } \tau. \\ e_i = v_i - \bar{v} = 0 \ (i = m+1, \dots, n): & \textit{sustaining bifurcation point}. \end{cases}$$

A sustaining bifurcation occurs at a point, where $v_i - \bar{v} = 0$ and $\lambda_i = 0$ are satisfied simultaneously. A *sustain point* is the sustaining bifurcation point at which the sustainability changes.

## 9.4 Hierarchical Symmetry-Breaking Bifurcation

Section 3.4 investigated the mechanism of the direct bifurcation at the uniform state

$$\boldsymbol{\lambda} = \boldsymbol{\lambda}^* = \frac{1}{n}(1, \dots, 1)^\top$$

for $n$ even. We extend this mechanism to encompass the case of $n$ odd and secondary and further bifurcations (Ikeda & Murota, 2019, Chapter 9). This mechanism is independent of the economic modeling and dynamics employed.

### *9.4.1 Bifurcation from the Uniform State*

The eigenvalues of the Jacobian matrix $J = \partial \boldsymbol{F}/\partial \boldsymbol{\lambda}$ are related to the irreducible representations of the group $\mathrm{D}_n$ and are given by extending (3.16)–(3.18) for $n$ even to encompass $n$ odd as follows:

$$\begin{cases} e^{(+)},\ e^{(-)} & \text{for } n = 2, \\ e^{(+)},\ e^{(-)},\ e^{(1)},\ e^{(1)},\ \dots,\ e^{(j_{\max})},\ e^{(j_{\max})} & \text{for } n = 2q+2 \ \ (q = 1, 2, \dots), \\ e^{(+)},\ e^{(1)},\ e^{(1)},\ \dots,\ e^{(j_{\max})},\ e^{(j_{\max})} & \text{for } n = 2q+1 \ \ (q = 1, 2, \dots). \end{cases} \tag{9.8}$$

Here, $e^{(+)}$ and $e^{(-)}$ are simple eigenvalues that are repeated once; $e^{(1)}, \dots, e^{(j_{\max})}$ are double eigenvalues repeated twice; and

$$j_{\max} = \begin{cases} n/2 - 1 & (n \text{ even}), \\ (n-1)/2 & (n \text{ odd}). \end{cases}$$

Critical points at the uniform state are classified in (3.19) as

$$\begin{cases} e^{(+)} = 0: & \text{stationary } (\textit{limit}) \text{ point of } \tau \ (M = 1). \\ e^{(-)} = 0: & \textit{simple bifurcation point } (M = 1). \\ e^{(j)} = 0: & \textit{double bifurcation point } (M = 2). \end{cases}$$

Here, $M$ is the *multiplicity* of a critical point that is defined as the number of zero eigenvalues of $J$.

At a simple bifurcation point with $e^{(-)} = 0$, which exists for $n$ even, we encounter symmetry breaking expressed by

$$\mathrm{D}_n \longrightarrow \mathrm{D}_{n/2},$$

where $\longrightarrow$ denotes an occurrence of bifurcation. This bifurcation doubles the *spatial period*: $T_n = 2\pi/n \longrightarrow T_{n/2} = 2 \times 2\pi/n$. The bifurcation is symmetric in the sense that the eigenvectors $\boldsymbol{\eta}^{(-)}$ and $-\boldsymbol{\eta}^{(-)}$ are *conjugate* (identical up to geometrical transformation $T(r)\boldsymbol{\eta}^{(-)} = -\boldsymbol{\eta}^{(-)}$ for $r \in \mathrm{D}_n$).

At a double bifurcation point with $e^{(j)} = 0$, which exists for $n \geq 3$, we encounter a reduction of symmetry expressed by

$$\mathrm{D}_n \longrightarrow \mathrm{D}_{n/\hat{n}}. \tag{9.9}$$

Here,

$$\hat{n} = n/\gcd(n, j) \quad (3 \leq \hat{n} \leq n) \tag{9.10}$$

is an important index for the double bifurcation point, where $\gcd(n, j)$ is the greatest common divisor of $n$ and $j$. This bifurcation extends the spatial period $\hat{n}$ times: $T_n = 2\pi/n \longrightarrow T_{n/\hat{n}} = \hat{n} \times 2\pi/n$. The double bifurcation is *symmetric* when $\hat{n}$ is even and is *asymmetric* when $\hat{n}$ is odd.

**Proposition 9.5** *When $n$ is a prime number greater than 2, there are only double bifurcations that are asymmetric and lead to the reduction of symmetry* $\mathrm{D}_n \longrightarrow \mathrm{D}_1$.

***Proof*** The prime number $n$ greater than 2 is odd. There is no simple bifurcation because $e^{(-)}$ for this bifurcation is absent for odd $n$ $(= 2q + 1)$ in (9.8). This bifurcation is asymmetric since the index of a double bifurcation is $\hat{n} = n/\gcd(n, j) = n$ $(j = 1, \ldots, j_{\max})$ and is odd. Since $n/\hat{n} = 1$, (9.9) gives $\mathrm{D}_n \longrightarrow \mathrm{D}_1$. □

**Proposition 9.6** *When $n$ is divisible by $m$ $(1 \leq m \leq n/2)$, there are bifurcations that lead to the reduction of symmetry* $\mathrm{D}_n \longrightarrow \mathrm{D}_m$.

***Proof*** The case of $n/m = 2$ corresponds to a simple bifurcation point. Otherwise, we have $3 \leq n/m \leq n$ and set $j = m$ $(1 \leq j < n/2)$ that gives $\hat{n} = n/\gcd(n, j) = n/m$. Thus, there is a double bifurcation associated with $j = m$. □

## 9.4.2 Hierarchical Bifurcation

We show the dependence of the bifurcation mechanism on the number $n$ of places for (1) prime number, (2) composite number $n = 2^k$, and (3) other composite numbers.

A prime number $p$ has the simplest hierarchical bifurcation structure (see Proposition 9.5):

$$\mathrm{D}_p \longrightarrow \mathrm{D}_1 \longrightarrow \mathrm{C}_1.$$

When $n$ is a composite number, the mechanism of bifurcation depends on the factorization of the number $n$, which, in general, takes the form of

$$n = 2^{n_2} 3^{n_3} 5^{n_5} \cdots$$

for some nonnegative integer $n_q \geq 0$ $(q = 2, 3, 5, \ldots)$. By Proposition 9.6, the existence of the factors $q^{n_q}$ $(n_q \geq 1)$ indicates that spatial period $q$ times bifurcations can occur $n_q$ times at most.

For $n = 2^k$ $(k = 1, 2, \ldots)$ places, simple bifurcations can take place repeatedly to engender a hierarchy of subgroups

$$\mathrm{D}_n \longrightarrow \mathrm{D}_{n/2} \longrightarrow \cdots \longrightarrow \mathrm{D}_2 \longrightarrow \mathrm{D}_1 \tag{9.11}$$

and lead to a *spatial period-doubling cascade*:[2]

$$T_n = 2\pi/n \longrightarrow 2 \times 2\pi/n \longrightarrow \cdots \longrightarrow \pi \longrightarrow 2\pi.$$

Figure 9.5 depicts this cascade for $n = 8 = 2^3$ places.

Based on the theory presented above, we can analyze agglomeration behaviors for a specific number $n$ of places, as conducted for $n = 4$ below.

**Example 9.3** By Proposition 9.3, four places have several invariant patterns: the uniform state $\boldsymbol{\lambda} = (1/4, 1/4, 1/4, 1/4)^\top$, regular twin places $(1/2, 0, 1/2, 0)^\top$, skewed twin places $(1/2, 1/2, 0, 0)^\top$, and a full agglomeration $(1, 0, 0, 0)^\top$. At the uniform state, there is a simple bifurcation and a double bifurcation with $\hat{n} = 4$. The

[2] A repeated doubling of the period through bifurcations takes place in many physical systems (e.g., Feigenbaum, 1978) and is called the period-doubling cascade.

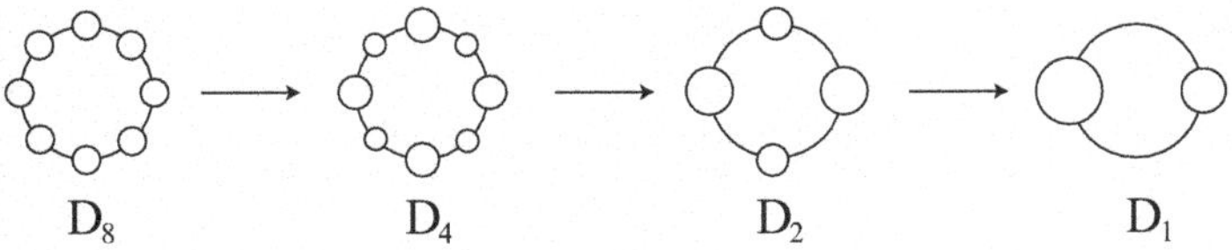


**Fig. 9.5** Spatial period-doubling cascade for $n = 8$ (the arrow denotes the occurrence of a simple bifurcation)

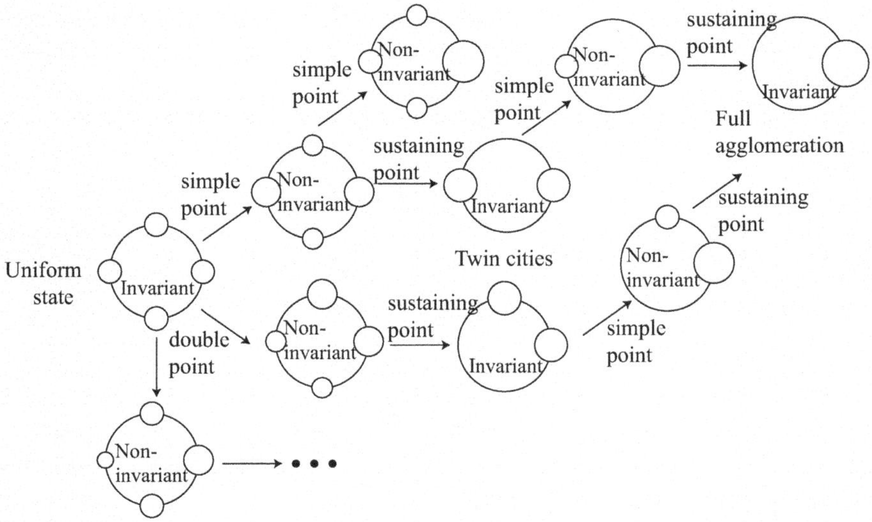


**Fig. 9.6** Theoretical hierarchical agglomeration mechanism for $n = 4$ places

simple bifurcation engenders a spatial period-doubling, non-invariant pattern:

$$\boldsymbol{\lambda} = (1/4 + a, 1/4 - a, 1/4 + a, 1/4 - a)^\top \quad (-1/4 < a < 1/4), \tag{9.12}$$

which reaches the invariant pattern $(1/2, 0, 1/2, 0)^\top$ at a sustaining bifurcation point ($a = 1/4$). Another simple bifurcation of the non-invariant pattern in (9.12) or of the invariant pattern $(1/2, 0, 1/2, 0)^\top$ leads to a further spatial period doubling. Figure 9.6 illustrates this bifurcation mechanism, comprising simple, double, and sustaining bifurcations. □

## 9.5 Bifurcation Behavior in Economic Geography Models

Agglomerations in the racetrack economy are investigated through the numerical analysis of the replicator dynamics in the Krugman model (Fujita et al., 1999) and

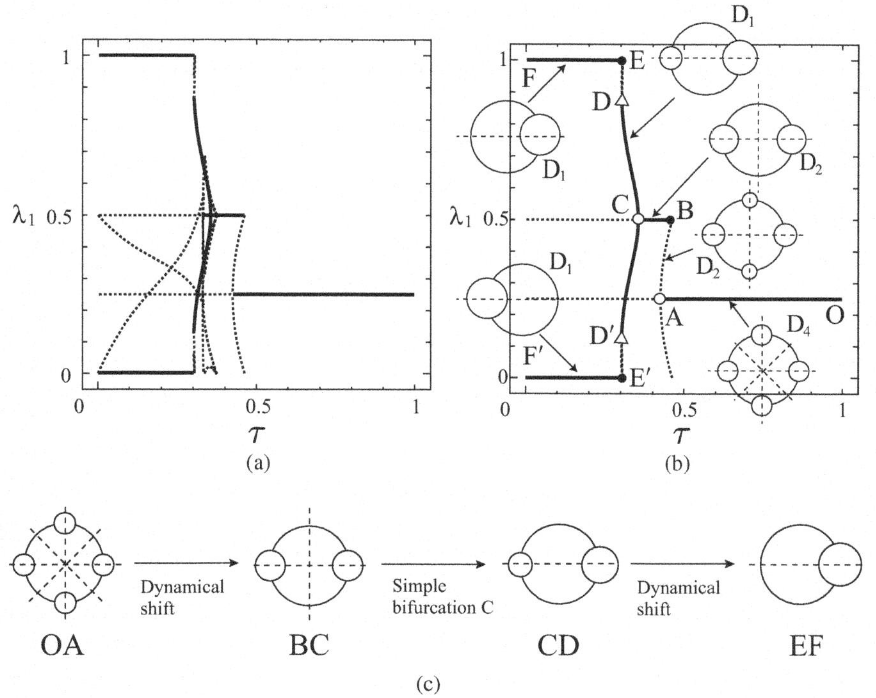


**Fig. 9.7** Solution curves of the four places ($n = 4$) and a predicted shift of stable equilibria in association with the decrease in transportation cost (Ikeda et al., 2012) (solid curve: stable; dashed curve: unstable; (○): simple bifurcation point; (△): limit point; (•): sustain point). (**a**) All solution curves. (**b**) Stable curves of invariant solutions and associated curves. (**c**) Predicted shift of stable equilibria as $\tau$ decreases

the Forslid and Ottaviano model (2003) (Sect. 2.2). We set $(\sigma, \mu) = (10.0, 0.4)$, which satisfies the so-called *no-black-hole condition*: $(\sigma - 1)/\sigma = 0.9 > \mu = 0.4$ in the former model and $\sigma - 1 = 9.0 > \mu = 0.4$ in (8.9) in the latter model.

## 9.5.1 Krugman Model

We demonstrate the successive and gradual transition of spatial agglomerations caused by a spatial period-doubling cascade for $n = 4$ and 16 places in the Krugman model.

Figure 9.7a shows the solution curves for the four places ($n = 4$). The horizontal lines at $\lambda_1 = 0$, 1/4, 1/2, and 1 are solution curves for $D_m$-symmetric ($m = 1, 2, 4$) invariant patterns. Non-horizontal bifurcation paths connect these lines. The bifurcation mechanism observed below is in complete agreement with the theoretical prediction in Fig. 9.6.

The stable curves for invariant patterns and associated paths shown in Fig. 9.7b are highlighted as those most likely to occur among the curves in (a); distributions of populations are portrayed for several equilibrium points. The stable curves of the invariant patterns are:

- OA: uniform state $\boldsymbol{\lambda} = (1/4, 1/4, 1/4, 1/4)^\top$ ($D_4$-symmetry)
- BC: period-doubling state $\boldsymbol{\lambda} = (1/2, 0, 1/2, 0)^\top$ ($D_2$-symmetry)
- EF: full agglomeration $\boldsymbol{\lambda} = (1, 0, 0, 0)^\top$ ($D_1$-symmetry)
- E′F′: the other full agglomeration $\boldsymbol{\lambda} = (0, 0, 1, 0)^\top$ ($D_1$-symmetry) (EF and E′F′ are conjugate solutions with the same economic meaning)

Bifurcation points on these invariant solutions are *pitchfork* and *sustaining bifurcation points*. The spatial period doubling occurs recurrently at pitchfork bifurcation point A for $D_4 \longrightarrow D_2$ and point C for $D_2 \longrightarrow D_1$ both denoted as (∘). We can predict a possible course that comprises four stable stages: OA, BC, CD, and EF. As $\tau$ decreases, dynamical shifts[3] are assumed between OA and BC and between CD and EF. Starting from the uniform state $\boldsymbol{\lambda} = (1/4, 1/4, 1/4, 1/4)^\top$, through bifurcations and dynamical shifts, we arrive at the complete concentration $\boldsymbol{\lambda} = (1, 0, 0, 0)^\top$. A spatial period-doubling cascade, $D_4 \longrightarrow D_2 \longrightarrow D_1$, occurs in this manner (see (9.11)).

Symmetries are preserved at the sustain points denoted as (•), where a solution curve for an invariant pattern intersects with one for a non-invariant pattern. Sustain point B has $D_2$-symmetry, and E and E′ have $D_1$-symmetry. Among the bifurcated paths, CD and CD′ are stable. The stable path CD becomes unstable at the limit (maximum) point of $\tau$ at D denoted by (△).

For 16 places, a spatial period-doubling cascade, $D_{16} \longrightarrow D_8 \longrightarrow D_4 \longrightarrow D_2 \longrightarrow D_1$, occurs as shown in Fig. 9.8. For the four and 16 places, the direct bifurcation is always *subcritical* pitchfork (tomahawk) and the secondary bifurcation is *supercritical* pitchfork.

## 9.5.2 Forslid and Ottaviano Model

We obtained spatial agglomeration patterns for various numbers $n$ of places in the Forslid and Ottaviano model (2003) (Sect. 2.2).

First, we numerically investigated the stability of invariant patterns. A pattern is called *stable* herein if it is stable for some range of $\tau$, and is called *unstable* if it is unstable for any $\tau$. For prime numbers $n = 3$, 5, and 7, Fig. 9.9 depicts the range of $\tau$ for stable invariant patterns. For $n = 3$, stable invariant patterns exist for any $\tau$, and there are distinct ranges of $\tau$ for the stable uniform state and the full agglomeration. For $n = 5$ and 7, there are three distinct ranges of $\tau$ in which the

[3] When a stable solution curve becomes unstable at a critical point where a stable bifurcated curve is nonexistent, a "dynamical shift" would take place to jump into another stable curve.

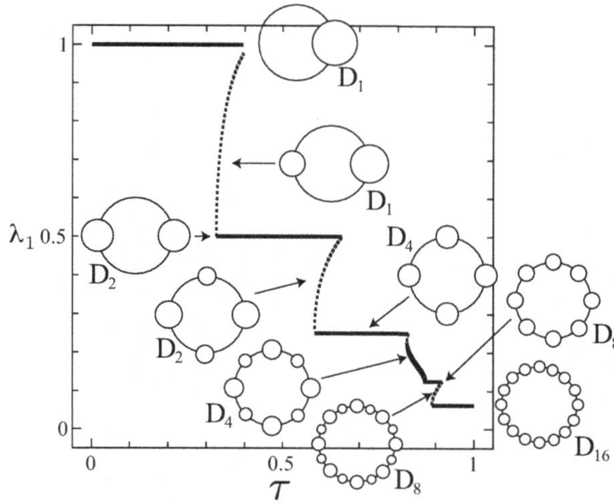


**Fig. 9.8** Solution curves of 16 places (Ikeda et al., 2012) (solid curve: stable; dashed curve: unstable)

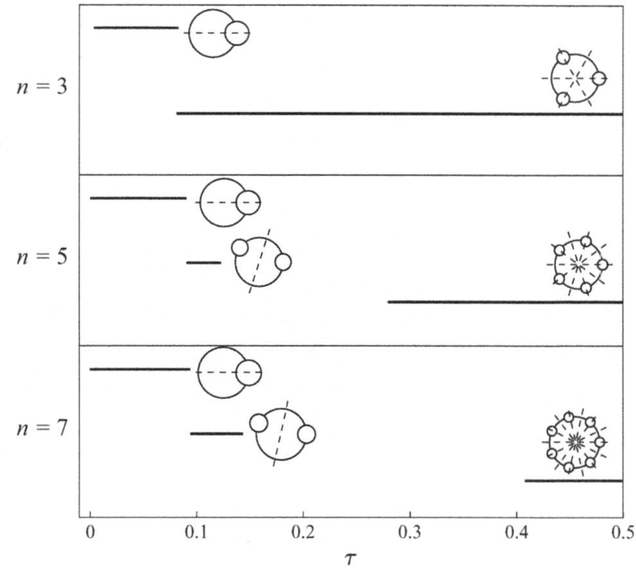


**Fig. 9.9** Ranges of $\tau$ for stable invariant patterns for $n = 3$, 5, and 7 (Ikeda et al., 2019) (solid curve: stable; dashed curve: unstable)

uniform state, skewed twin places, and full agglomeration are stable; there are gaps of $\tau$ in which stable invariant patterns are absent.

Next, the stability of invariant patterns for various numbers $n$ of places is classified in Table 9.2. All the regular $m$ places are stable. On the other hand, not all invariant patterns of skewed $2m$ places are stable, and stable ones have shapes closer to regular twin or regular four places.

**Table 9.2** Stability of invariant patterns for various numbers $n$ of places

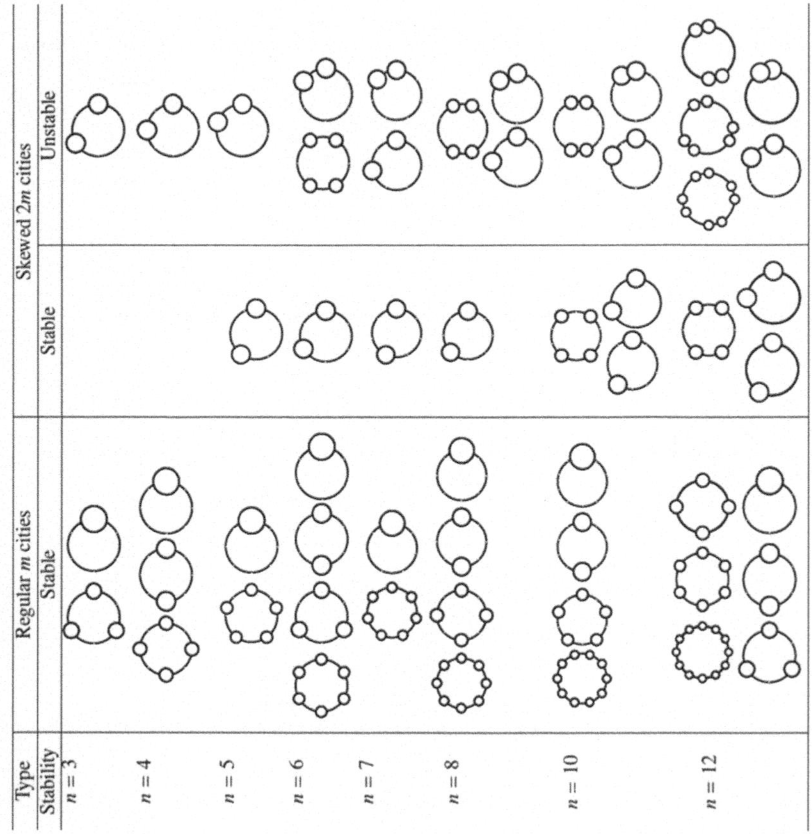

**Fig. 9.10** Solution curves for $n = 3$, 5, and 6 (Ikeda et al., 2019) (solid curve: stable; dashed curve: unstable; $\lambda_{\max} = \max(\lambda_1, \ldots, \lambda_n)$; (○): simple bifurcation; (△): double bifurcation; (●): sustain point). (**a**) $n = 3$. (**b**) $n = 5$. (**c**) $n = 6$

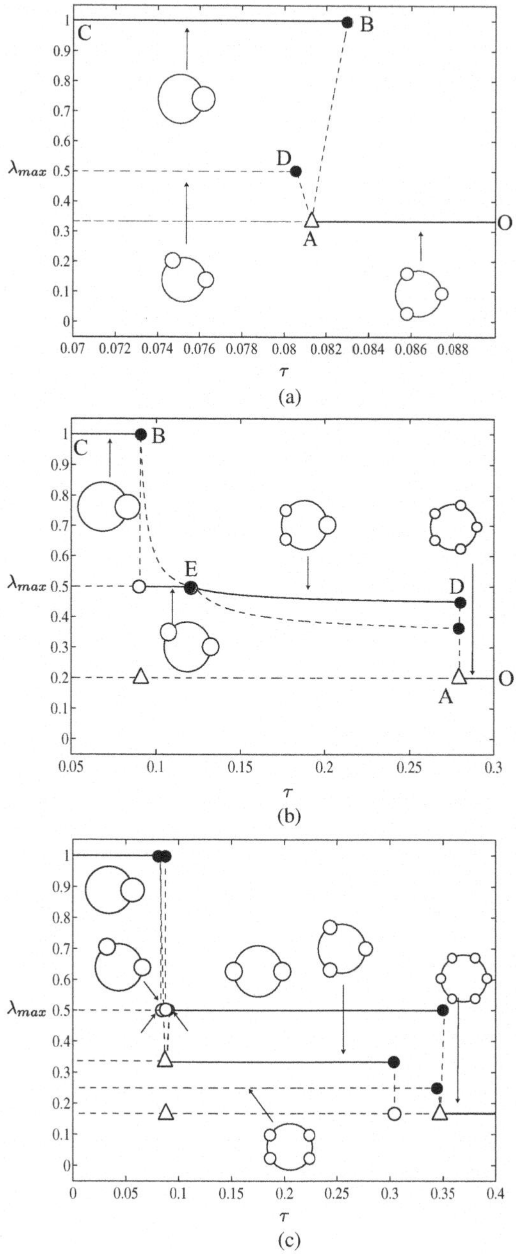


Last, we obtained solution curves for $n = 3$, 5, and 6 in Fig. 9.10. Invariant patterns correspond to the horizontal lines. No stable, non-invariant patterns exist for $n = 3$ in panel (a). For $n = 5$ in (b), the stable non-invariant pattern DE resides on a wide range of $\tau$ and fills the gap of $\tau$ of the stable invariant patterns, unlike

$n = 3$. For a composite number $n = 6 = 2 \cdot 3$ in (c), there are multiple stable solutions for the same value of $\tau$ due to the coexistence of spatial period doubling and tripling.

## References

Akamatsu, T., Takayama, Y., & Ikeda, K. (2012). Spatial discounting, Fourier, and racetrack economy: A recipe for the analysis of spatial agglomeration models. *Journal of Economic Dynamics and Control, 36*(11), 1729–1759.

Barbero, J., & Zofío, J. L. (2016). The multiregional core-periphery model: The role of the spatial topology. *Networks and Spatial Economics, 16*(2), 469–496.

Feigenbaum, M. J. (1978). Qualitative universality for a class of nonlinear transformations. *Journal of Statistical Physics, 19,* 25–52.

Forslid, R., & Ottaviano, G. I. P. (2003). An analytically solvable core-periphery model. *Journal of Economic Geography, 3,* 229–240.

Fujita, M., Krugman, P., & Venables, A. J. (1999). *The spatial economy: Cities, regions, and international trade*. MIT Press.

Hofbauer, J., & Sigmund, K. (1998). *Evolutionary games and population dynamics*. Cambridge University Press.

Ikeda, K., Akamatsu, T., & Kono, T. (2012). Spatial period-doubling agglomeration of a core-periphery model with a system of cities. *Journal of Economic Dynamics and Control, 36*(5), 754–778.

Ikeda, K., & Murota, K. (2019). *Imperfect bifurcation in structures and materials* (3rd ed.). Springer.

Ikeda, K., Murota, K., Akamatsu, T., & Takayama, Y. (2017). Agglomeration patterns in a long narrow economy of a new economic geography model: Analogy to a racetrack economy. *International Journal of Economic Theory, 13,* 113–145.

Ikeda, K., Onda, M., & Takayama, Y. (2018). Spatial period doubling, invariant pattern, and break point in economic agglomeration in two dimensions. *Journal of Economic Dynamics & Control, 92,* 129–152.

Ikeda, K., Onda, M., & Takayama, Y. (2019). Bifurcation theory of a racetrack economy in a spatial economy model. *Networks and Spatial Economics, 19*(1), 57–82.

Mossay, P., & Picard, P. M. (2011). On spatial equilibria in a social interaction model. *Journal of Economic Theory, 146*(6), 2455–2477.

Osawa, M., Akamatsu, T., & Takayama, Y. (2017). Harris and Wilson (1978) model revisited: The spatial period-doubling cascade in an urban retail model. *Journal of Regional Science, 57*(3), 442–466.

Picard, P. M., & Tabuchi, T. (2010). Self-organized agglomerations and transport costs. *Economic Theory, 42,* 565–589.

Tabuchi, T., & Thisse, J. F. (2011). A new economic geography model of central places. *Journal of Urban Economics, 69,* 240–252.

# Chapter 10
# Invariant Patterns on Square and Hexagonal Lattices

**Abstract** This chapter presents a study of invariant patterns in replicator dynamics on square and hexagonal lattices with periodic boundary conditions. Invariant patterns are solutions to the governing equation for any value of the bifurcation parameter. We propose a theoretical procedure that can exhaustively identify invariant patterns, including uniform, monocentric (full agglomeration), and polycentric distributions. We introduce many theoretically possible invariant patterns and identify those that are stable in an economic geography model. Solution curves of invariant patterns are connected by those of non-invariant patterns to form a complex mesh-like structure, akin to warp and weft threads. Chapter 2 provides the foundation for the economic geography model. Chapters 4 and 5 are prerequisites for the direct bifurcation of the lattices.

**Keywords** Bifurcation · Economic geography model · Group-theoretic bifurcation analysis · Hexagonal lattice · Invariant pattern · Replicator dynamics · Square lattice

## 10.1 Introduction

Square and hexagonal lattices with periodic boundary conditions exhibit many direct and further bifurcating curves from a uniform state. These curves form a complex network of solution curves (see Fig. 4.8 for a square lattice).[1] This chapter undertakes an important scientific mission of elucidating the mechanism of this complicated network. For this purpose, we combine the mechanisms of invariant patterns and symmetry-breaking bifurcations from the uniform state (Chaps. 4 and 5).

[1] For the square lattice, see Ikeda et al. (2018b) and Kogure and Ikeda (2022). For the hexagonal lattice, see Ikeda and Murota (2014) and Ikeda et al. (2012, 2018a).

 
K. Ikeda, Y. Takayama, *Bifurcation Theory of Pattern Formation in Economic Geography*, Interdisciplinary Applied Mathematics 61,
https://doi.org/10.1007/978-3-031-71686-7_10

Invariant patterns are solutions to the governing equation, irrespective of the value of the bifurcation (transportation cost) parameter (Sect. 9.2). This chapter carries out a theoretical study of invariant patterns on square and hexagonal lattices in replicator dynamics that accommodates many invariant patterns.[2] We present a group-theoretic condition for the existence of invariant patterns and exhaustively identify invariant patterns on these lattices using this condition.

In the numerical bifurcation analysis of a symmetric system, it is customary to obtain a uniform solution and direct, secondary, and tertiary bifurcating solutions successively, and, in turn, find a family of stable solutions. Given the existence of invariant patterns, we propose the following innovative bifurcation analysis procedure to find stable solution curves:

- Obtain all invariant patterns and investigate their stability to find the curves (zones of $\tau$) for stable invariant patterns.
- Obtain bifurcating solution curves connecting stable invariant patterns and investigate their stability to find stable bifurcating curves.

We apply this procedure to an economic geography model (Forslid & Ottaviano, 2003) on the lattices to observe bifurcation diagrams with horizontal lines for invariant patterns and non-horizontal curves for bifurcating solutions. These lines and curves form a structure like warp and weft.

We propose a conjecture on the connectivity between the uniform state and invariant patterns: When a bifurcating path branches from the uniform state in the direction of its critical eigenvector (for the zero eigenvalue of the Jacobian matrix $J = \partial \boldsymbol{F}/\partial \boldsymbol{\lambda}$), the population tends to be agglomerated to places with large positive components of the eigenvector, and then the path arrives at an invariant pattern. A numerical analysis of a square lattice assesses this conjecture.

This chapter is organized as follows: A theory of invariant patterns is introduced in Sect. 10.2. In Sect. 10.3, stability and sustainability of invariant patterns are investigated. Section 10.4 advances the numerical bifurcation and stability analysis for an economic geography model on the lattices.

## 10.2 Invariant Patterns on Square and Hexagonal Lattices

We consider an $n \times n$ *square lattice* and a rhombic, $n \times n$ *hexagonal lattice* both with periodic boundary conditions (Fig. 10.1). Examples of these lattices for $n = 2$ are shown by the dashed lines.

[2] Invariant patterns in the replicator dynamics in the racetrack economy are studied theoretically by Ikeda et al. (2012, 2019b) and have come to be used in the analysis of lattices (Ikeda et al., 2018b, 2019a and Kogure & Ikeda, 2022).

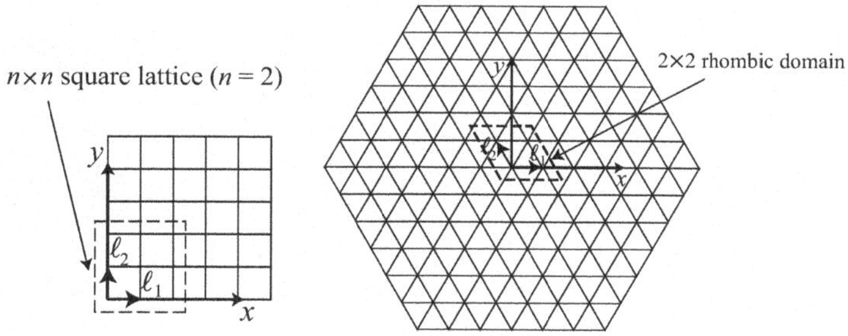


(a) (b)

**Fig. 10.1** Two types of lattices. (**a**) Square lattice. (**b**) Hexagonal lattice

The *symmetry group* of these lattices is

$$G = \langle r, s, p_1, p_2 \rangle.$$

For the square lattice, the elements of $G$ are defined as follows (see Sect. 4.2):

- $r$ : counterclockwise *rotation* about the origin at an angle of $\pi/2$
- $s$ : *reflection* $y \mapsto -y$
- $p_1$ : periodic *translation* along the vector $\boldsymbol{\ell}_1 = \tilde{L}(1,0)^\top$ by a length $\tilde{L}$
- $p_2$ : periodic *translation* along the vector $\boldsymbol{\ell}_2 = \tilde{L}(0,1)^\top$ by a length $\tilde{L}$

For the hexagonal lattice, the elements of $G$ are defined as follows (see Sect. 5.2):

- $r$ : counterclockwise *rotation* about the origin at an angle of $\pi/3$
- $s$ : *reflection* $y \mapsto -y$
- $p_1$ : periodic *translation* along the vector $\boldsymbol{\ell}_1 = \tilde{L}(1,0)^\top$ by a length $\tilde{L}$
- $p_2$ : periodic *translation* along the vector $\boldsymbol{\ell}_2 = \tilde{L}(-1/2, \sqrt{3}/2)^\top$ by a length $\tilde{L}$

### *10.2.1 Theory of Invariant Patterns in Replicator Dynamics*

We consider the governing equation $\boldsymbol{F} = (F_i(\boldsymbol{\lambda}, \phi) \mid i = 1, \dots, N) = \boldsymbol{0}$ in the replicator dynamics (1.7):

$$F_i(\boldsymbol{\lambda}, \phi) = (v_i(\boldsymbol{\lambda}, \phi) - \bar{v}(\boldsymbol{\lambda}, \phi))\lambda_i \quad (i = 1, \dots, N).$$

Here, $F_i(\boldsymbol{\lambda}, \phi)$ $(i = 1, \dots, N)$ is a sufficiently smooth function, $\boldsymbol{\lambda}$ is the independent variable vector, and $\bar{v} = \sum_{i=1}^{N} \lambda_i v_i$.

We consider a corner solution (a pattern $\boldsymbol{\lambda}$ with zero components) of this dynamics and rearrange the order of its components to arrive at

$$\hat{\boldsymbol{\lambda}} = (\boldsymbol{\lambda}_m^+, \boldsymbol{0}_{N-m}) \qquad (1 \le m < N) \tag{10.1}$$

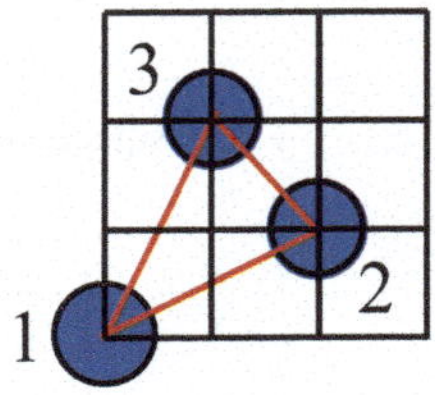


**Fig. 10.2** A non-invariant pattern (the area of a circle denotes a population size)

with $\boldsymbol{\lambda}_m^+ = \{\lambda_i > 0, \ \sum_{1=1}^m \lambda_i = 1 \mid 1 \le i \le m\}$ and the $(N-m)$-dimensional zero vector $\mathbf{0}_{N-m}$. The governing equation $\boldsymbol{F}(\boldsymbol{\lambda}, \tau)$ can be rearranged as

$$\hat{\boldsymbol{F}} = \{\boldsymbol{F}^+(\boldsymbol{\lambda}_m^+, \mathbf{0}_{N-m}, \tau), \boldsymbol{F}^0(\boldsymbol{\lambda}_m^+, \mathbf{0}_{N-m}, \tau)\}. \tag{10.2}$$

As candidates for invariant patterns, we consider the uniform state $\boldsymbol{\lambda}^* = \frac{1}{N}\mathbf{1}_N = \frac{1}{N}(1, \ldots, 1)$ and the *core–periphery pattern*:

$$\boldsymbol{\lambda}_m^{\mathrm{CP}} = \left(\frac{1}{m}\mathbf{1}_m, \mathbf{0}_{N-m}\right) \quad (1 \le m < N). \tag{10.3}$$

This pattern has a two-level hierarchy: $m$ core places with the same positive population $1/m$, and other peripheral places with no population.

In studying invariant patterns, we revisit Lemma 9.1:

**Lemma 10.1** *The core–periphery pattern* $\boldsymbol{\lambda}_m^{\mathrm{CP}} = (\frac{1}{m}\mathbf{1}_m, \mathbf{0}_{N-m})$ *is an invariant pattern if the set of points for* $\frac{1}{m}\mathbf{1}_m$ *belongs to an orbit of a subgroup* $G'$ *of* $G$.

The following example demonstrates the use of this lemma.

**Example 10.1** The pattern in Fig. 10.2 is a non-invariant pattern although it has the form $\boldsymbol{\lambda}_3^{\mathrm{CP}} = (\frac{1}{3}\mathbf{1}_3, \mathbf{0}_{13})$ in (10.3) with $m = 3$. This pattern is invariant under reflection across the diagonal line. However, the place 1 cannot be permuted with the other two places, 2 and 3, by this reflection. Since it is not possible to find the subgroup $G'$ in Lemma 10.1, the pattern is not an invariant pattern. □

In particular, the patterns for $m = 1$, $m = 2$, and $m = n^2$ are *full agglomeration*, *twin places*, and the *uniform state*, respectively, and they play pivotal roles throughout this chapter. We have the following proposition for these patterns.

**Proposition 10.1** *A full agglomeration, twin places, and the uniform state are invariant patterns on the square and hexagonal lattices.*

***Proof*** (i) A full agglomeration is an invariant pattern by Proposition 9.1. (ii) For twin places, we consider two nodes $(n_1, n_2)$ and $(n_1', n_2')$. For a square lattice, we have

$$r^2 {p_1}^i {p_2}^j \cdot (n_1, n_2) \equiv (-n_1 - i, -n_2 - j) \mod n.$$

Hence, we see the exchangeability between any pair of $(n_1, n_2)$ and $(n'_1, n'_2)$:

$$g \cdot (n_1, n_2) \equiv (n'_1, n'_2), \quad g \cdot (n'_1, n'_2) \equiv (n_1, n_2) \mod n$$

by $g = r^2 {p_1}^i {p_2}^j$ with $i = -n_1 - n'_1$ and $j = -n_2 - n'_2$. By choosing $G' = \langle r^2 {p_1}^i {p_2}^j \rangle$, we see that a twin ($m = 2$) at any places is an invariant pattern by Lemma 10.1. For a hexagonal lattice, the exchangeability is demonstrated in Ikeda and Murota (2014, Remark 7.1). (iii) We can prove that the uniform state is an invariant pattern by extending the proof for the twins to any pair of points. □

We can search for invariant patterns by finding a set of $m$ nodal points and a *subgroup* $G'$ of the group $G$, satisfying Lemma 10.1. We propose the following procedure to obtain all invariant patterns on the hexagonal lattice with given $n$ ($\geq 2$) and $m$ ($1 \leq m \leq n^2 = N$):

- Choose a set of $m$ nodal points among a total of $n^2$ nodal points.
- Find a subgroup $G'$ that keeps the set of points invariant.
- If the elements of $G'$ permute any two of the $m$ nodal points, we choose this group as $G'$ in Lemma 10.1 to ensure that the set of points gives the nonzero components $\frac{1}{m}\mathbf{1}_m$ in (10.3) of an invariant pattern.

In this procedure, we already know that the full agglomerations ($m = 1$), the twins ($m = 2$), and the uniform state ($m = n^2$) are invariant patterns by Proposition 10.1; therefore, we focus on $3 \leq m < n^2$. Moreover, the number $m$ is not arbitrary but depends on the lattice size $n$, as explained in Proposition 10.2. Table 10.1 lists possible values of $m$ for the hexagonal lattice, e.g., for $n = 3$, 4, 5, 6.

**Table 10.1** Possible values of the number $m$ for several sizes $n$ for the hexagonal lattice ($\mathbb{Z}$ is a set of integer numbers)

| $n$ | $12n^2$ | $m$ | $12n^2/m \in \mathbb{Z}$ |
|---|---|---|---|
| 3 | 108 | 1 | 108 |
| | | 2 | 54 |
| | | 3 | 36 |
| | | 4 | 27 |
| | | 6 | 18 |
| | | 9 | 12 |
| 4 | 192 | 1 | 192 |
| | | 2 | 96 |
| | | 3 | 64 |
| | | 4 | 48 |
| | | 6 | 32 |
| | | 8 | 24 |
| | | 12 | 16 |
| | | 16 | 12 |

| $n$ | $12n^2$ | $m$ | $12n^2/m \in \mathbb{Z}$ |
|---|---|---|---|
| 5 | 300 | 1 | 300 |
| | | 2 | 150 |
| | | 3 | 100 |
| | | 4 | 75 |
| | | 5 | 60 |
| | | 6 | 50 |
| | | 10 | 30 |
| | | 25 | 12 |

| $n$ | $12n^2$ | $m$ | $12n^2/m \in \mathbb{Z}$ |
|---|---|---|---|
| 6 | 432 | 1 | 432 |
| | | 2 | 216 |
| | | 3 | 144 |
| | | 4 | 108 |
| | | 6 | 72 |
| | | 8 | 54 |
| | | 9 | 48 |
| | | 12 | 36 |
| | | 18 | 24 |
| | | 24 | 18 |
| | | 27 | 16 |
| | | 36 | 12 |

**Proposition 10.2** *The number* $m$ $(1 \leq m \leq n^2)$ *divides* $|G|$ $(= 8n^2$ *for the square lattice and* $12n^2$ *for the hexagonal one).*

***Proof*** Since $G'$ with $|G'|$ elements is a subgroup of $G$ with $|G|$ elements, $|G'|$ divides $|G|$ by Lagrange's theorem. The number $m$ of elements of an orbit divides $|G'|$ (Kochendörffer, 1970, §3.1.2). Hence, $|G|$ is divisible by $m$. □

### *10.2.2 Examples of Invariant Patterns*

By using the procedure presented in Sect. 10.2.1, we find invariant patterns on the square and hexagonal lattices.

Examples of invariant patterns on a $6 \times 6$ *square lattice* are depicted in Fig. 10.3. Each invariant pattern is connected by an arrow $\rightarrow$ with the associated eigenvector of $J = \partial \boldsymbol{F}/\partial \boldsymbol{\lambda}$ for the bifurcating solution of the uniform state (cf. Figs. 4.4 and 4.7). Each pair connected by $\rightarrow$ displays similar geometrical patterns. In the numerical bifurcation analysis to be conducted in Sect. 10.4, populations in places with positive components of the eigenvectors tend to increase along a bifurcating curve from the uniform state; by contrast, the populations in places with negative ones tend to decrease. For example, a full agglomeration can emerge from a critical point associated with $\boldsymbol{q}_1^{(4;1,0)} + \boldsymbol{q}_3^{(4;1,0)}$ (see the top-left of Fig. 10.3a). Based on this tendency, we propose the following conjecture:

*Conjecture 10.1* In the Krugman model and its variants (cf. Remark 2.1 in Sect. 2.2), places with large positive components of the eigenvectors for the bifurcating solutions tend to gather populations after bifurcation.

A remark is on the symmetry/asymmetry of the bifurcating solutions. When the solutions in the positive and negative directions from the bifurcation point are *conjugate*, these solutions are called symmetric and can arrive at conjugate invariant patterns (see Fig. 10.3a). When the two solutions are non-conjugate, they are asymmetric and can arrive at two different invariant patterns (see Fig. 10.3b).

For the hexagonal lattice, invariant patterns that exist for any $n$ are presented in Fig. 10.4 for the lattice size $n = 3$ and 6. This figure encompasses all invariant patterns on the $3 \times 3$ *hexagonal lattice*. These patterns include regular triangle, racetrack, and stripe, in addition to full agglomeration, twin, and uniform (cf. Proposition 10.1). Other invariant patterns exist only for specific value(s) of $n$. Figure 10.5 depicts such patterns for $n = 6$.

## 10.3 Stability and Sustainability

The Jacobian matrix of the rearranged governing equation (10.2) in the replicator dynamics takes the form in (9.6):

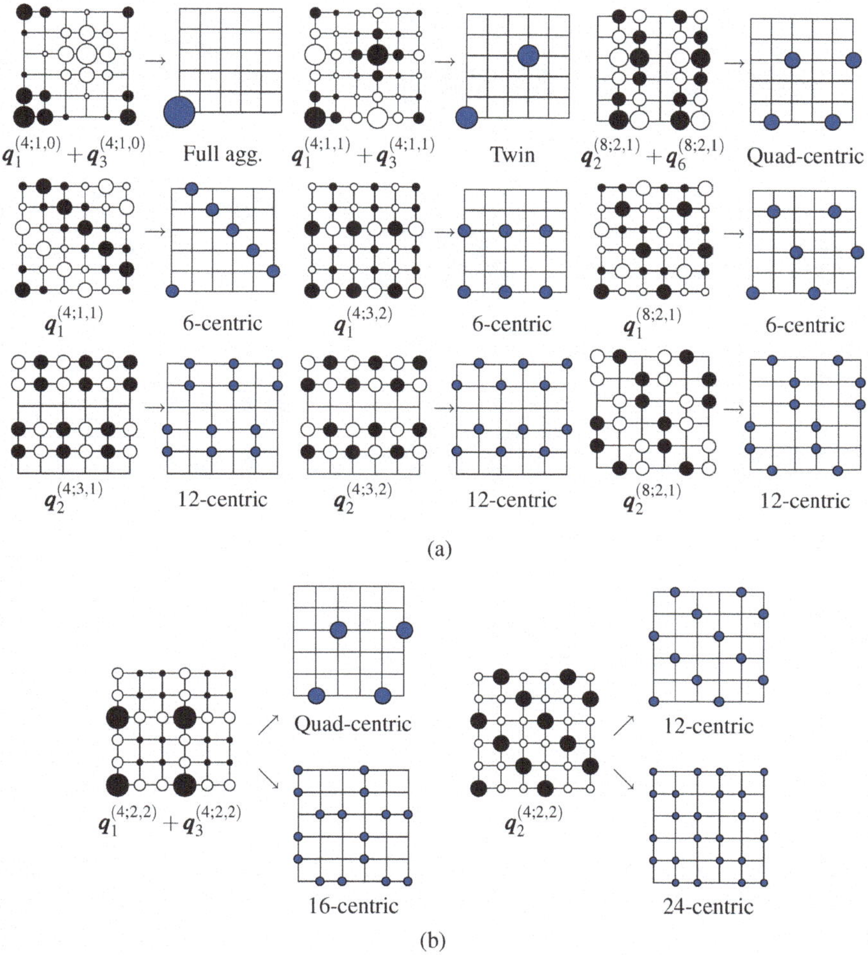


**Fig. 10.3** Examples of invariant patterns that emerge from the uniform state and the associated eigenvectors on the $6 \times 6$ square lattice (Kogure et al., 2021) (an arrow $\rightarrow$ connects the eigenvector of $J$ for a bifurcating solution on the left and the corresponding invariant pattern on the right). (**a**) Symmetric bifurcating solutions. (**b**) Asymmetric bifurcating solutions

$$\hat{J} = \begin{bmatrix} J^{+} & J^{+0} \\ O & J^{0} \end{bmatrix} \quad \text{with} \quad J^{0} = \mathrm{diag}(v_{m+1} - \bar{v}, \ldots, v_N - \bar{v}). \tag{10.4}$$

We can classify the critical points as

$$\begin{cases} J^{+} \text{ has a zero eigenvalue :} & \textit{breaking bifurcation point} \text{ or } \textit{stationary point} \text{ of } \tau. \\ e_i = v_i - \bar{v} = 0 \ (\mathrm{i} = \mathrm{m}+1, \ldots, \mathrm{n}) : & \textit{sustaining bifurcation point.} \end{cases}$$

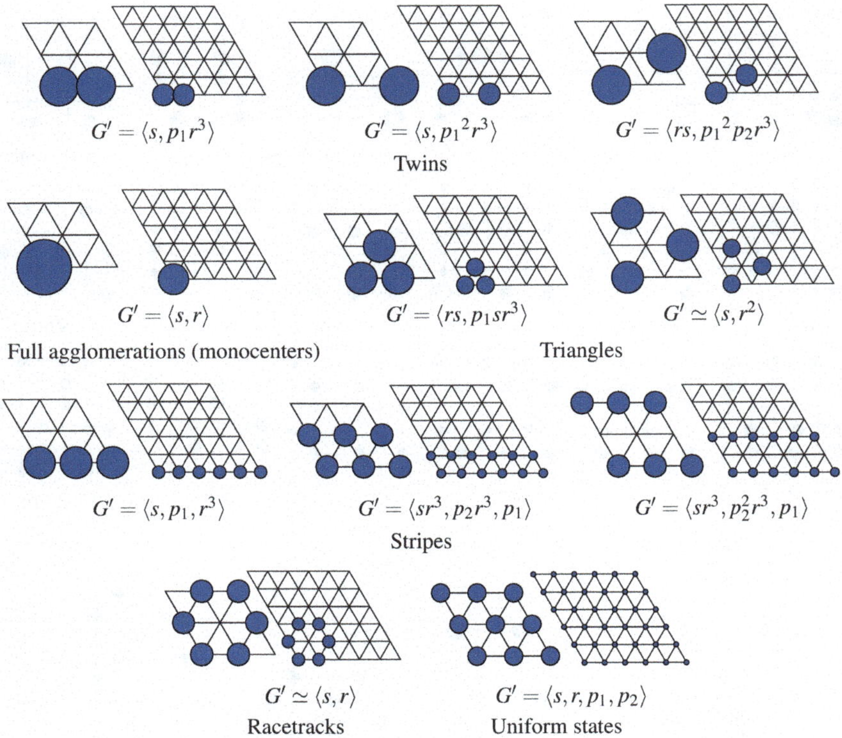


**Fig. 10.4** Invariant patterns that exist for any $n$ ($n \geq 3$)

A solution curve $(\bar{\boldsymbol{\lambda}}, \tau)$ for an invariant pattern $\boldsymbol{\lambda} = \bar{\boldsymbol{\lambda}}$ has sustaining bifurcation points (cf. Fig. 10.6 in Sect. 10.4). A sustain point $\tau_S$ is defined as the value of $\tau$ at a sustaining bifurcation point where the sustainability changes. Then, we have the following proposition.

**Proposition 10.3** *The solution curve of the uniform state has no sustaining bifurcation point, and that for the full agglomeration has no breaking one. Curves of other invariant patterns possibly have both breaking and sustaining bifurcation points.*

***Proof*** The curve of the uniform state has no sustain point due to the absence of $J^0$ in (10.4). The curve for the full agglomeration has no break point because $J^+(1, \mathbf{0}_{N-1}, \tau) = -\lambda_1 v_1$ is always negative. Curves of other invariant patterns possibly have both breaking and sustaining bifurcation points due to the presence of both $J^+$ and $J^0$. □

The coexistence of the break point $\tau_B$ and the sustain point $\tau_S$ for a solution curve for an invariant pattern is observed for the square and hexagonal lattices (cf. Sect. 10.4); by contrast, this feature is absent in the two-place economy (see Fig. 1.5 in Sect. 1.6.1). We introduce the following assumption based on the numerical

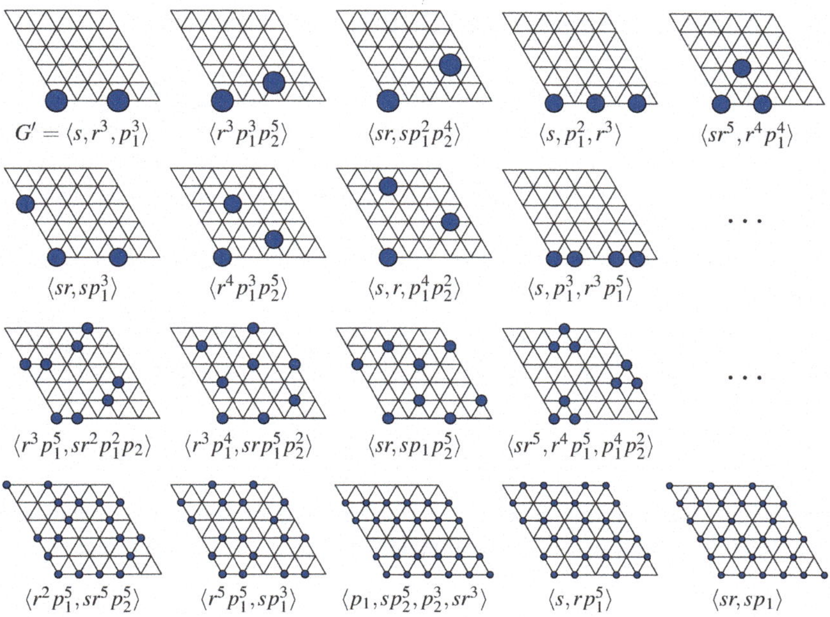


**Fig. 10.5** Examples of invariant patterns for $n = 6$ that do not exist for all $n$ (see Kogure et al., 2021 for the rest of patterns)

behavior in Sect. 10.4. This assumption is expected to hold in the Krugman model and its variants, whereas there are other models that behave differently (cf. Remark 2.1 in Sect. 2.2).

**Assumption 10.1** *On the solution curve of an invariant pattern other than the uniform state and the full agglomeration,* $\tau_B$ *and* $\tau_S$ *exist such that the solutions for the pattern are stable for* $\tau > \tau_B$ *and are sustainable for* $\tau < \tau_S$.

We introduce the following classification:

$$\begin{cases} \text{Well-posed invariant pattern:} & \tau_B < \tau_S. \\ \text{Ill-posed invariant pattern:} & \tau_B > \tau_S. \end{cases} \tag{10.5}$$

**Proposition 10.4** *Under Assumption 10.1, a well-posed invariant pattern is stable in the range* $\tau_B < \tau < \tau_S$. *An ill-posed one is unstable for any* $\tau$.

***Proof*** A stable spatial equilibrium satisfies the stability condition $\tau > \tau_B$ and the sustainability condition $\tau < \tau_S$ simultaneously. For a well-posed invariant pattern with $\tau_B < \tau_S$, there exists an interval $\tau_B < \tau < \tau_S$ which satisfies these two conditions. For an ill-posed one with $\tau_B > \tau_S$, no interval of $\tau$ satisfies $\tau > \tau_B$ and $\tau < \tau_S$ simultaneously. □

## 10.4 Numerical Stability and Bifurcation Analysis

We present a numerical stability and bifurcation analysis of the replicator dynamics on the square and hexagonal lattices. We employ the *trade freeness* $\phi$, defined in (2.11), as the bifurcation parameter. We use the Forslid and Ottaviano (2003) model (Sect. 2.2) and set $(\sigma, \mu) = (6.0, 0.4)$ for the square lattice and $(\sigma, \mu) = (8.0, 0.4)$ for the hexagonal lattice.

To search for stable solution curves, we employ the following procedure that exploits the existence of invariant patterns:

- Obtain all invariant patterns and investigate their stability through the eigenvalue analysis of the Jacobian matrix for $\phi \in (0, 1)$ to find stable curves (zones of $\phi$) for invariant patterns.
- Obtain bifurcating solution curves connecting these stable curves for invariant patterns and investigate their stability to find stable bifurcating curves.

This procedure is convenient as we can start the bifurcation analysis from any stable invariant pattern, whereas it is conventional to start from the uniform state.

**Square Lattice**

We demonstrate the emergence of bifurcating solutions and investigate the connectivity of these solutions to invariant patterns on the $6 \times 6$ *square lattice*, which accommodates a wide variety of invariant patterns (see Fig. 10.3).

Figure 10.6a shows the solution curves. When the trade freeness $\phi$ increases from zero, the uniform state loses its stability at bifurcation point A for $\mu = (1; -)$, and the bifurcating curve is engendered (see panel (b) for its bifurcation mode), and this curve is connected to the 18-centric invariant pattern. In accordance with Conjecture 10.1 in Sect. 10.2.2, the population is agglomerated completely to places with large positive components of the eigenvector of $J$ of the bifurcating solution. Such is the case also for other solution curves (see Table 4.1 in Sect. 4.5 for the list of bifurcating solutions). Thus, Conjecture 10.1 helps predict invariant patterns that emerge through bifurcations in the present model.

There are bifurcating solution curves for non-invariant patterns connecting the uniform distribution to the invariant patterns. So far as these solution curves are concerned, stable equilibria are always invariant patterns, and non-invariant patterns are all unstable just after bifurcation but often regain stability by arriving at stable invariant patterns. We have observed a mesh-like structure of the horizontal curves for stable invariant patterns and non-horizontal curves of unstable, non-invariant ones. Thus, knowledge of invariant patterns helps us understand the mechanism of the mesh-like structure.

The uniform state has only the break point $\tau_{\mathrm{B}}$ at point A, and the full agglomeration has only the sustain point $\tau_{\mathrm{S}}$ at point H (cf. Proposition 10.3). Other invariant patterns have both $\tau_{\mathrm{B}}$ and $\tau_{\mathrm{S}}$ and are all well-posed satisfying $\tau_{\mathrm{B}} < \tau_{\mathrm{S}}$ in (10.5) and are stable in the range $\tau_{\mathrm{B}} < \tau < \tau_{\mathrm{S}}$ in line with Assumption 10.1 and Proposition 10.4.

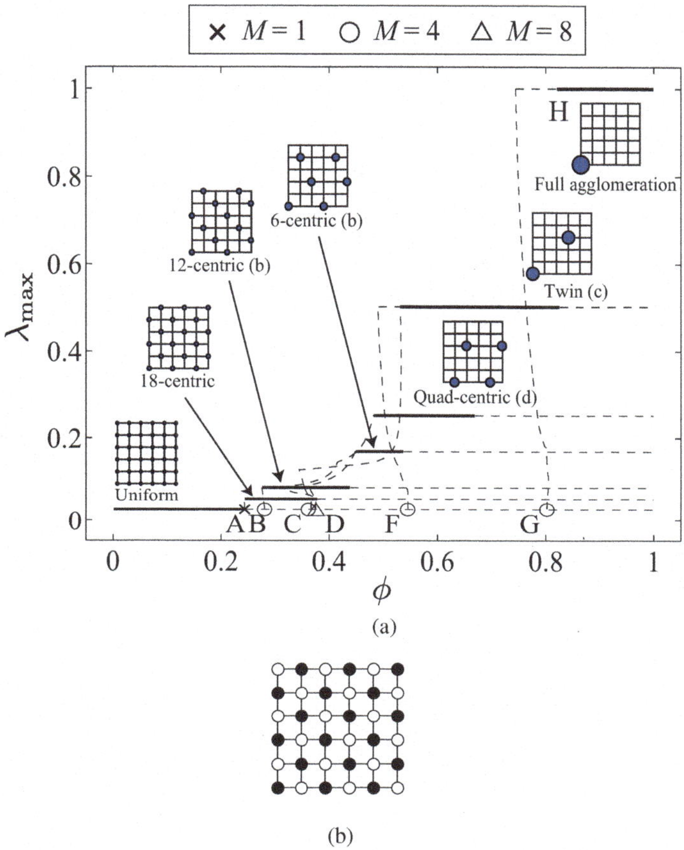


**Fig. 10.6** Stable invariant patterns engendered by direct bifurcation on the 6 × 6 square lattice (Kogure et al., 2021) (the vertical axis shows $\lambda_{\max} = \max(\lambda_1, \ldots, \lambda_N)$; $M$: multiplicity of the critical point; solid curve: stable; dashed one: unstable; the label of each invariant pattern corresponds to Fig. 10.7). (**a**) Solution curves. (**b**) Bifurcation mode at point A

We investigate the stability of all invariant patterns on the 6 × 6 square lattice in $\phi \in (0, 1)$, and the range of $\phi$ where the associated patterns are stable is depicted by solid lines on the top of Fig. 10.7. There are as many as 22 invariant patterns that are stable for some range of $\phi$. We see a tendency that when the trade freeness $\phi$ increases from a small value, the number of places with positive population decreases. Most of these invariant patterns are not connected directly to the uniform state and are consequences of secondary and further bifurcations.

**Hexagonal Lattice**

For the 6 × 6 *hexagonal lattice*, we investigated the stability of all invariant patterns (Figs. 10.4 and 10.5). Invariant patterns that are stable for some $\phi$ are shown in Fig. 10.8. There are as many as 19 stable invariant patterns. As shown in Table 10.2,

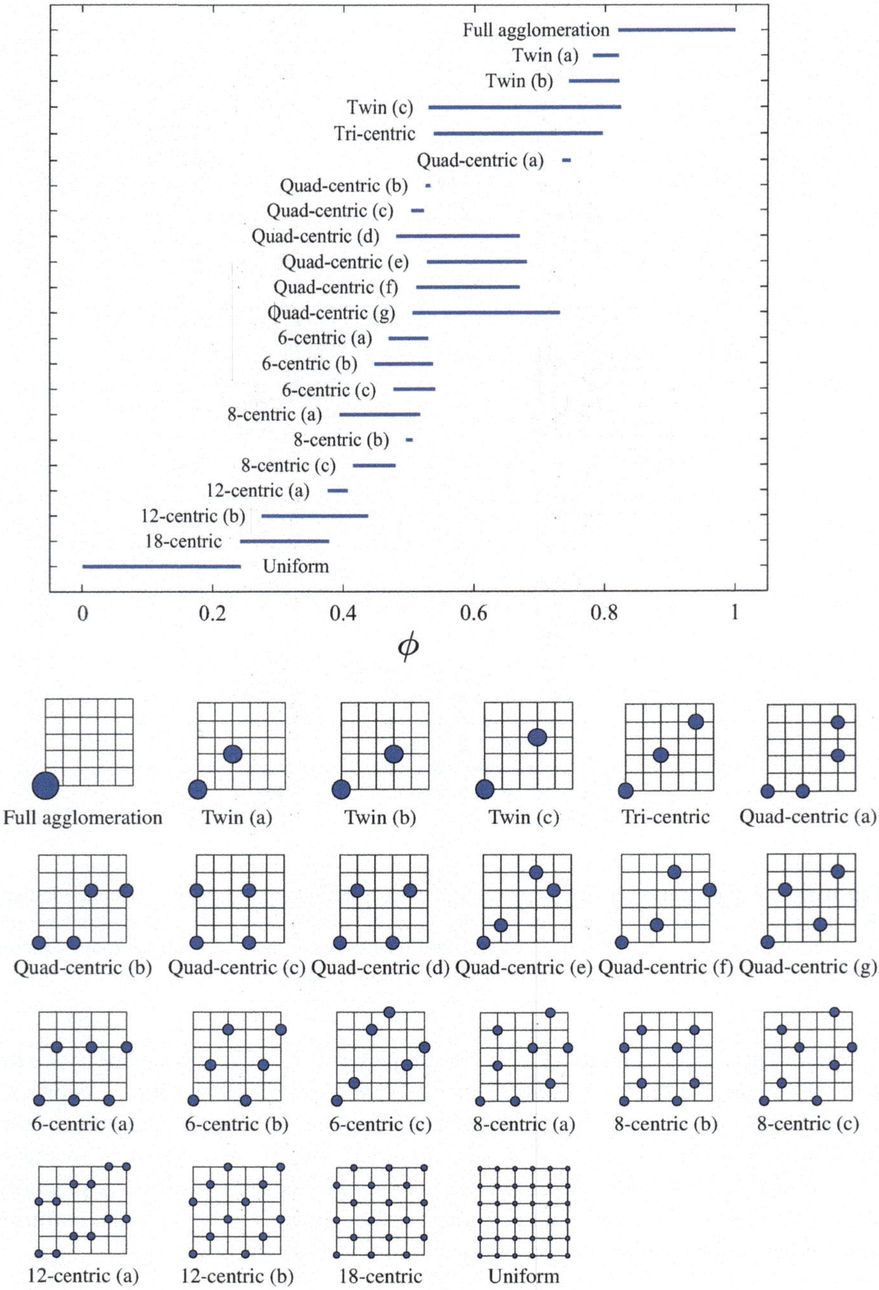


**Fig. 10.7** The ranges of $\phi$ for stable invariant patterns on the $6 \times 6$ square lattice (Kogure et al., 2021)

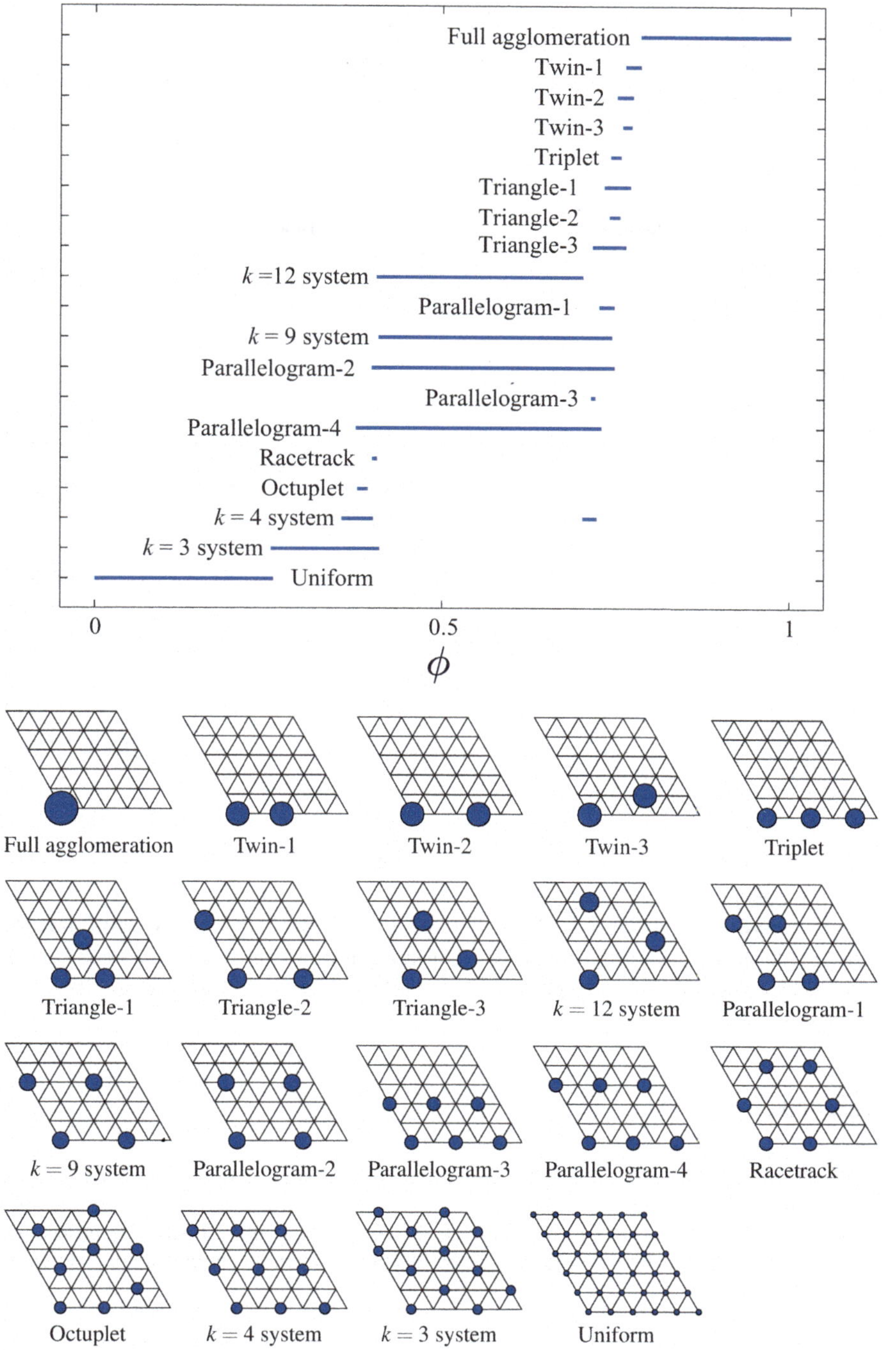


**Fig. 10.8** The ranges of $\phi$ for stable invariant patterns on the $6 \times 6$ hexagonal lattice (Ikeda et al., 2019a)

**Table 10.2** The number of stable and unstable invariant patterns classified with reference to the number $m$ of agglomerated places for the $6 \times 6$ hexagonal lattice

| $m$ | 1 | 2 | 3 | 4 | 6 | 8 | 9 | 12 | 16 | 18 | 24 | 27 | 36 |
|---|---|---|---|---|---|---|---|---|---|---|---|---|---|
| Stable | 1 | 3 | 5 | 3 | 3 | 1 | 1 | 1 | 0 | 0 | 0 | 0 | 1 |
| Unstable | 0 | 3 | 2 | 8 | 10 | 5 | 2 | 9 | 1 | 3 | 2 | 1 | 0 |
| Total | 1 | 6 | 7 | 11 | 13 | 6 | 3 | 10 | 1 | 3 | 2 | 1 | 1 |

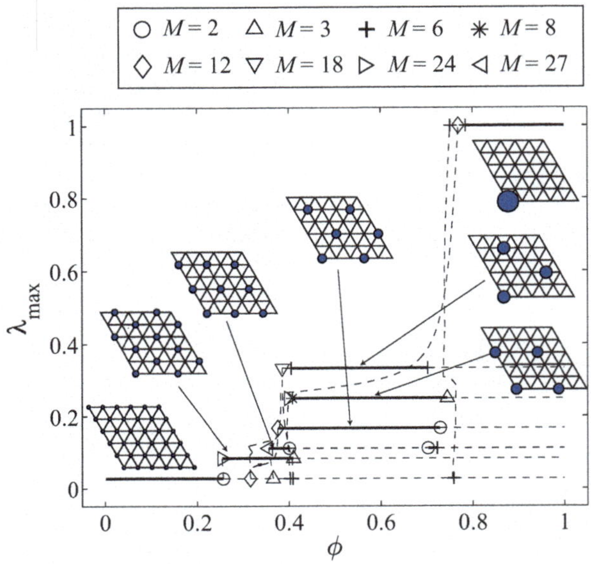


**Fig. 10.9** Solution curves of invariant and non-invariant patterns on the $6 \times 6$ hexagonal lattice (Ikeda et al., 2019a) ($\lambda_{\max} = \max(\lambda_1, \ldots, \lambda_N)$; solid curve: stable; dashed curve: unstable)

invariant patterns with large $m$ tend to become unstable (except for the uniform state with $m = 36$).

The solution curves in Fig. 10.9 display a mesh-like structure of equilibrium paths made up of horizontal lines for invariant patterns and non-horizontal curves for non-invariant patterns. Again, non-invariant patterns are inferior in stability.

## References

Forslid, R., & Ottaviano, G. I. P. (2003). An analytically solvable core-periphery model. *Journal of Economic Geography, 3,* 229–240.

Ikeda, K., Aizawa, H., Kogure, Y., & Takayama, Y. (2018a). Stability of bifurcating patterns of spatial economy models on a hexagonal lattice. *International Journal of Bifurcation and Chaos, 28*(11), 1850138-1–30.

Ikeda, K., Akamatsu, T., & Kono, T. (2012). Spatial period-doubling agglomeration of a core-periphery model with a system of cities. *Journal of Economic Dynamics and Control, 36,* 754–778.

Ikeda, K., Kogure, Y., Aizawa, H., & Takayama, Y. (2019a). Invariant patterns for replicator dynamics on a hexagonal lattice. *International Journal of Bifurcation and Chaos, 29,* 1930014.

Ikeda, K., & Murota, K. (2014). *Bifurcation theory for hexagonal agglomeration in economic geography*. Springer.

Ikeda, K., Murota, K., & Akamatsu, T. (2012). Self-organization of Lösch's hexagons in economic agglomeration for core-periphery models. *International Journal of Bifurcation and Chaos, 22,* 1230026-1–29.

Ikeda, K., Onda, M., & Takayama, Y. (2018b). Spatial period doubling, invariant pattern, and break point in economic agglomeration in two dimensions. *Journal of Economic Dynamics and Control, 92,* 129–152.

Ikeda, K., Onda, M., & Takayama, Y. (2019b). Bifurcation theory of a racetrack economy in a spatial economy model. *Networks and Spatial Economics, 19*(1), 57–82.

Kochendörffer, R. (1970). *Group theory* McGraw Hill.

Kogure, Y., & Ikeda, K. (2022). Group-theoretic study of economic agglomerations on a square lattice. *MPRA Paper* 112842.

Kogure, Y., Ikeda, K., & Aizawa, H. (2021). Group-theoretic bifurcation mechanism of economic agglomerations on a square lattice. *International Journal of Bifurcation and Chaos, 31*(13), 2130040.

# Chapter 11
# Global–Local Spatial Network

**Abstract** This chapter develops a global–local spatial platform to represent the hierarchical structure of cities, towns, villages, and so on. A global system describes the geographical distribution of a system of cities, whereas each local square lattice models the population size and local geography of each city. Through group-theoretic bifurcation analysis, we elucidate the bifurcation mechanism of the monocentric distribution on a square lattice using the replicator dynamics. This bifurcation explains how satellite places appear around a large city. As a realistic example, we utilize an air network of cities with the seven largest hub airports in the USA. We model this network as a series of local square lattices of various sizes connected by an equidistant economy of a global network. This system simulates the competition among local square lattices. As an example of this competition, we observe successive sustaining bifurcations leading to the gradual emergence of satellite places around a large city in the analysis of an economic geography model. Chapter 9 is a prerequisite for sustaining bifurcations.

**Keywords** Air network · Economic geography model · Equidistant economy · Global network · Group-theoretic bifurcation analysis · Local network · Satellite city · Square lattice · Sustaining bifurcation

## 11.1 Introduction

We can observe a hierarchical spatial structure of economic agglomeration worldwide, comprising countries, cities, towns, and so on. Central place theory describes the geometric mechanism of the self-organization of this hierarchical structure (Christaller, 1933). Chapter 2 introduced an attempt to simulate such a hierarchy in economic geography models. Quantitative Spatial Economics (QSE) has been developed as a realistic spatial platform with many places (Redding & Rossi-Hansberg, 2017); however, it does not necessarily provide insight into the bifurcation mechanism.

 
K. Ikeda, Y. Takayama, *Bifurcation Theory of Pattern Formation in Economic Geography*, Interdisciplinary Applied Mathematics 61,
https://doi.org/10.1007/978-3-031-71686-7_11

This chapter presents a global–local system as a large-scale spatial platform for economic agglomerations. This system is capable of expressing a hierarchical structure of economic agglomeration and is endowed with a bifurcation mechanism. A global network comprises a system of cities and expresses the geographical distribution and interaction of cities. Each city has a microstructure comprising a system of local places. The size of a local system characterizes the population of a city, and the distribution of places in this system expresses its geographical properties.

Diverse network systems of economic activities can be modeled by appropriately choosing an economic geography model, a global system, and a series of local systems. The global–local system applies to various problems, such as (1) air networks employed in this chapter, (2) express-train networks, and (3) a hierarchy of countries, cities, towns, and so on. Although the global–local system is applied here to a two-level hierarchy, this system can be extended to a higher level hierarchy.

Figure 11.1 depicts candidates for global and/or local spatial platforms. The two-place economy in (a) is popular but does not have much spatial structure. The line segment in (b) expresses a chain of cities. The *equidistant economy* in (c) represents a system of cities connected, e.g., by airplanes, where the term "equidistant" means that the transportation costs between any two cities are identical. The radial network in (d) is found in many traditional cities like Paris. The square lattice in (e) and the hexagonal lattice in (f) are suitable for modeling densely and regularly distributed places.

As a local platform, we use a two-dimensional square lattice in this chapter. Square lattice-like road networks exist worldwide. For example, Chicago (USA) and Kyoto (Japan) are well known for accommodating such square networks (see Fig. 4.1). There is a trade-off: An irregular-shaped lattice has more flexibility to express the geometry of a local system (city) but relies too heavily on numerical

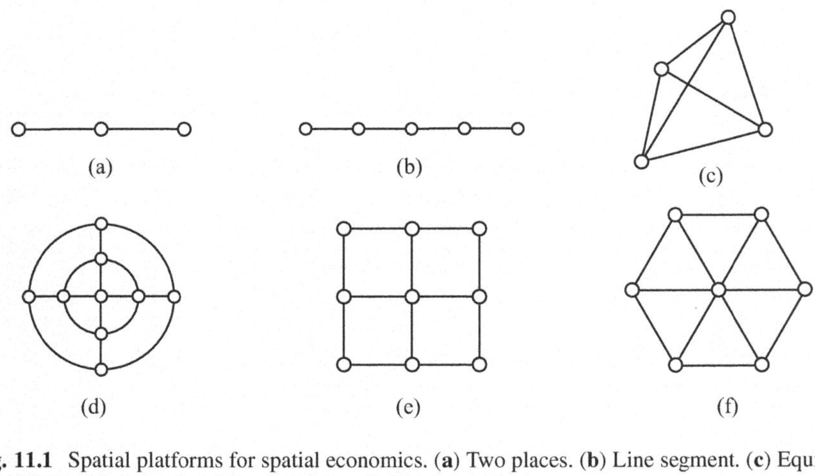


**Fig. 11.1** Spatial platforms for spatial economics. (**a**) Two places. (**b**) Line segment. (**c**) Equidistant. (**d**) Radial network. (**e**) Square lattice. (**f**) Hexagonal lattice

analysis, whereas a regular lattice, such as a square lattice, is endowed with the analytical tractability due to the bifurcation mechanism.

We propose a global–local spatial platform suitable for expressing a hierarchical structure of spatial economic agglomerations. As a realistic example, we employ an air network of cities with the seven largest hub airports in the USA and model it by an equidistant economy with a series of local square lattices with different sizes. We simulate economic agglomeration in this network in an economic geography model (Forslid & Ottaviano, 2003). We show successive sustaining bifurcations and gradual emergence of satellite places around a large central city.

This chapter highlights sustaining bifurcation as a catalyst for engendering satellite places around the full agglomeration (monocenter) at the center of a local square lattice network. Sustaining bifurcation is not given much attention in economic geography.

This chapter is organized as follows: In Sect. 11.2, we introduce a method for the construction of the global–local spatial platform. The mechanism of sustaining bifurcation on a square lattice is addressed in Sect. 11.3. Section 11.4 presents a global–local system of the air transportation network in the USA.

## 11.2 Construction of Global–Local Spatial Platform

We explain how to construct a global–local spatial platform.

### *11.2.1 Construction of Local System*

As a platform for a *local system*, we employ a square lattice. There are various square lattices with square symmetry, as depicted in Fig. 11.2. The simplest one is a cross-shaped economy shown on the left, while a doughnut-shaped domain on the right represents a city with a lake or a castle at the center.

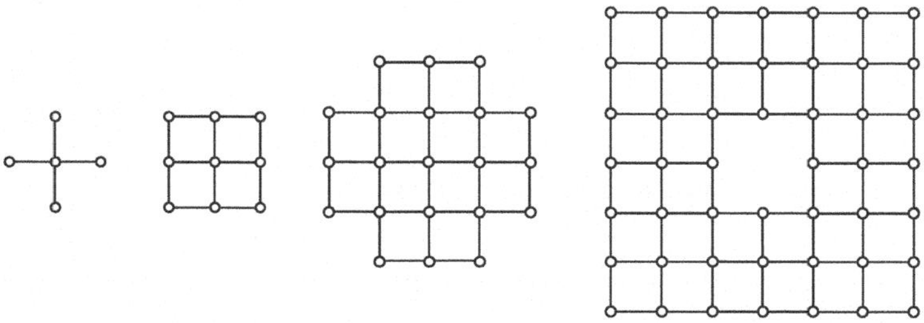

**Fig. 11.2** Local systems with square symmetry

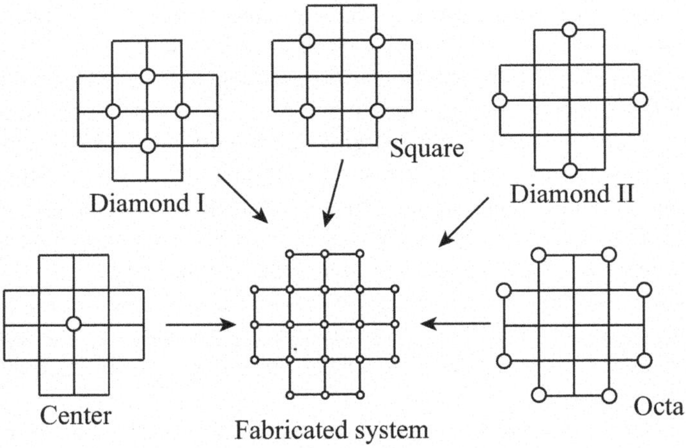


**Fig. 11.3** Construction of a local system through the assemblage of orbits with square symmetry

We choose each local system so that it has a square symmetry expressed by the dihedral group $D_4$. Figure 11.3 illustrates an example of the construction of a square lattice system by an assemblage of $D_4$-*symmetric orbits*. Such local $D_4$-symmetry is exploited in the description of sustaining bifurcation in Sect. 11.3.

In preparation for the bifurcation analysis, we carry out an *orbit decomposition* of places on a square domain with respect to the group $D_4$. Each square lattice with square symmetry can be expressed as an assemblage of *orbits*; each orbit comprises a series of points with square symmetry. We denote these orbits by

$$P = \{1\} \cup (\alpha_1 \cup \ldots \cup \alpha_{n_1}) \cup (\beta_1 \cup \ldots \cup \beta_{n_2}), \tag{11.1}$$

where $\{1\}$ represents an orbit comprising only the place at the center, $\alpha_i$ ($i = 1, \ldots, n_1$) represents an orbit with square- or diamond-shaped four places, and $\beta_i$ ($i = 1, \ldots, n_2$) represents an orbit with octagonal eight places.

For example, a system of $5 \times 5$ places in Fig. 11.4a has six orbits comprising one, four, or eight places ($n_1 = 4$ and $n_2 = 1$). Patterns with square symmetry on a square lattice are made up of some or all of the orbits in (11.1). For example, square patterns for $\{1\} \cup \alpha_1$, $\{1\} \cup \alpha_4$, and $\{1\} \cup \alpha_1 \cup \beta_1$ are depicted in Fig. 11.4b.

### *11.2.2 Construction of Global System*

We can construct a global system by choosing a series of cities and modeling the transportation network among the cities. Candidates of a global system are two-place economy, line segment economy, equidistant economy, radial network, and square and hexagonal lattices depicted in Fig. 11.1.

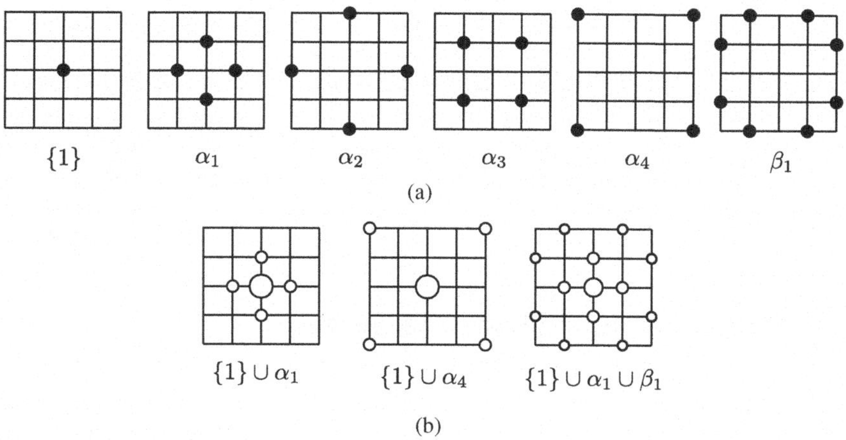


**Fig. 11.4** Orbits with respect to $D_4$ for a $5 \times 5$ square lattice. (**a**) Orbits with respect to $D_4$. (**b**) Spatial patterns constructed by assembling $D_4$-symmetric orbits

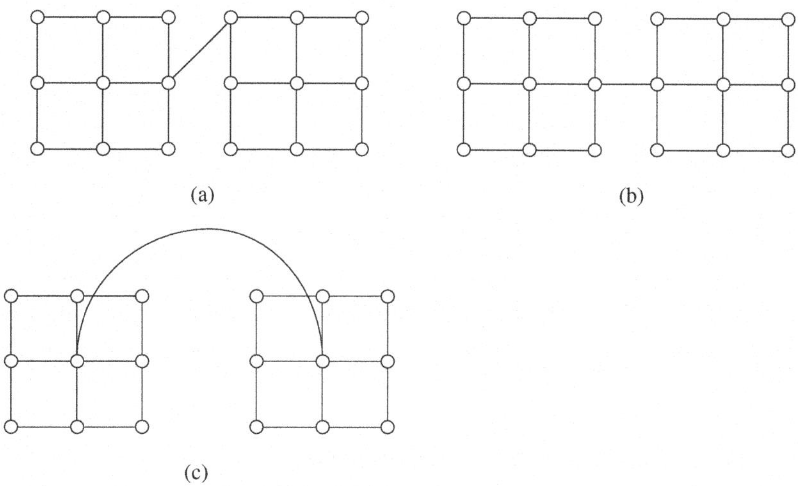


**Fig. 11.5** Two types of connections of two local systems. (**a**) Side-by-side connection (No symmetry). (**b**) Side-by-side connection ($D_2$-symmetry). (**c**) Center-to-center connection ($(D_4 \times D_4)\dot{+}S_2$-symmetry)

Among several possible ways to connect local platforms, we consider (1) side-by-side connection and (2) center-to-center connection. It is vital for the success of the analysis of the global–local system to connect local platforms pertinently. For example, Fig. 11.5 depicts these two types of connections. The side-by-side connection in (a) has no symmetry and its bifurcation structure is completely lost. By contrast, the connection in (b) has $D_2$-symmetry and retains some bifurcation structure. We recommend the center-to-center connection in (c) because this global–

local system has $(D_4 \times D_4)\dot{+}S_2$-symmetry and is endowed with a rich bifurcation mechanism, where $S_2$ is a symmetric group of degree 2.

The global system undergoes breaking bifurcation due to the loss of symmetry in the global system. Sustaining bifurcation occurs due to the loss of population in some places. The usefulness of such a rich bifurcation mechanism in the investigation of economic agglomeration behavior is demonstrated in Sect. 11.4 using a realistic example.

## 11.3 Sustaining Bifurcation of Monocenter

We investigate the mechanism of sustaining bifurcations (Sect. 11.3.1) and present an exhaustive list of bifurcating solutions of a local square lattice. These bifurcating solutions are observed in Sect. 11.3.2 in the numerical analysis of the Forslid and Ottaviano (2003) model (the FO model presented in Sect. 2.2).

As the bifurcation parameter, we employ the *trade freeness* in (2.11):

$$\phi = \exp[-\tau(\sigma - 1)\tilde{L}] \quad (0 < \phi < 1),$$

where $\tau$ is the transportation cost parameter and $\tilde{L}$ is the distance unit (the distance between two neighboring places). We consider the governing equation $\boldsymbol{F} = (F_i(\boldsymbol{\lambda}, \phi) \mid i = 1, \ldots, N) = \boldsymbol{0}$ in the replicator dynamics (1.7):

$$F_i(\boldsymbol{\lambda}, \phi) = (v_i(\boldsymbol{\lambda}, \phi) - \bar{v}(\boldsymbol{\lambda}, \phi))\lambda_i \quad (i = 1, \ldots, N).$$

Here, $F_i(\boldsymbol{\lambda}, \phi)$ $(i = 1, \ldots, N)$ is a sufficiently smooth function, $\boldsymbol{\lambda}$ is the independent variable vector, and $\bar{v} = \sum_{i=1}^{N} \lambda_i v_i$.

### 11.3.1 Theory of Sustaining Bifurcation

We consider *sustaining bifurcations* for the monocentric distribution (full agglomeration) $\boldsymbol{\lambda}^{\mathrm{FA}} = (1, 0, \ldots, 0)$ at the center of the square lattice. Recall that the monocentric distribution at any place is an invariant pattern (Proposition 9.1). The eigenvalues of the Jacobian matrix $J = \partial \boldsymbol{F}/\partial \boldsymbol{\lambda}$ for the monocentric distribution $\boldsymbol{\lambda}^{\mathrm{FA}}$ are evaluated to $-v_1(< 0)$, $v_2 - v_1, \ldots, v_N - v_1$ by Lemma 9.2 (the first eigenvalue can be evaluated from (9.7)). Then, the sustaining bifurcation condition is expressed as $v_k - v_1 = 0$ for some $k$ $(2 \leq k \leq N)$. (It is straightforward to generalize this condition to other square distributions.) By symmetry, the indirect utility $v_i$ in the same orbit (cf. (11.1)) takes the same value, denoted as

$$v_{\alpha_1}, \ldots, v_{\alpha_{n_1}} \text{ for } \alpha_i \ (i = 1, \ldots, n_1),$$
$$v_{\beta_1}, \ldots, v_{\beta_{n_2}} \text{ for } \beta_i \ (i = 1, \ldots, n_2).$$

Then, we have the following proposition for sustaining bifurcation.

**Proposition 11.1** *For the monocentric distribution, a sustaining bifurcation occurs if one of the following conditions is satisfied:*

$$v_{\alpha_i} - \bar{v} = 0 \text{ for some } \alpha_i \ (i = 1, \ldots, n_1),$$
$$v_{\beta_i} - \bar{v} = 0 \text{ for some } \beta_i \ (i = 1, \ldots, n_2).$$

The bifurcation mechanism of the monocentric distribution $\boldsymbol{\lambda}^{\mathrm{FA}}$ is described below (see Kogure & Ikeda, 2022a). Let $\phi_{\mathrm{c}}^{k}$ be the value of trade freeness that satisfies $v_k - v_1 = 0$ $(2 \le k \le N)$. In the analysis of bifurcating solutions from a critical point $(\boldsymbol{\lambda}^{\mathrm{FA}}, \phi_{\mathrm{c}}^{k})$, we employ incremental variables $\psi = \phi - \phi_{\mathrm{c}}^{k}$ and $\boldsymbol{x} = (x_1, \ldots, x_4) = \{\lambda_j \mid j \in \alpha_i\}$ for $\alpha_i$ and $\boldsymbol{x} = (x_1, \ldots, x_8) = \{\lambda_j \mid j \in \beta_i\}$ for $\beta_i$ (see the left of Fig. 11.6 for the definition of $\boldsymbol{x}$). Then, the bifurcation equation takes the following form:

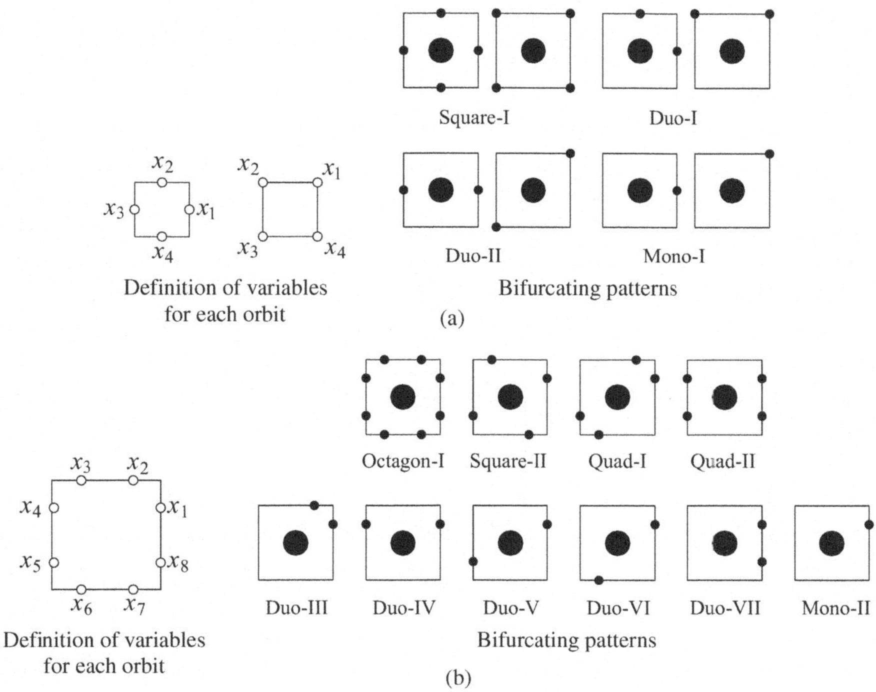


**Fig. 11.6** Definition of variables and bifurcating patterns for orbits $\alpha_i$ and $\beta_i$. (**a**) Type $\alpha_i$ orbits. (**b**) Type $\beta_i$ orbits

**Lemma 11.1**

*(i) For Type $\alpha_i$ orbit, the bifurcation equation is four-dimensional and is expressed as*

$$
\begin{aligned}
\tilde{F}_1(x_1, x_2, x_3, x_4, \psi) &= x_1 R(x_1, x_2, x_3, x_4, \psi) = 0,\\
\tilde{F}_2(x_1, x_2, x_3, x_4, \psi) &= x_2 R(x_2, x_3, x_4, x_1, \psi) = 0,\\
\tilde{F}_3(x_1, x_2, x_3, x_4, \psi) &= x_3 R(x_3, x_4, x_1, x_2, \psi) = 0,\\
\tilde{F}_4(x_1, x_2, x_3, x_4, \psi) &= x_4 R(x_4, x_1, x_2, x_3, \psi) = 0
\end{aligned}
$$

*with a smooth function $R(\boldsymbol{x}, \psi)$ satisfying $R(x_1, x_2, x_3, x_4, \psi) = R(x_1, x_4, x_3, x_2, \psi)$.*

*(ii) For Type $\beta_i$ orbit, the bifurcation equation is eight-dimensional and is given as*

$$
\begin{aligned}
\tilde{F}_1(\boldsymbol{x}, \psi) &= x_1 R(x_1, x_2, x_3, x_4, x_5, x_6, x_7, x_8, \psi) = 0,\\
\tilde{F}_2(\boldsymbol{x}, \psi) &= x_2 R(x_2, x_1, x_8, x_7, x_6, x_5, x_4, x_3, \psi) = 0,\\
\tilde{F}_3(\boldsymbol{x}, \psi) &= x_3 R(x_3, x_4, x_5, x_6, x_7, x_8, x_1, x_2, \psi) = 0,\\
\tilde{F}_4(\boldsymbol{x}, \psi) &= x_4 R(x_4, x_3, x_2, x_1, x_8, x_7, x_6, x_5, \psi) = 0,\\
\tilde{F}_5(\boldsymbol{x}, \psi) &= x_5 R(x_5, x_6, x_7, x_8, x_1, x_2, x_3, x_4, \psi) = 0,\\
\tilde{F}_6(\boldsymbol{x}, \psi) &= x_6 R(x_6, x_5, x_4, x_3, x_2, x_1, x_8, x_7, \psi) = 0,\\
\tilde{F}_7(\boldsymbol{x}, \psi) &= x_7 R(x_7, x_8, x_1, x_2, x_3, x_4, x_5, x_6, \psi) = 0,\\
\tilde{F}_8(\boldsymbol{x}, \psi) &= x_8 R(x_8, x_7, x_6, x_5, x_4, x_3, x_2, x_1, \psi) = 0
\end{aligned}
$$

*with a smooth function $R(x_1, x_2, x_3, x_4, x_5, x_6, x_7, x_8, \psi)$.*

Solving the bifurcation equation for $\alpha_i$ and that for $\beta_i$, we obtain the bifurcating solutions in Proposition 11.2. Figure 11.6 depicts these bifurcating patterns. The stability of all the bifurcating solutions depends on the case.

**Proposition 11.2**

*(i) A bifurcation point for Type $\alpha_i$ orbit has the following bifurcating solutions:*

$$
\boldsymbol{x} = \begin{cases}
w(1, 1, 1, 1): & \text{Square-I},\\
w(1, 1, 0, 0): & \text{Duo-I},\\
w(1, 0, 1, 0): & \text{Duo-II},\\
w(1, 0, 0, 0): & \text{Mono-I}.
\end{cases}
$$

*(ii) A bifurcation point for Type $\beta_i$ orbit has the following bifurcating solutions:*

$$
\boldsymbol{x} = \begin{cases}
w(1, 1, 1, 1, 1, 1, 1, 1): & \text{Octagon-I},\\
w(1, 0, 1, 0, 1, 0, 1, 0): & \text{Square-II},\\
w(1, 1, 0, 0, 1, 1, 0, 0): & \text{Quad-I},\\
w(1, 0, 0, 1, 1, 0, 0, 1): & \text{Quad-II},
\end{cases}
$$

$$\boldsymbol{x} = \begin{cases} w(1,1,0,0,0,0,0,0): & \text{Duo-III}, \\ w(1,0,0,1,0,0,0,0): & \text{Duo-IV}, \\ w(1,0,0,0,1,0,0,0): & \text{Duo-V}, \\ w(1,0,0,0,0,1,0,0): & \text{Duo-VI}, \\ w(1,0,0,0,0,0,0,1): & \text{Duo-VII}, \\ w(1,0,0,0,0,0,0,0): & \text{Mono-II}. \end{cases}$$

*Here,* $w > 0$ *is a constant.*

### 11.3.2 Bifurcation Behavior of an Economic Geography Model

We consider the $5 \times 5$ *square lattice* in the FO model, which has two economic model parameters: $\sigma > 1$ and $\mu \in (0, 1)$. The $5 \times 5$ places are decomposed into six orbits (cf. Fig. 11.4a):

$$\begin{cases} \{1\}: & \text{a place at the center}, \\ \alpha_1, \ldots, \alpha_4: & \text{4 places}, \\ \beta_1: & \text{8 places}. \end{cases}$$

For the monocentric distribution $\boldsymbol{\lambda}^{\text{FA}}$ at the center of the square lattice, there are five bifurcation points corresponding to $\alpha_1, \ldots, \alpha_4$ and $\beta_1$ orbits.

Figure 11.7 shows the stability areas (colored blue) of the monocentric distribution $\boldsymbol{\lambda}^{\text{FA}}$ in the space of $(\phi, \mu) \in (0, 1) \times (0, 1)$ for $\sigma = 6.0$ and $\sigma = 3.0$. For each $\mu$, the monocentric distribution loses its stability at a sustaining bifurcation point at the leftward border of the stability area when $\phi$ decreases from 1. An increase of $\mu$ or a decrease of $\sigma$ expands the stability area leftward.

We investigate the emergence of bifurcating solutions from the stable monocentric distribution referring to the solution curves for $\sigma = 6.0$ and 3.0 ($\mu = 0.4$) in Fig. 11.8. When $\phi$ decreases from 1, a sustaining bifurcation occurs at point E. The stable bifurcating path EA shows that the stable agglomeration pattern shifts from the monocentric distribution to the Square-I pattern (Fig. 11.6a) for Type $\alpha_3$ orbit at $\sigma = 6.0$ and Type $\alpha_4$ orbit at $\sigma = 3.0$. Thus, the type of stable bifurcating solution depends on the value of $\sigma$. Other bifurcating paths, EB, EC, and ED, are all unstable.

We further investigate the influence of the parameters $\sigma$ and $\mu$ on the sustaining bifurcation. Figure 11.9 shows the change in bifurcating solution in the space of $(1/\sigma, \mu)$ for $5 \times 5$ and $7 \times 7$ square lattices. We distinguish different bifurcating solutions using different colors. No bifurcation occurs in the upper-right white zone, in which the no-black-hole condition $\mu < \sigma - 1$ in (8.9) is violated. The location of satellite places engendered through the sustaining bifurcation shifts outward as $\sigma$ decreases and $\mu$ increases.

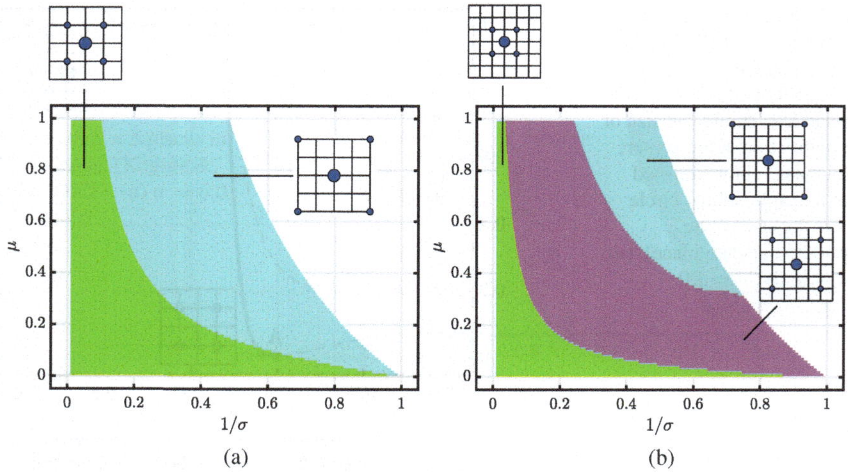


**Fig. 11.9** The dependence of the types of bifurcating solutions on the values of $(1/\sigma, \mu) \in (0, 1) \times (0, 1)$ in the FO model on square lattices (Kogure & Ikeda, 2022b) (a blue circle shows the size of the population at each place). (**a**) 5 × 5 places. (**b**) 7 × 7 places

**Table 11.1** Seven largest hub airports in the USA in 2019 by United States Department of Transportation (https://www.bts.gov/content/passengers-boarded-top-50-us-airports) and the population of the cities for these airports in 2021 by World Population Review (https://worldpopulationreview.com/us-cities) ($\lambda^* = 330{,}000$)

| City | Airport | Total enplaned passengers (rank) | Population of city $\Lambda_i$ | Normalized population $\Lambda_i/\lambda^*$ | Lattice size |
|---|---|---|---|---|---|
| Atlanta | Hartsfield-Jackson Atlanta Int. | 53,505,357 (1) | 524,067 | 1.6 | 2 |
| Los Angeles | Los Angeles Int. | 42,965,731 (2) | 3,983,540 | 12.0 | 3 × 3 |
| Chicago | Chicago O'Hare Int. | 40,887,890 (3) | 2,679,080 | 8.1 | 3 × 3 |
| Dallas | Dallas/Fort Worth Int. | 35,785,318 (4) | 1,347,120 | 4.1 | 1 + 4 |
| Denver | Denver Int. | 33,592,645 (5) | 749,103 | 2.3 | 2 |
| New York | John F. Kennedy Int. | 31,123,436 (6) | 8,230,290 | 24.9 | 5 × 5 |
| San Francisco | San Francisco Int. | 27,715,305 (7) | 883,255 | 2.7 | 2 |

- Determine a square lattice with the number of places close to $\Lambda_i/\lambda^*$.

We model the local networks of these hub cities by either square lattices with several sizes or two places as follows:

- 5 × 5 square lattice: New York
- 3 × 3 square lattice: Los Angeles and Chicago
- Star economy with 1 + 4 places: Dallas
- Two places: San Francisco, Denver, and Atlanta

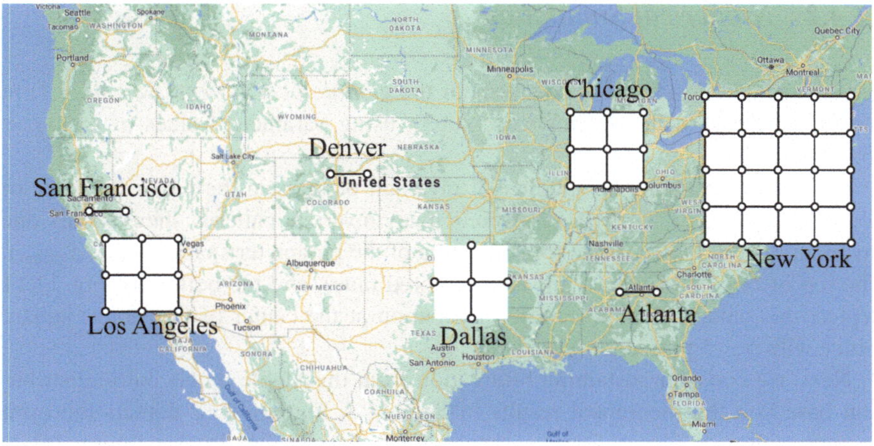


**Fig. 11.10** Global–local model of seven largest airport hub cities in the USA

Figure 11.10 depicts these local transportation networks on the map of the USA. We consider a global–local system where the centers of these local platforms (the left place for the two-place economy) are connected equidistantly by an air transportation network.

We consider the transportation between two places, $i$ and $j$, belonging to a global–local system and introduce two types of transportation cost parameters:

- $\tau_{\text{local}}$ for *intra-city transportation cost*
- $\tau_{\text{global}}$ for *inter-city transportation cost*

When the two places $i$ and $j$ belong to the same local square lattice, there is no inter-city transportation cost, and the transportation cost is obtained by replacing $\tau$ by $\tau_{\text{local}}$ in (2.10). That is, $\tau_{ij} = \exp(L_{ij}\tau_{\text{local}})$, where $L_{ij}$ is the road distance between places $i$ and $j$; when a unit of goods is transported from place $i$, only $1/\tau_{ij}$ arrives at place $j$.

When the two places $i$ and $j$ belong to different local square lattices, there are both intra- and inter-city transportation costs. Denoting by $i^*$ the center of the lattice to which $i$ belongs and $j^*$ that $j$ belongs, we can express the transportation cost between the two places $i$ and $j$ as

$$\tau_{ij} = \exp[(L_{i,i^*} + L_{j,j^*})\tau_{\text{local}} + \tau_{\text{global}}].$$

We employ, as bifurcation parameters, the local and global *trade freeness* parameters:

$$\phi_{\text{local}} = \exp[-(\sigma - 1)\tau_{\text{local}}], \quad \phi_{\text{global}} = \exp[-(\sigma - 1)\tau_{\text{global}}].$$

### 11.4.2 Numerical Simulation

We advance the results of the analysis of the *global–local system* (Sect. 11.4.1) in the FO model (Forslid & Ottaviano, 2003), using parameter values of $(\sigma, \mu) = (6.0, 0.4)$.

First, we investigate the stability of invariant patterns in the parameter zone of $(\phi_{\text{local}}, \phi_{\text{global}}) \in (0, 1) \times (0, 1)$. Among many invariant patterns, there is only one invariant pattern that is stable for some area of the parameter space $(\phi_{\text{local}}, \phi_{\text{global}})$. The top of Fig. 11.11 shows this area (colored blue), and the bottom of Fig. 11.11 depicts this pattern: a monocentric distribution at the center of the largest square lattice with $5 \times 5$ places for New York.

Next, we consider sustaining bifurcation from this stable monocentric distribution based on the theoretical results in Sect. 11.3. Figure 11.12 depicts the orbit decomposition of places for this monocentric distribution. In this figure, $\{1\}$ is not associated with bifurcation, and the sustaining bifurcation mechanisms of $\alpha_i$ $(i = 1, 2, 3, 4, 7)$ and $\beta_1$ are given in Sect. 11.3.1; those for other orbits can be consulted with Remark 11.1 below.

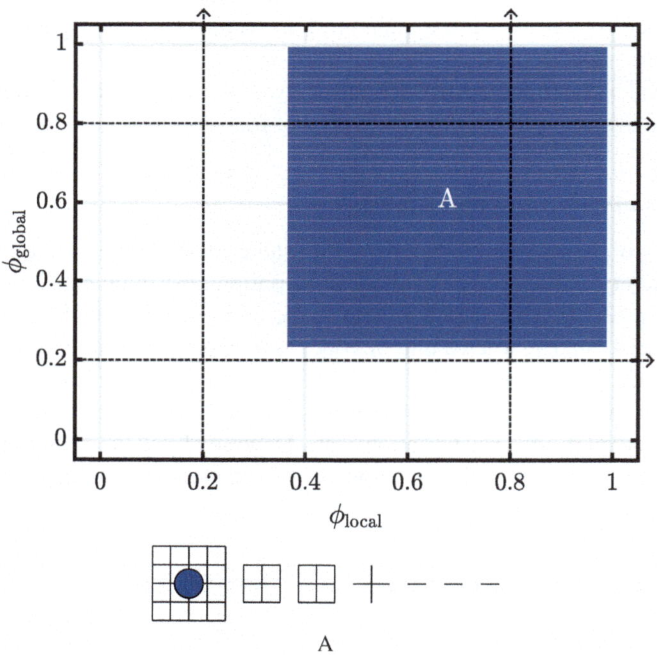


**Fig. 11.11** The stability area $(\phi_{\text{local}}, \phi_{\text{global}})$ of the monocentric distribution at the center of the square lattice with $5 \times 5$ places (Kogure & Ikeda, 2022b) (the two vertical dotted lines and the two horizontal ones correspond to the analysis in Fig. 11.13

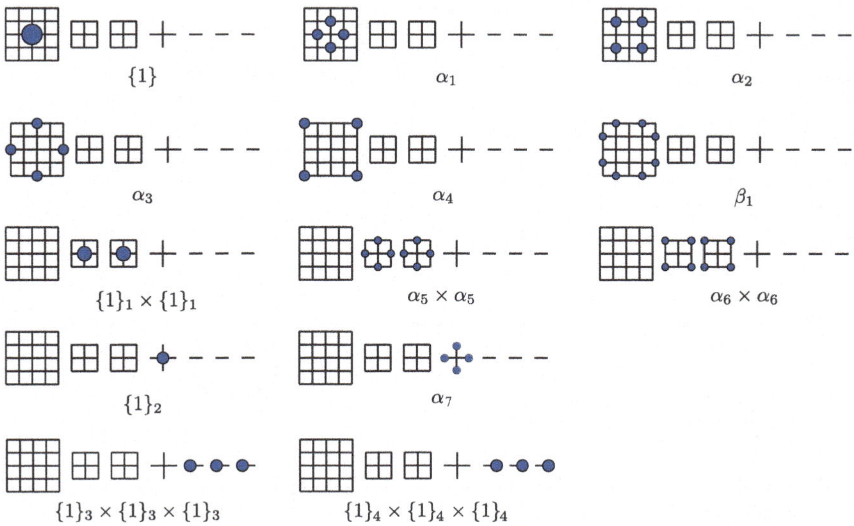


**Fig. 11.12** Orbit decomposition of the monocenter (full agglomeration to the center) of the $5 \times 5$ square lattice

We first fix the local trade freeness to be $\phi_{\text{local}} = 0.2$ (corresponding to the left vertical dotted line in Fig. 11.11).[1] Figure 11.13a plots the stable solution curves in the space of $(\phi_{\text{global}}, \lambda_{\text{max}})$. The monocentric distribution is not stable for any value of global trade freeness $\phi_{\text{global}}$, although there is a stable state close to this distribution. This stable state encounters the sustaining bifurcation at point A, from which a stable bifurcating path branches engendering population at the vertices of the $3 \times 3$ lattices shown by $(\times)$. After that, there are successive sustaining bifurcations at points B, C, D, E, F, and G occurring as many as six times, through which the population is distributed to many other places.

We next fix the value of global trade freeness $\phi_{\text{global}}$ and examine the transition of stable equilibria. Figure 11.13b shows the solution curve for a small global trade freeness of $\phi_{\text{global}} = 0.2$.[2] At the sustaining bifurcation points A, B, and C, satellite places appear successively on the $5 \times 5$ lattice. There is a population at the center of the $3 \times 3$ lattices for any value of $\phi_{\text{local}}$, and this population grows as $\phi_{\text{local}}$ decreases.

**Remark 11.1** For a bifurcation point associated with $\{1\}_1 \times \{1\}_1$, there are two independent bifurcation equations of the same form: $\tilde{F}_i(x_i, \psi) = x_i R(x_i, \psi) = 0 \ (i = 1, 2)$ for some function $R(x_i, \psi)$. There are nonzero $x_i$ satisfying $R(x_i, \psi) = 0 \ (i = 1, 2)$ and two types of bifurcating solutions: (1) $x_1 = 0$ and $x_2 \neq 0$ and (2) $x_1 = x_2 \neq 0$. For a bifurcation point associated with $\alpha_5 \times \alpha_5$ and

[1] A case of a large local trade freeness of $\phi_{\text{local}} = 0.8$ (corresponding to the right vertical dotted line in Fig. 11.11) is available in Kogure and Ikeda (2022b).

[2] A case of a large global trade freeness of $\phi_{\text{global}} = 0.8$ is given in Kogure and Ikeda (2022b).

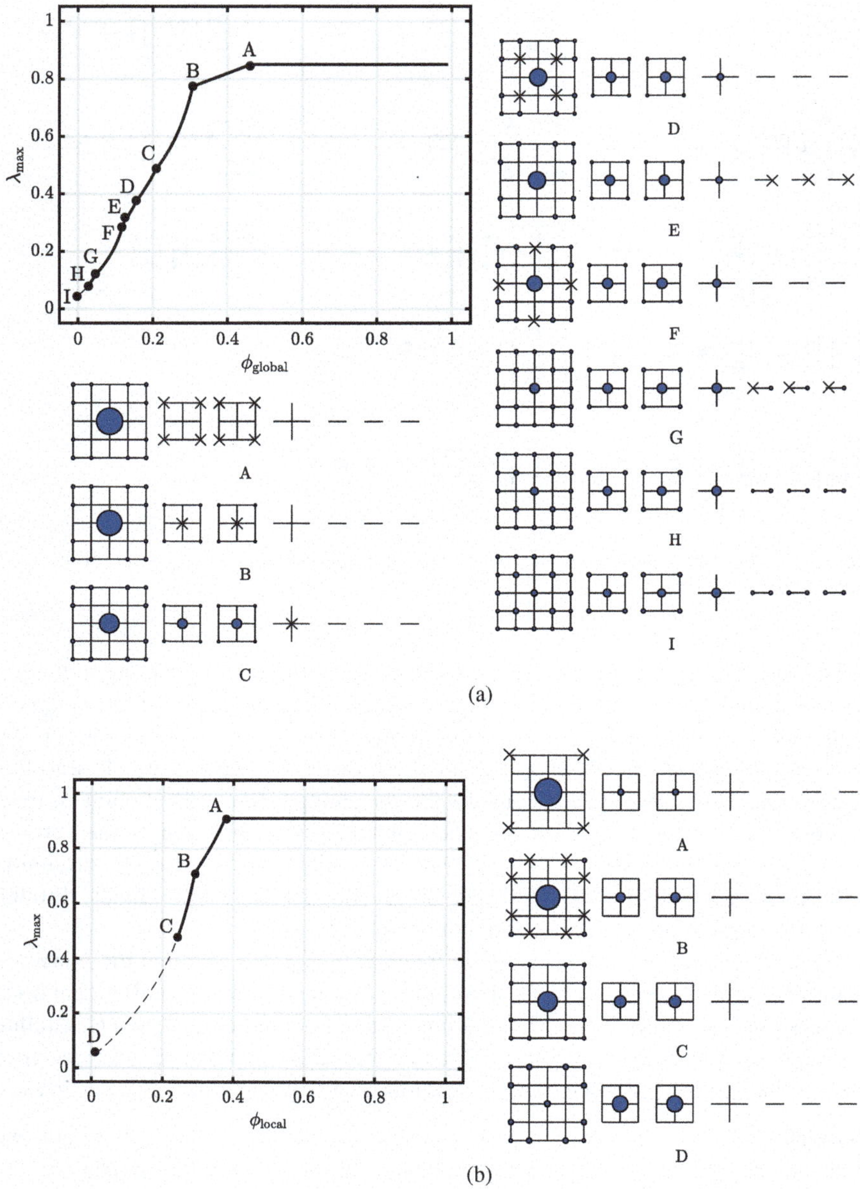


**Fig. 11.13** Solution curves obtained by varying $\phi_{\text{global}}$ or $\phi_{\text{local}}$ (Kogure & Ikeda, 2022b) (the vertical axis shows $\lambda_{\max} = \max(\lambda_1, \ldots, \lambda_{54})$; solid curve: stable; dashed one: stable; ($\times$) indicates the place where population emerges at a sustaining bifurcation point). (**a**) $\phi_{\text{local}} = 0.2$ and $\phi_{\text{global}} \in (0, 1)$. (**b**) $\phi_{\text{global}} = 0.2$ and $\phi_{\text{local}} \in (0, 1)$

$\alpha_6 \times \alpha_6$, there are two sets of bifurcation equations of the same form, and each set has the same bifurcation mechanism for $\alpha_i$ presented in Proposition 11.2(i). □

## References


Christaller, W. (1933). *Die zentralen Orte in Süddeutschland*. Gustav Fischer. English Translation: *Central places in Southern Germany*. Prentice Hall (1966).

Forslid, R., & Ottaviano, G. I. P. (2003). An analytically solvable core–periphery model. *Journal of Economic Geography, 3*, 229–240.

Kogure, Y., & Ikeda, K. (2022a). Group-theoretic study of economic agglomerations on a square lattice. *MPRA Paper* 112842.

Kogure, Y., & Ikeda, K. (2022b). Global–local spatial platform for economic geography: Mechanism for sustaining bifurcation. *International Journal of Bifurcation and Chaos, 32*(11), 2230026.

Redding, S. J., & Rossi-Hansberg, E. (2017). Quantitative spatial economics. *Annual Review of Economics* 9, 21–58.

# Chapter 12
# Spatial Period-Doubling Cascade in Long-Narrow Economy

**Abstract** The mechanism of the self-organization of spatial agglomeration patterns in a long-narrow economy is elucidated in light of an analogy to a racetrack economy. A spatial period-doubling bifurcation cascade in the racetrack economy characterizes the transition of spatial agglomerations in the long-narrow economy. Formulas for break points in an economic geography model of the racetrack economy are developed to predict the occurrence of spatial period doubling in the long-narrow economy. There are several prerequisites: Chap. 8 provides the analytical analysis of break points, and Chaps. 3 and 9 cover the bifurcation of the racetrack economy.

**Keywords** Bifurcation · Break point · Economic geography model · Long-narrow economy · Pattern formation · Racetrack analogy · Racetrack economy · Spatial period doubling

## 12.1 Introduction

Narrow industrial belts support megalopolises worldwide. A chain of cities stretches from Boston to Washington, D.C., in a closed, narrow corridor between the Atlantic Ocean and the Appalachian Mountains. New York prospers as the core of this chain.

This chapter highlights a long-narrow economy (Fig. 12.1a) as a spatial platform for a chain of cities and elucidates its spatial agglomeration mechanism, based on Ikeda et al. (2017). Spatial agglomeration in this platform proceeds gradually without undergoing bifurcation.[1] Other spatial platforms in this book, such as the racetrack economy (Fig. 12.1b), have homogeneity among places (all places are initially identical) and undergo various types of bifurcations (cf. Chaps. 3 and 9).

---

[1] Although there is a bifurcation due to the loss of bilateral symmetry, this bifurcation does not occur in the study in this chapter.

 
K. Ikeda, Y. Takayama, *Bifurcation Theory of Pattern Formation in Economic Geography*, Interdisciplinary Applied Mathematics 61,
https://doi.org/10.1007/978-3-031-71686-7_12

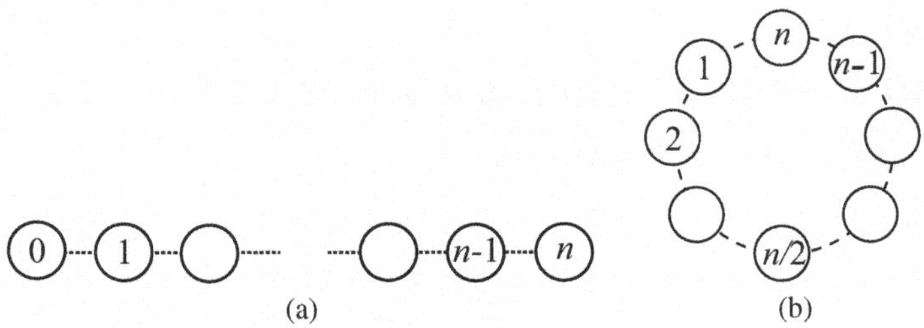


**Fig. 12.1** Spatial platforms of economic activities. (**a**) Long-narrow economy. (**b**) Racetrack economy

Despite this difference, the long-narrow economy and the racetrack economy have similar geometries: chains of places arranged side by side. The basic strategy of this chapter, called *racetrack economy analogy*, is to interpret and describe the mechanism of economic agglomerations in the long-narrow economy based on theoretical information about the racetrack economy. The analogy agrees with the underlying intuitive belief that both economies would tend to behave similarly as the number of places increases. The racetrack economy undergoes a sequence of recurrent bifurcations, known as a spatial period-doubling cascade.[2] This cascade helps describe how agglomerations proceed in the long-narrow economy.

We employ the model by Forslid and Ottaviano (2003) due to its analytical solvability. In this model with replicator dynamics, we describe the break point for the racetrack economy by combining the formulas for the direct bifurcation in Chap. 8 with those for the secondary bifurcation (Ikeda et al., 2017). We demonstrate that these formulas can index the transition of spatial patterns in the long-narrow economy.

There are several studies on the long-narrow economy and its continuous-space counterpart. A megalopolis consisting of large core cities connected to *an industrial belt*, i.e., *a continuum of cities*, associated with lower transportation costs, is observed (Mori, 1997). Fujita and Mori (1997) found a stable full agglomeration and a discretized, highly regular central place system *a la* Christaller (1933) and Lösch (1940). Mossay and Picard (2011) conducted a comparative study of the long-narrow economy in a continuous space and the racetrack economy.

This chapter is organized as follows: Sect. 12.2 advances the mechanism of a spatial period-doubling cascade in the racetrack economy and presents the formulas for the break point in this cascade. Spatial agglomeration in the long-narrow economy is investigated based on the racetrack economy analogy in Sect. 12.3.

[2] See Sect. 9.4.2 and, e.g., Tabuchi and Thisse (2011), Ikeda et al. (2012), and Akamatsu et al. (2012, 2016).

## 12.2 Racetrack Economy: Idealized Spatial Platform

As an idealized spatial platform of a long-narrow economy, we introduce the *racetrack economy*, comprising a chain of $n$ places (labeled $i = 1, \ldots, n$) spread equally on the circumference of a circle (Fig. 12.1b).

### 12.2.1 Spatial Period Doubling Bifurcation Cascade

In the racetrack economy, the uniform state $\boldsymbol{\lambda}^* = \frac{1}{n}(1, \ldots, 1)^\top$ is an invariant pattern that is a solution to the governing equation $\boldsymbol{F}(\boldsymbol{\lambda}, \tau) = \boldsymbol{0}$ for any $\tau$ (Sect. 9.2.3). In this economy, with $n = 2^k$ ($k = 2, 3, \ldots$) places, the bifurcation from this state proceeds as explained in Sect. 9.4. At a *pitchfork* bifurcation point at the uniform state, bifurcating equilibria branch in the direction of an eigenvector $\boldsymbol{\eta}^* = (1, -1, \ldots, 1, -1)^\top$ of the Jacobian matrix $J(\boldsymbol{\lambda}^*, \tau) = \frac{\partial \boldsymbol{F}}{\partial \boldsymbol{\lambda}}(\boldsymbol{\lambda}^*, \tau)$, satisfying $J(\boldsymbol{\lambda}^*, \tau)\boldsymbol{\eta}^* = \boldsymbol{0}$. After that, pitchfork bifurcations can occur recurrently until reaching the *full agglomeration*, i.e., complete agglomeration to a single core place. It is the *spatial period-doubling cascade* in (9.11):

$$\mathrm{D}_n \to \mathrm{D}_{n/2} \to \cdots \to \mathrm{D}_2 \to \mathrm{D}_1. \tag{12.1}$$

Figure 12.2 depicts this cascade for $n = 8$ places. Akamatsu et al. (2012) observed this cascade in the Pflüger (2004). The pitchfork bifurcation is either subcritical or supercritical. In the study of a long-narrow economy in Sect. 12.3.2, we have supercritical ones.

### 12.2.2 Formulas for Break points

We obtain the formulas for the break point in the racetrack economy by combining the results for the direct bifurcation in Chap. 8 with those for the secondary bifurcation developed in Ikeda et al. (2017). We employ the analytically solvable model by Forslid and Ottaviano (2003) with the replicator dynamics in (1.7). The

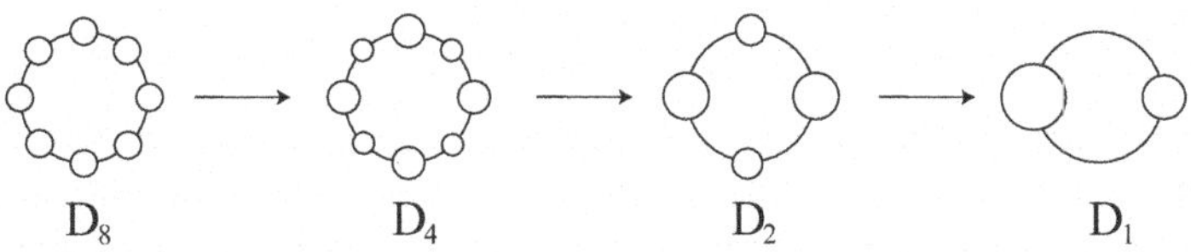


**Fig. 12.2** Spatial period-doubling bifurcation cascade in the racetrack economy with $n = 8$ places (the size of an area represents population size)

total length $\mathscr{L}$ of the road along the circumference is chosen as $\mathscr{L} = 1$, and neighboring places are connected by an inter-place road of length $\tilde{L} = 1/n$.

The *break point* for the $m$th period doubling is denoted by $\tau = \tau_m$, where $\tau$ is the transportation cost parameter defined by (2.10). We introduce the following assumption to describe this cascade.

**Assumption 12.1** *The number of places is $n = 2^k$ $(k \geq m + 1)$, and the secondary bifurcation occurs from the state of $\hat{\boldsymbol{\lambda}} = \frac{2}{n}(1, 0, \ldots, 1, 0)^\top$.*

**Lemma 12.1** *The break point $\tau_m$ exists under a generalized no-black-hole condition:*

$$\frac{\mu}{\sigma - 1} < \frac{1}{m}, \quad m = 1, 2. \tag{12.2}$$

***Proof*** See (8.9) for the proof of $m = 1$ and Ikeda et al. (2017) for $m = 2$. □

The following proposition advances the formula for break points up to the second doubling of the spatial period.

**Proposition 12.1** *The break point is given by*

$$\frac{\tau_m}{n} = \frac{1}{m(\sigma - 1)} \log \left( \frac{1 + \epsilon_m^{1/2}}{1 - \epsilon_m^{1/2}} \right); \quad m = 1, 2; \;\; n = 2^k \; (k \geq m + 1); \tag{12.3}$$

$$\epsilon_1 = \frac{\tilde{\mu} + \mu'}{\tilde{\mu}\mu' + 1}, \quad \epsilon_2 = \frac{\tilde{\mu} + \mu'}{\tilde{\mu}\mu' + (1 + \tilde{\mu})/2}; \quad \tilde{\mu} = \frac{\mu}{\sigma}, \quad \mu' = \frac{\mu}{\sigma - 1}.$$

***Proof*** See (8.20) and (8.24) for $m = 1$ and Ikeda et al. (2017) for $m = 2$. □

The break point $\tau_m$ is proportional to $n = 1/\tilde{L}$. A shorter $\tilde{L}$ or larger $n$, therefore, hastens the occurrence of a breaking bifurcation by increasing the break point $\tau_m$ and intensifying the competition between neighboring places. The effects of the parameters $\sigma$ and $\tau$ are not explicit in the formula (12.3) for the break point $\tau_m$. We introduce the following approximate formula to express their effects explicitly.

**Corollary 12.1** *Under the condition $\sigma \gg 1$, we have an approximate formula:*

$$\frac{\tau_m}{n} \approx \frac{\hat{\tau}_m}{n} = \sqrt{\frac{8}{m!}} \frac{\mu^{1/2}}{(\sigma - 1)^{3/2}}; \quad m = 1, 2; \;\; n = 2^k \; (k \geq m + 1). \tag{12.4}$$

***Proof*** See (8.25) for $m = 1$ and Ikeda et al. (2017) for $m = 2$. □

A lower $\sigma$ and/or a higher $\mu$ with a larger agglomeration force tends to hasten the onset of a breaking bifurcation. This tendency concurs with economic intuition in the literature that there are a few large agglomerations when $\sigma$ is low and/or $\mu$ is high (e.g., Fujita et al., 1999).

Note that the formulas (12.3) and (12.4) cover only $m = 1$ and 2. Nonetheless, it will be observed in the numerical analyses (Table 12.1(b) in Sect. 12.3.2) that these formulas have extended applicability to $m = 3$ and 4. Then, the formula (12.4) indicates that pitchfork bifurcations occur recurrently in the decreasing order of $\tau$ at the rate of $1/\sqrt{m!}$, which quickly converges to 0 as $m \longrightarrow +\infty$. We pose a conjecture that "as the recurrent bifurcations proceed, the relative change in the break point decreases."

## 12.3 Agglomeration Behaviors of Long-Narrow Economy

We consider a *long-narrow economy* in Fig. 12.1a. There are $n + 1$ places (labeled $i = 0, 1, \ldots, n$) evenly spread on a line segment with the unit length, and neighboring places are connected by an inter-place road of length $\tilde{L} = 1/n$. This economy has the same length and the same number $n$ of inter-place roads as the racetrack economy in Sect. 12.2. The economy, unlike the racetrack economy, has spatial heterogeneity because a border place ($i = 0$ or $n$) has a single neighboring place, but inside places have two.

We use the same model as that employed in Sect. 12.2 and use parameter values of $(\sigma, \mu) = (10.0, 0.4)$, satisfying the no-black-hole condition (12.2) for the existence of direct and secondary bifurcations.

### *12.3.1 Transition of Agglomeration Patterns*

We observe the transition of spatial agglomeration patterns in the long-narrow economy with 17 places. Figure 12.3a shows the stable solution curve expressed by the relation between the population $\lambda_8$ of the place at the center and the transportation cost parameter $\tau$. At the points denoted by ($\circ$) on this curve, the transition of population distributions as $\tau$ decreases is depicted in Fig. 12.3b using the bar charts.

An *almost uniformly distributed state* prevails for large transportation costs ($\tau \approx 5.0$) as local firms operating in a small market area enjoy the benefit of reduced transportation distance of goods at the expense of small economies of scale. At $\tau \approx 1.5$, border effects start to take effect and the populations at the border places 0 and 16 with smaller market areas decrease, and those in the neighboring places 1 and 15 increase. As $\tau$ decreases, the number of agglomerated places decreases recurrently as $17 \rightarrow 7 \rightarrow 5 \rightarrow 3 \rightarrow 1$ en route to the full agglomeration at the center, as explained below.

*Spatial Period-Doubling State I* ($1.0 \geq \tau \geq 0.8$) The populations at odd-numbered places 1, 3, ..., 15 almost disappear, and those at even-numbered ones increase. Core and peripheral places alternate spatially to engender a chain of spatially repeated core–periphery patterns. The spatial period between the core places is doubled to $L/\tilde{L} = 2$. Since two peripheral places surround a core place, this pattern

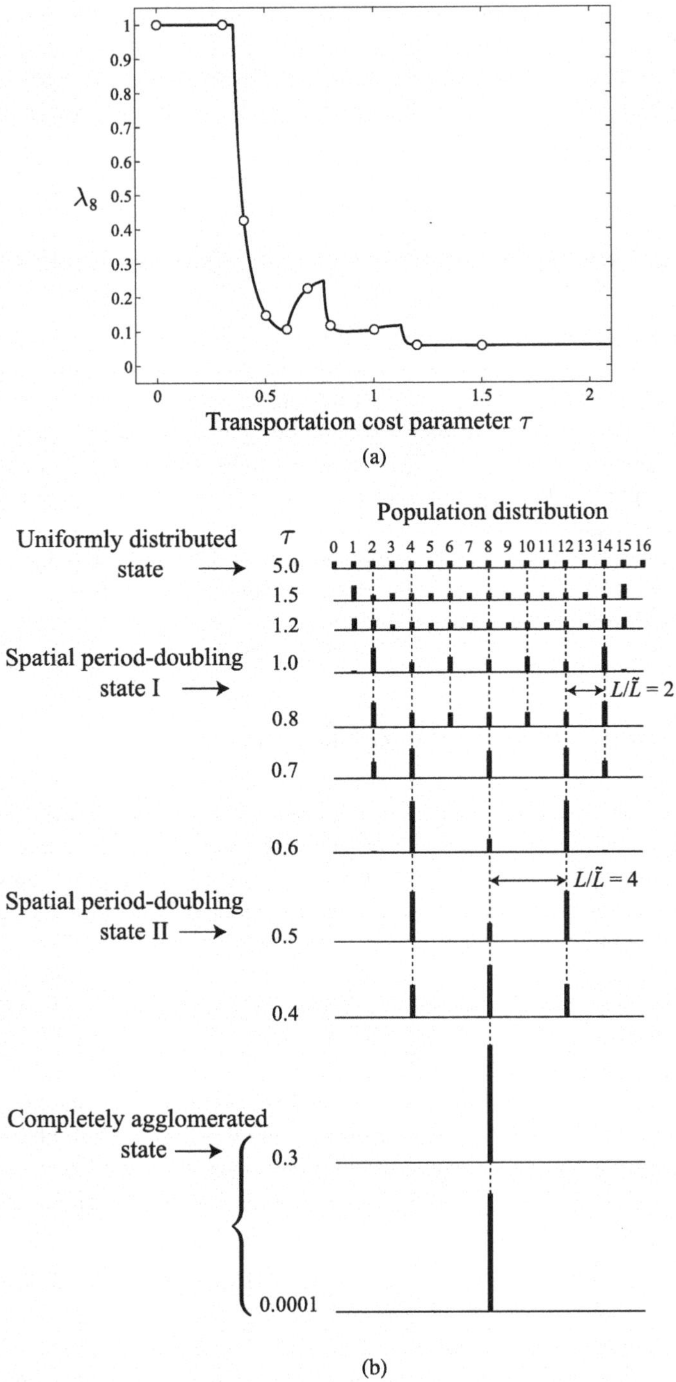


**Fig. 12.3** Spatial agglomeration in the long-narrow economy with $n = 17$ places observed as the transportation cost $\tau$ decreases (Ikeda et al., 2017). (**a**) Relation between $\lambda_8$ and $\tau$. (**b**) Population bar charts

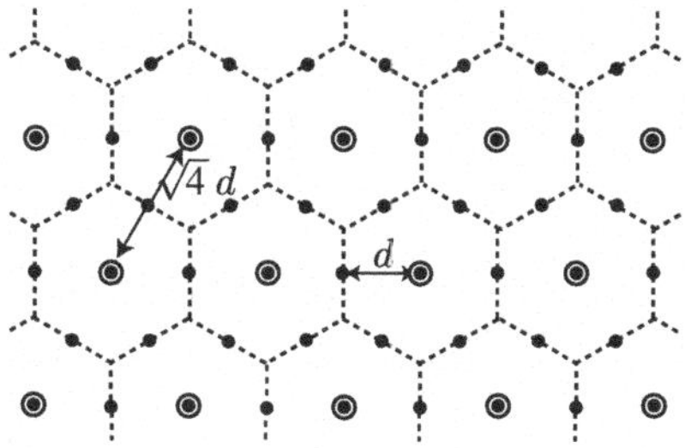


**Fig. 12.4** $k = 4$ system of Christaller (1933)

looks like a one-dimensional counterpart of Christaller's $k = 4$ system in Fig. 12.4 (Christaller, 1933). The chain of spatially repeated core–periphery patterns observed above is in line with "a highly regular system *a la* Christaller" found in a different economic model with increasing population size by Fujita and Mori (1997).

*Spatial Period-Doubling State II* ($0.6 \geq \tau \geq 0.4$) Three places, 4, 8, and 12, gather all the population, thereby displaying a core–periphery pattern with enlarged peripheral zones and a further doubled spatial period $L/\tilde{L} = 4$.

The existence of full agglomeration to a place at the center ($\tau \leq 0.353$) is robust against the number of places as it exists in $\tau \leq 0.356$ for 33 places and in $\tau \leq 0.358$ for 65 places.

A question on the spatial agglomeration set forth above is "How and when does a chain of core–periphery patterns emerge?" We answer this question in Sect. 12.3.2 in light of the racetrack economy analogy and the formulas for break points in Sect. 12.2.2.

### 12.3.2 Racetrack Economy Analogy

The geometrical configurations of the long-narrow economy and the racetrack economy are similar in that places are located equidistantly. We conduct a comparative study of these two economies to demonstrate the similarities in their spatial agglomeration patterns.

Figure 12.5 shows solution curves for the long-narrow economy with 65 places and the racetrack economy with 64 places. Spatial agglomerations of these two economies are illustrated comparatively in this figure for several values of $\tau$. In the racetrack economy, the curve OACDEGH in Fig. 12.5 displays a course of agglomeration through *spatial period-doubling cascade* occurring at *pitchfork* bifurcation points A, D, and G (cf. (12.1)). A dynamic shift takes place from a stable state at A to another stable state at C along an unstable curve AC shown by a dashed curve.

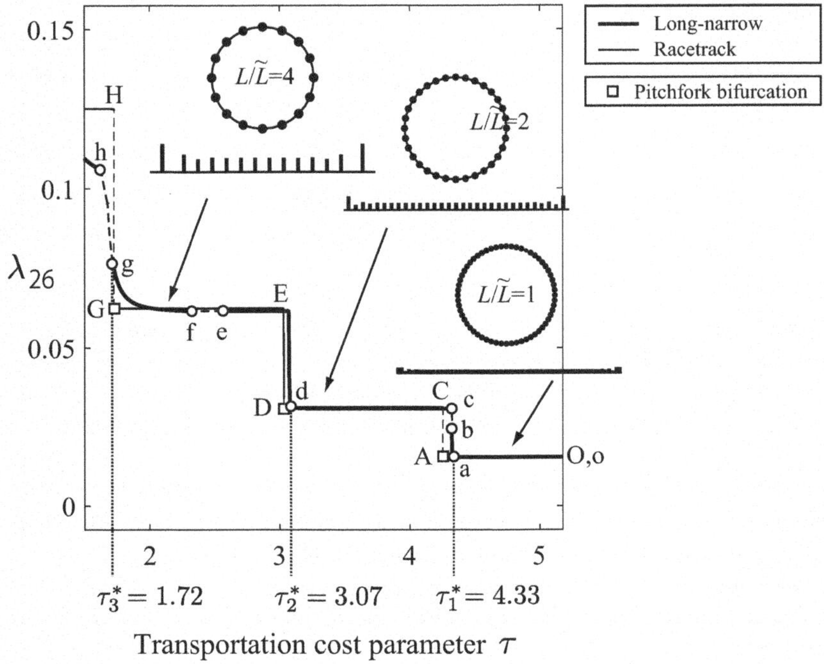


**Fig. 12.5** Comparison of the spatial agglomerations in the long-narrow economy with 65 places and those in the racetrack economy with 64 places (Ikeda et al., 2017) (bold curve: long-narrow economy; thin curve: racetrack economy; solid curve: stable; dashed curve: unstable)

In the long-narrow economy, most of the curves are stable (shown by bold solid curves); several short unstable ones (shown by bold dashed ones) are enclosed by a pair of critical points (such as b–c, e–f, and g–h).[3]

The curve oabcdef of the long-narrow economy closely follows the curve OACDEG of the racetrack economy. Thus, the cascade can be seen clearly in the present case with 65 places, as border effects are weakened compared to the case with 17 places in Sect. 12.3.1. Moreover, population distributions depicted in Fig. 12.5 in both economies, especially near the middle, display an amazing resemblance: recurrent emergence of chains of spatially repeated *core–periphery patterns* with increasing peripheral zones as $\tau$ decreases. This demonstrates the validity of the *racetrack economy analogy*.

Break points $\tau_1^* = 4.33$, $\tau_2^* = 3.07$, $\tau_3^* = 1.72$, and $\tau_4^* = 0.94$, as read from Fig. 12.5, indicate changes in the agglomeration pattern. The formulas presented in Sect. 12.2 are of assistance in an analytical prediction of these break points to some measures of accuracy as listed in Table 12.1.

[3] Point b is a maximum point of $\tau$, at which stability changes. Bifurcating curves (e.g., from bifurcation points e and f) are suppressed in Fig. 12.5 because they are not related to the discussion on the racetrack economy analogy.

**Table 12.1** Break points in the long-narrow and the racetrack economies obtained computationally and analytically

(a) Computed $\tau_m^*$ in the long-narrow economy

| $n$ | 17 | 33 | 65 |
|---|---|---|---|
| $\tau_1^*$ | 1.15 | 2.21 | 4.33 |
| $\tau_2^*$ | 0.81 | 1.55 | 3.07 |
| $\tau_3^*$ | – | 0.90 | 1.72 |
| $\tau_4^*$ | – | – | 0.94 |

(b) Computed $\tau_m^*$ in the racetrack economy

| $n$ | 16 | 32 | 64 |
|---|---|---|---|
| $\tau_1^*$ | 1.06 | 2.13 | 4.25 |
| $\tau_2^*$ | 0.76 | 1.52 | 3.03 |
| $\tau_3^*$ | 0.43 | 0.86 | 1.73 |
| $\tau_4^*$ | 0.10 | 0.45 | 0.89 |

(c) Analytical $\tau_m$ by (12.3)

| $n$ | 16 | 32 | 64 |
|---|---|---|---|
| $\tau_1$ | 1.06 | 2.13 | 4.25 |
| $\tau_2$ | 0.76 | 1.52 | 3.03 |
| $\tau_3$ | – | – | – |
| $\tau_4$ | – | – | – |

(d) Approximate $\hat{\tau}_m$ by (12.4)

| $n$ | 16 | 32 | 64 |
|---|---|---|---|
| $\hat{\tau}_1$ | 1.06 | 2.12 | 4.24 |
| $\hat{\tau}_2$ | 0.75 | 1.50 | 3.00 |
| $\hat{\tau}_3$ | 0.43 | 0.87 | 1.73 |
| $\hat{\tau}_4$ | – | 0.43 | 0.87 |

## References


Akamatsu, T., Mori, T., & Takayama, Y. (2016). Agglomerations in a multi-region economy: Polycentric versus monocentric patterns. *DPRIETI Discussion Paper Series* 16-E-009.

Akamatsu, T., Takayama, Y., & Ikeda, K. (2012). Spatial discounting, Fourier, and racetrack economy: A recipe for the analysis of spatial agglomeration models. *Journal of Economic Dynamics and Control, 36*(11), 1729–1759.

Christaller, W. (1933). *Die zentralen Orte in Süddeutschland*. Gustav Fischer (1933). English translation: Central Places in Southern Germany. Prentice Hall (1966).

Forslid, R., & Ottaviano, G. I. P. (2003). An analytically solvable core-periphery model. *Journal of Economic Geography, 3,* 229–340.

Fujita, M., Krugman, P., & Venables, A. J. (1999). *The spatial economy: Cities, regions, and international trade*. MIT Press.

Fujita, M., & Mori, T. (1997). Structural stability and the evolution of urban systems. *Regional Science and Urban Economics, 42,* 399–442.

Ikeda, K., Akamatsu, T., & Kono, T. (2012). Spatial period-doubling agglomeration of a core-periphery model with a system of cities. *Journal of Economic Dynamics and Control, 36,* 754–778.

Ikeda, K., Murota, K., Akamatsu, T., & Takayama, Y. (2017). Agglomeration patterns in a long narrow economy of a new economic geography model: Analogy to a racetrack economy. *International Journal of Economic Theory, 13*(1), 113–145.

Lösch, A. (1940). *Die räumliche Ordnung der Wirtschaft*. Gustav Fischer. English translation: *The Economics of Location*. Yale University Press (1954).

Mori, T. (1997). A modeling of megalopolis formation: the maturing of city systems. *Journal of Urban Economics, 42,* 133–157.

Mossay, P., & Picard, P. M. (2011). On spatial equilibria in a social interaction model. *Journal of Economic Theory, 146*(6), 2455–2477.

Pflüger, M. (2004). A simple, analytically solvable, Chamberlinian agglomeration model. *Regional Science and Urban Economics, 34*(5), 565–573.

Tabuchi, T., & Thisse, J.-F. (2011). A new economic geography model of central places. *Journal of Urban Economics, 69,* 240–252.

# Chapter 13
# Spatial Period-Doubling Cascade on Square Lattice

**Abstract** We investigate the self-organization of patterns on a square lattice, highlighting the spatial period-doubling cascade among invariant patterns. A group-theoretic bifurcation analysis is conducted in light of the racetrack economy analogy. We have found the existence of two types of cascades in the square lattice, whereas the racetrack economy has only one type of cascade. In the numerical analysis of an economic geography model, we observe the transition of stable invariant patterns as the transportation cost decreases. Several prerequisites include Chaps. 3 and 4 for the direct bifurcation from the uniform state, Chap. 9 for the hierarchical bifurcation in the racetrack economy, and Chaps. 9 and 10 for invariant patterns.

**Keywords** Bifurcation · Economic geography model · Group-theoretic bifurcation analysis · Pattern formation · Racetrack analogy · Racetrack economy · Replicator dynamics · Spatial period doubling · Square lattice · Sustaining bifurcation

## 13.1 Introduction

Chapter 8 proposed formulas for break points that can synthetically describe the bifurcations of two spatial platforms: a racetrack economy and a square lattice with periodic boundary conditions. This implies the commonality between these two platforms, referred to as the *racetrack economy analogy*. This analogy describes how spatial agglomeration on the square lattice proceeds as the transportation cost decreases. The racetrack economy, with the dihedral group symmetry, has a simple, analytically tractable spatial structure (Chaps. 3 and 9). Moreover, this economy undergoes a *spatial period-doubling bifurcation cascade* that engenders a chain of alternating core and periphery places, as presented in Fig. 13.1a.

We theoretically assess the existence of this cascade on the square lattice (Fig. 13.1b) as an extension of the bifurcation theory on the direct bifurcation from the uniform state presented in Chap. 4. A square lattice has two types of spatial



K. Ikeda, Y. Takayama, *Bifurcation Theory of Pattern Formation in Economic Geography*, Interdisciplinary Applied Mathematics 61,
https://doi.org/10.1007/978-3-031-71686-7_13

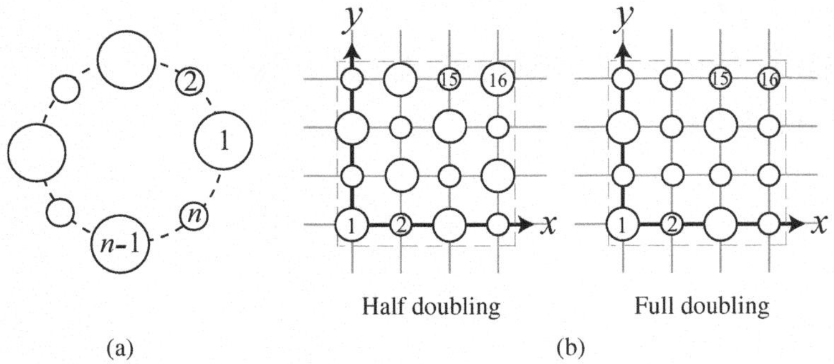


**Fig. 13.1** Two spatial platforms in the state of spatial period doubling. (**a**) Racetrack economy. (**b**) Square lattice

period doubling: half doubling and full doubling, whereas the racetrack economy has only one type of doubling. We highlight the spatial period-doubling cascade among invariant patterns as the mechanism for the transition of these patterns. The cascade plays a pivotal role in investigating the bifurcation mechanism of the square lattice that has many possible spatial patterns, as studied in Chap. 10 (Kogure et al., 2021).

This chapter is organized as follows: Sect. 13.2 presents a group-theoretic analysis procedure. In Sect. 13.3, the spatial period-doubling cascade in the racetrack economy is revisited. Section 13.4 advances two types of cascades on the square lattice. Numerical examples are presented in Sect. 13.5.

## 13.2 Group-Theoretic Analysis Procedure

We revisit and extend the group-theoretic analysis procedure described in Sect. 4.3.1 (Ikeda & Murota, 2014). The governing equation is

$$\boldsymbol{F}(\boldsymbol{\lambda}, \tau) = \boldsymbol{0}. \tag{13.1}$$

Here, $\boldsymbol{F}: \mathbb{R}^N \times \mathbb{R} \to \mathbb{R}^N$ is a sufficiently smooth nonlinear function in the vector $\boldsymbol{\lambda} \in \mathbb{R}^N$ and the bifurcation parameter $\tau \in \mathbb{R}$. The symmetry of the governing equation is formulated through the *equivariance* condition:

$$T(g)\boldsymbol{F}(\boldsymbol{\lambda}, \tau) = \boldsymbol{F}(T(g)\boldsymbol{\lambda}, \tau), \quad g \in G \tag{13.2}$$

in terms of an $N \times N$ orthogonal *matrix representation* $T$ of the group $G$.

We consider a *group-theoretic critical point* $(\boldsymbol{\lambda}^*, \tau_c)$ (cf. Footnote 3 in Sect. 3.3) with the multiplicity $M (\geq 1)$. The Jacobian matrix $J_c = \partial \boldsymbol{F}/\partial \boldsymbol{\lambda}(\boldsymbol{\lambda}^*, \tau_c)$ at this point has $M$ zero eigenvalues. Let $(\boldsymbol{\eta}_i \mid i = 1, \ldots, N)$ be an orthonormal basis of $\mathbb{R}^N$ such that $J\boldsymbol{\eta}_i = \mathbf{0}$ $(i = 1, \ldots, M)$.

At the critical point, we define incremental variables $\boldsymbol{w} = (w_i) \in \mathbb{R}^M$ of $\boldsymbol{\lambda}$ and $\tilde{\tau}$ of $\tau$, respectively, by

$$\boldsymbol{\lambda} = \boldsymbol{\lambda}^* + \sum_{i=1}^{M} w_i \boldsymbol{\eta}_i, \qquad \tau = \tau_c + \tilde{\tau}.$$

The full system of equations $\boldsymbol{F}(\boldsymbol{\lambda}, \tau) = \mathbf{0}$ in (13.1) is reduced, in a neighborhood of $(\boldsymbol{\lambda}_c, \tau_c)$, to the *bifurcation equation* in the incremental variables as

$$\tilde{\boldsymbol{F}}(\boldsymbol{w}, \tilde{\tau}) = \mathbf{0} \tag{13.3}$$

for some function $\tilde{\boldsymbol{F}}$. This equation inherits the equivariance in (13.2) as

$$\tilde{T}(g)\tilde{\boldsymbol{F}}(\boldsymbol{w}, \tilde{\tau}) = \tilde{\boldsymbol{F}}(\tilde{T}(g)\boldsymbol{w}, \tilde{\tau}), \qquad g \in G,$$

where $\tilde{T}$ is a sub-representation of $T$ in $M$-dimensional kernel space of the Jacobian matrix.

We employ the *equivariant branching lemma* to find bifurcating patterns in view of the symmetry group of these patterns.[1] The symmetry of a vector $\boldsymbol{w}$ is described by its isotropy subgroup

$$\Sigma(\boldsymbol{w}) = \Sigma(\boldsymbol{w}; G, \tilde{T}) = \{g \in G \mid \tilde{T}(g)\boldsymbol{w} = \boldsymbol{w}\}.$$

A subgroup $\Sigma$ of $G$ is said to be an *isotropy subgroup* (with respect to $\tilde{T}$) if $\Sigma = \Sigma(\boldsymbol{w})$ for some $\boldsymbol{w} \neq \mathbf{0}$. For a subgroup $\Sigma$ of $G$, its *fixed-point subspace* is defined as follows:

$$\text{Fix}(\Sigma) = \{\boldsymbol{w} \mid \tilde{T}(g)\boldsymbol{w} = \boldsymbol{w} \text{ for all } g \in \Sigma\}. \tag{13.4}$$

For an isotropy subgroup $\Sigma$, we have $\dim \text{Fix}(\Sigma) \geq 1$. An isotropy subgroup $\Sigma$ with $\dim \text{Fix}(\Sigma) = 1$ is called an *axial subgroup*. When $\tilde{T}$ is an absolutely irreducible representation different from the unit representation and (13.3) is generic, a unique solution branch exists for the axial subgroup.

The *symmetry group* of the *racetrack economy*, located at the origin of the $xy$-plane, is the dihedral group:

$$G = \mathrm{D}_n = \langle s, r \rangle.$$

[1] For details of this lemma, see, e.g., Vanderbauwhede (1980) and Ikeda and Murota (2014, §2.5.4) and (2019, §8.7).

This group is generated by $s$ : the *reflection* $y \mapsto -y$ and $r$ : $2\pi/n$ counterclockwise *rotation* around the origin. Partial symmetry of $G = \mathrm{D}_n$ is described by a *subgroup* $G'$ of $G$, such as $G' = \mathrm{D}_m = \langle s, r^{n/m} \rangle$, where $m$ is an integer dividing $n$ $(1 \leq m < n)$.

The symmetry group of the square lattice is

$$G = \langle r, s, p_1, p_2 \rangle,$$

which is generated by $r$ : counterclockwise *rotation* about the origin at an angle of $\pi/2$, $s$ : *reflection* $y \mapsto -y$, $p_1$ : $x$-directional periodic *translation* by a length $\tilde{L}$, and $p_2$ : $y$-directional one by a length $\tilde{L}$ ($\tilde{L}$ is the distance unit between neighboring places).

## 13.3 Spatial Period Doubling on Racetrack Economy

The bifurcation mechanism of the racetrack economy in Chap. 9 is revisited in preparation for the racetrack economic analogy to be employed in the investigation of a square lattice in the next section. The racetrack economy with $n = 2^k$ $(k = 2, 3, \ldots)$ places accommodates a *spatial period-doubling cascade* (Ikeda et al., 2012), expressed by a chain of groups in (9.11):

$$\mathrm{D}_n \rightarrow \mathrm{D}_{n/2} \rightarrow \cdots \rightarrow \mathrm{D}_1, \tag{13.5}$$

where ($\rightarrow$) denotes a simple pitchfork bifurcation. For example, Fig. 13.2 depicts this cascade for $n = 16$ places. The core (agglomerated) places shown by (○) are located equidistantly, and the normalized spatial period $\tilde{T}$ between these places is doubled repeatedly as the number of the agglomerated places decreases recurrently from 16, 8, 4, 2, to 1. A chain of regular polygonal symmetries starts from the uniform state and ends up with a full agglomeration.

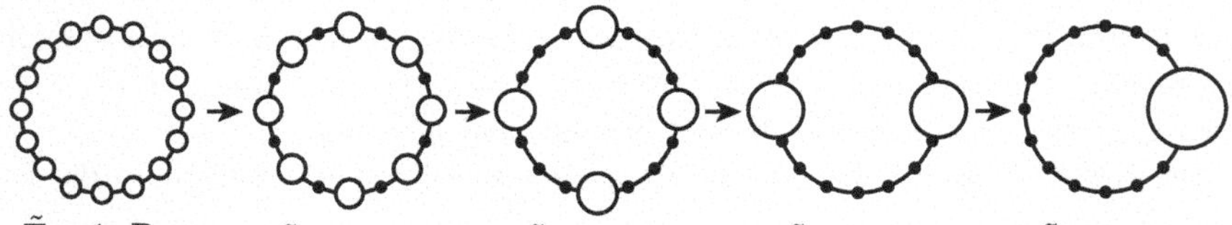


**Fig. 13.2** Regular polygonal invariant patterns on a racetrack economy and spatial period-doubling cascade between these patterns ($n = 16$; $\tilde{T} = T/\tilde{L}$) (the area of a circle denotes a population size; (○): agglomerated place; (•): place with no population)

**Direct Bifurcation**

We consider a simple *pitchfork* bifurcation point on the $\mathrm{D}_n$-invariant uniform state $\boldsymbol{\lambda}_{\mathrm{R}0} \equiv \frac{1}{n}\mathbf{1} \in \mathbb{R}^n$. The *critical eigenvector* for the zero eigenvalue of the Jacobian matrix $J$ at this point is $\mathrm{D}_{n/2}$-invariant vector

$$\boldsymbol{\eta}_{\mathrm{R}1} = (1, -1, \ldots, 1, -1) = (\hat{\boldsymbol{e}}_1, \ldots, \hat{\boldsymbol{e}}_1) \in \mathbb{R}^n \tag{13.6}$$

with $\hat{\boldsymbol{e}}_1 = (1, -1)$. This eigenvector represents alternating core and periphery places. The population distribution of the bifurcating solution is given by

$$\boldsymbol{\lambda} = \frac{1}{n}\mathbf{1} + a\, \boldsymbol{\eta}_{\mathrm{R}1} \quad (0 < |a| \le 1/n). \tag{13.7}$$

At a sustaining bifurcation point ($a = 1/n$ or $-1/n$ in (13.7)), we arrive at $\mathrm{D}_{n/2}$-symmetric *regular polygonal invariant patterns*:

$$\boldsymbol{\lambda}_{\mathrm{R}1} = \frac{2}{n}(\underbrace{\boldsymbol{e}_1, \ldots, \boldsymbol{e}_1}_{n/2 \text{ times}}), \quad \boldsymbol{\lambda}^*_{\mathrm{R}1} = \frac{2}{n}(\underbrace{\boldsymbol{e}^*_1, \ldots, \boldsymbol{e}^*_1}_{n/2 \text{ times}}) \tag{13.8}$$

with $\boldsymbol{e}_1 = (1, 0)$ and $\boldsymbol{e}^*_1 = (0, 1)$. For later use, we introduce notations:

$$\boldsymbol{e}_\ell = (1, 0, \ldots, 0) = (\boldsymbol{e}_{\ell-1}, \mathbf{0}) \in \mathbb{R}^{2^\ell}, \quad \boldsymbol{e}^*_\ell = (\mathbf{0}, \boldsymbol{e}_{\ell-1}) \in \mathbb{R}^{2^\ell};$$
$$\hat{\boldsymbol{e}}_\ell = (\boldsymbol{e}_{\ell-1}, -\boldsymbol{e}_{\ell-1}) \in \mathbb{R}^{2^\ell}.$$

**Secondary and Further Bifurcations**

We investigate secondary and further bifurcations in the cascade (13.5). We introduce regular polygonal patterns $\boldsymbol{\lambda}_{\mathrm{R}\ell}$ and $\boldsymbol{\lambda}^*_{\mathrm{R}\ell} \in \mathbb{R}^n$ ($\ell = 0, \ldots, k$) that appear in the chain of groups in (13.5) and have a two-level hierarchy without populations at periphery places.

The critical eigenvector of the secondary simple pitchfork bifurcation is

$$\boldsymbol{\eta}_{\mathrm{R}2} = (\boldsymbol{e}_1, -\boldsymbol{e}_1, \ldots, \boldsymbol{e}_1, -\boldsymbol{e}_1) = (\hat{\boldsymbol{e}}_2, \ldots, \hat{\boldsymbol{e}}_2).$$

A sustaining bifurcation follows subsequently to produce regular polygonal patterns:

$$\boldsymbol{\lambda}_{\mathrm{R}2} = \frac{2^2}{n}(\boldsymbol{e}_1,\ \mathbf{0}, \ldots, \boldsymbol{e}_1,\ \mathbf{0}) = \frac{2^2}{n}(\boldsymbol{e}_2, \ldots, \boldsymbol{e}_2),$$
$$\boldsymbol{\lambda}^*_{\mathrm{R}2} = \frac{2^2}{n}(\mathbf{0},\ \boldsymbol{e}_1, \ldots, \mathbf{0},\ \boldsymbol{e}_1) = \frac{2^2}{n}(\boldsymbol{e}^*_2, \ldots, \boldsymbol{e}^*_2)$$

with $\boldsymbol{e}_2 = (1, 0, 0, 0)$ and $\boldsymbol{e}_2^* = (0, 0, 1, 0)$. After that, a further simple bifurcation and a sustaining bifurcation can occur recurrently until reaching a full agglomeration. This recurrence engenders

$$\eta_{\mathrm{R}\ell} = (\boldsymbol{e}_{\ell-1}, -\boldsymbol{e}_{\ell-1}, \ldots, \boldsymbol{e}_{\ell-1}, -\boldsymbol{e}_{\ell-1}) = (\underbrace{\hat{\boldsymbol{e}}_\ell, \ldots, \hat{\boldsymbol{e}}_\ell}_{n/2^\ell \text{ times}}), \quad \ell = 1, \ldots, k;$$

$$\boldsymbol{\lambda}_{\mathrm{R}\ell} = \frac{2^\ell}{n}(\boldsymbol{e}_{\ell-1}, \boldsymbol{0}, \ldots, \boldsymbol{e}_{\ell-1}, \boldsymbol{0}) = \frac{2^\ell}{n}(\underbrace{\boldsymbol{e}_\ell, \ldots, \boldsymbol{e}_\ell}_{n/2^\ell \text{ times}}),$$

$$\boldsymbol{\lambda}_{\mathrm{R}\ell}^* = \frac{2^\ell}{n}(\boldsymbol{0}, \boldsymbol{e}_{\ell-1}, \ldots, \boldsymbol{0}, \boldsymbol{e}_{\ell-1}) = \frac{2^\ell}{n}(\underbrace{\boldsymbol{e}_\ell^*, \ldots, \boldsymbol{e}_\ell^*}_{n/2^\ell \text{ times}}), \quad \boldsymbol{0} \in \mathbb{R}^{2^{\ell-1}}, \ \ell = 1, \ldots, k.$$

The regular polygonal patterns in this recurrence are invariant patterns by Proposition 9.3 and are stationary points for any $\tau$.

## 13.4 Spatial Period Doublings on Square Lattice

The mechanism of the transition of spatial patterns on a square lattice is advanced and is meshed consistently with the spatial period-doubling cascade in the racetrack economy in the previous section. Spatial patterns engendered in this transition are invariant patterns (Fig. 13.3). We consider an $n \times n$ square lattice with $n = 2^k$ ($k = 2, 3, \ldots$) places and employ the *equivariant branching lemma* referring to the symmetry group of the patterns. By contrast, the bifurcation equation is solved in Sect. 4.3.3 to obtain the solution.

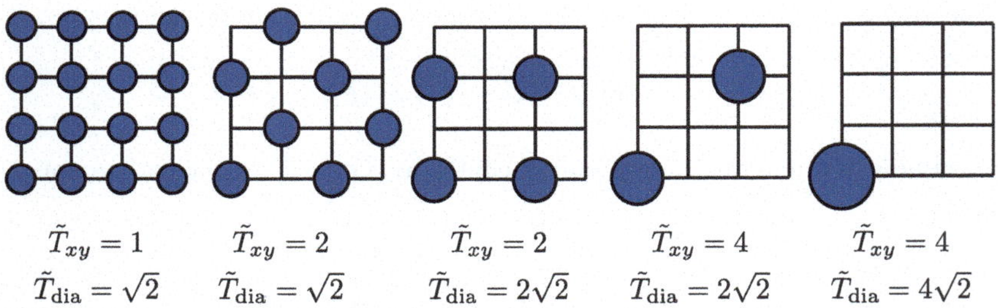


**Fig. 13.3** Invariant patterns appearing in spatial period doubling on a square lattice ($n = 4$; $\tilde{T}_{xy} = T_{xy}/\tilde{L}$, $\tilde{T}_{\mathrm{dia}} = T_{\mathrm{dia}}/\tilde{L}$; the area of a circle denotes a population size; (○): agglomerated place; (•): place with no population for the racetrack economy)

### 13.4.1 Half Doubling of Spatial Period

To begin with, we introduce half doubling, whereas the other type of doubling, full doubling, is presented in Sect. 13.4.2.

**Direct Bifurcation**
When $n$ is even, there is a *simple bifurcation point* associated with the *one-dimensional irreducible representation* $\mu = (1; -)$, being defined by (4.6):

$$T^{\mu}(r) = 1, \quad T^{\mu}(s) = 1, \quad T^{\mu}(p_1) = -1, \quad T^{\mu}(p_2) = -1. \tag{13.9}$$

We assume that the variable $\boldsymbol{w} = w$ for the bifurcation equation (13.3) corresponds to the column vector (4.20):

$$\begin{aligned} \eta_{\mathrm{H}1} &= \{\cos(\pi(n_1 - n_2)) \mid n_1, n_2 = 0, 1, \ldots, n-1\} \\ &= (1, -1, \ldots, 1, -1; \ -1, 1, \ldots, -1, 1; \ \ldots\ldots)^{\top} \\ &= (\boldsymbol{h}_1, \ldots, \boldsymbol{h}_1)^{\top}. \end{aligned} \tag{13.10}$$

Here, $\boldsymbol{h}_1 = (\eta_{\mathrm{R}1}, -\eta_{R1})$ with $\eta_{R1}$ defined in (13.6). There is a bifurcating solution from this bifurcation point.

**Proposition 13.1** *The simple bifurcation point at the uniform state has the bifurcating path with the symmetry* $\langle r, s, p_1 p_2, p_1^{-1} p_2\rangle$ *in the direction of* $\eta_{\mathrm{H}1}$ *in* (13.10).

***Proof*** We replace $\tilde{T}(g)$ in (13.4) by $T^{\mu}(g)$ for $\mu = (1; -)$. Using $\boldsymbol{w} = w$ $(M = 1)$ and (13.9), we obtain

$$T^{\mu}(r)w = w, \quad T^{\mu}(s)w = w,$$

$$T^{\mu}(p_1 p_2)w = (-1)(-1)w = w, \quad T^{\mu}(p_1^{-1} p_2)w = (-1)(-1)w = w.$$

That is, $T^{\mu}(g)w = w$ holds for any element of the targeted symmetry $\Sigma = \langle r, s, p_1 p_2, p_1^{-1} p_2\rangle$. Thus, the fixed-point subspace $\mathrm{Fix}^{\mu}(\Sigma)$ in (13.4) is one-dimensional. The *equivariant branching lemma* generically guarantees the existence of a bifurcating path with the symmetry $\Sigma$. □

**Secondary and Further Bifurcations**
Secondary and further bifurcations on the lattice can be analyzed similarly to the direct bifurcation.

**Proposition 13.2** *Through hierarchical bifurcation, there emerges a series of patterns expressed by a chain of groups*

$$\langle r, s, p_1, p_2\rangle \rightarrow \langle r, s, p_1 p_2, p_1^{-1} p_2\rangle \rightarrow \langle r, s, p_1^2, p_2^2\rangle \rightarrow \cdots \rightarrow \mathrm{D}_2. \tag{13.11}$$

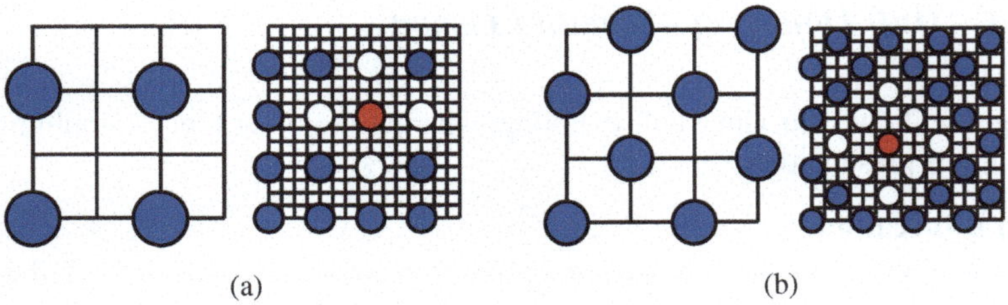

**Fig. 13.4** Classification of spatial period-doubling invariant patterns on square lattices (the area of a circle denotes the population size). (**a**) Foursquare patterns. (**b**) Oblique patterns

***Proof*** If we set $P_1 = p_1 p_2$ and $P_2 = p_1^{-1} p_2$, we get the relations

$$\langle r, s, p_1 p_2, p_1^{-1} p_2 \rangle = \langle r, s, P_1, P_2 \rangle, \quad \langle r, s, p_1^2, p_2^2 \rangle = \langle r, s, P_1 P_2, P_1^{-1} P_2 \rangle.$$

Like the direct bifurcation $\langle r, s, p_1, p_2 \rangle \to \langle r, s, p_1 p_2, p_1^{-1} p_2 \rangle$, there is a secondary bifurcation $\langle r, s, P_1, P_2 \rangle \to \langle r, s, P_1 P_2, P_1^{-1} P_2 \rangle$, which can be rewritten as $\langle r, s, p_1 p_2, p_1^{-1} p_2 \rangle \to \langle r, s, p_1^2, p_2^2 \rangle$. Other bifurcations can be dealt with similarly. □

We focus on a spatial period-doubling bifurcation cascade that engenders two-level hierarchical patterns without populations at peripheral places. The patterns can be classified into *foursquare* and *oblique* patterns, as depicted in Fig. 13.4. The oblique ones will turn out to be superior in stability in an economic geography model (Sect. 13.5). These are invariant patterns.

**Proposition 13.3** *Foursquare and oblique patterns are invariant patterns.*

***Proof*** The subgroup $G'$ in Lemma 10.1 is chosen as one of the groups in (13.11) to ensure that the patterns in these propositions are invariant patterns. □

**Remark 13.1** For foursquare patterns with sufficiently large $n$ (Fig. 13.4a), each first-level center (red circle) is surrounded by the four closest first-level centers (white circles). For oblique ones, each first-level center is surrounded by as many as eight closest first-level centers with the same road distance (Fig. 13.4b). Consequently, the first-level centers of the oblique ones are more densely distributed in comparison with those of the foursquare ones. □

A cascade of pitchfork and sustaining bifurcations emerging from the uniform state with $\boldsymbol{\lambda}_{\mathrm{R}0} = \frac{1}{n^2}\mathbf{1} \in \mathbb{R}^{n^2}$ produce a series of patterns:

$$\boldsymbol{\lambda}_{\mathrm{H}1} = \frac{1}{n}(\underbrace{\hat{\boldsymbol{h}}_1, \ldots, \hat{\boldsymbol{h}}_1}_{n/2 \text{ times}}) \quad \text{with } \hat{\boldsymbol{h}}_1 = (\boldsymbol{\lambda}_{\mathrm{R}1}, \boldsymbol{\lambda}_{R1}^*);$$

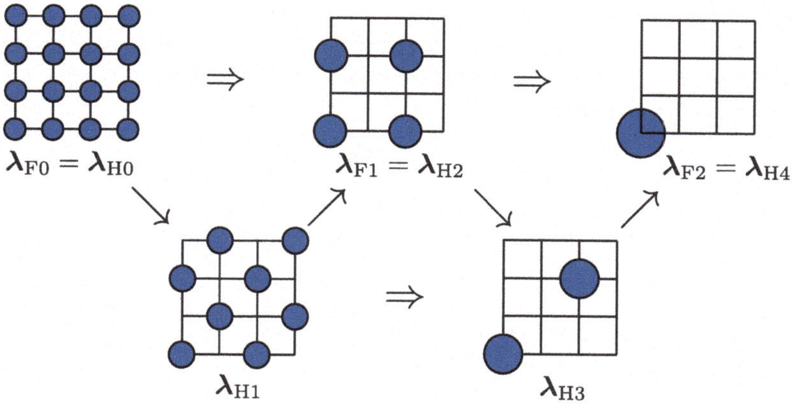


**Fig. 13.5** Spatial period-doubling cascades on a square lattice for ($n = 4$) ($\Rightarrow$: full doubling; $\searrow$ and $\nearrow$: half doubling; the area of a circle denotes a population size)

$$\lambda_{H2} = \frac{2}{n}(\underbrace{\hat{h}_2, \ldots, \hat{h}_2}_{n/2 \text{ times}}) \quad \text{with } \hat{h}_2 = (\lambda_{R1}, \mathbf{0}), \quad \mathbf{0} \in \mathbb{R}^n;$$

$$\lambda_{H3} = \frac{4}{n}(\underbrace{\hat{h}_3, \ldots, \hat{h}_3}_{n/4 \text{ times}}) \quad \text{with } \hat{h}_3 = (\lambda_{R2}, \mathbf{0}, \lambda^*_{R2}, \mathbf{0}), \quad \mathbf{0} \in \mathbb{R}^n,$$

and so on. Patterns $\lambda_{H\ell}$ ($\ell = 1, 2, \ldots$) are composed of an ensemble of sub-patterns given by the patterns $\lambda_{R\ell}$ and $\lambda^*_{R\ell}$ for the racetrack economy. This composition ensures the usefulness of the *racetrack economy analogy*.

$\lambda_{H2\ell}$ has a foursquare pattern, and $\lambda_{H(2\ell-1)}$ has an oblique one. These two types of patterns appear alternately in the cascade. In the simple bifurcation $\lambda_{H0} \searrow \lambda_{H1}$ depicted in Fig. 13.5, the spatial period $T_{xy}$ in the horizontal and the vertical directions are doubled, and another spatial period $T_{\text{dia}}$ in the two diagonal directions remains unchanged; we measure the diagonal spatial period $T_{\text{dia}}$ not by the road distance but by the Euclidean distance. The spatial period $T_{xy}$ is doubled when an oblique pattern appears, whereas $T_{\text{dia}}$ is doubled when a foursquare pattern appears. Such doubling is called herein *half doubling of spatial period*. This doubling in two types of spatial periods differs from the doubling in a single direction for the racetrack economy. Such difference shows the square lattice's importance in studying two-dimensional agglomerations.

### 13.4.2 Full Doubling of Spatial Period

The other type of doubling, which is called full doubling, emerges from a double bifurcation point on the uniform state.

**Direct Bifurcation**
We consider a *double bifurcation point* at the uniform state that is associated with the *two-dimensional irreducible representation* $\mu = (2)$, which exists only when $n$ is even, and is defined by (4.7):

$$T^{\mu}(r) = \begin{bmatrix} & 1 \\ 1 & \end{bmatrix}, \quad T^{\mu}(s) = \begin{bmatrix} 1 & \\ & 1 \end{bmatrix}, \quad T^{\mu}(p_1) = \begin{bmatrix} -1 & \\ & 1 \end{bmatrix}, \quad T^{\mu}(p_2) = \begin{bmatrix} 1 & \\ & -1 \end{bmatrix}. \tag{13.12}$$

By (4.21), we can assume that the variable $\boldsymbol{w} = (w_1, w_2)^{\top}$ for the bifurcation equation (13.3) corresponds to the two vectors:

$$\boldsymbol{q}_1 = \{\cos(\pi n_1) \mid n_1, n_2 = 0, 1, \ldots, n-1\},$$
$$\boldsymbol{q}_2 = \{\cos(\pi n_2) \mid n_1, n_2 = 0, 1, \ldots, n-1\}.$$

Then, we have the following proposition.

**Proposition 13.4** *Bifurcating solution of a square pattern associated with $\mu = (2)$ exists in the direction of $\boldsymbol{q}_1 + \boldsymbol{q}_2$ with the symmetry of $\langle r, s, p_1^2, p_2^2 \rangle$.*

***Proof*** We set $\Sigma = \langle r, s, p_1^2, p_2^2 \rangle$ and note

$$\mathrm{Fix}^{\mu}(\langle r, s, p_1^2, p_2^2 \rangle) = \mathrm{Fix}^{\mu}(\langle r \rangle) \cap \mathrm{Fix}^{\mu}(\langle s, p_1^2, p_2^2 \rangle).$$

Since $T^{\mu}(r)(w_1, w_2)^{\top} = (w_2, w_1)^{\top}$ by (13.12), $T^{\mu}(r)(w_1, w_2)^{\top} = (w_1, w_2)^{\top}$ holds when $(w_1, w_2) = w(1, 1)$. Hence, we get $\mathrm{Fix}^{\mu}(\langle r \rangle) = \{w(1, 1) \mid w \in \mathbb{R}\}$. We also have $\mathrm{Fix}^{\mu}(\langle s, p_1^2, p_2^2 \rangle) = \mathbb{R}^2$ since $T^{\mu}(s) = T^{\mu}(p_1^2) = T^{\mu}(p_2^2) = I$ by (13.12). Therefore, $\mathrm{Fix}^{\mu}(\Sigma) = \{w(1, 1)^{\top} \mid w \in \mathbb{R}\}$. Thus, the targeted symmetry $\Sigma$ is an isotropy subgroup with $\dim \mathrm{Fix}^{\mu}(\Sigma) = 1$. The *equivariant branching lemma* generically guarantees the existence of a bifurcating path with the symmetry $\Sigma$. □

The direction of the eigenvector for this bifurcation is given by

$$\begin{aligned} \boldsymbol{\eta}_{\mathrm{F1}} &= \left\{ \frac{1}{2}\cos(\pi n_1) + \frac{1}{2}\cos(\pi n_2) \mid n_1, n_2 = 0, 1, \ldots, n-1 \right\} \\ &= (1, 0, \ldots, 1, 0;\ 0, -1, \ldots, 0, -1;\ \ldots\ldots)^{\top} \\ &= (\boldsymbol{f}_1, \ldots, \boldsymbol{f}_1)^{\top} \end{aligned} \tag{13.13}$$

with $\boldsymbol{f}_1 = (\boldsymbol{\lambda}_{\mathrm{R1}}, -\boldsymbol{\lambda}^*_{\mathrm{R1}})$ and $\boldsymbol{\lambda}_{\mathrm{R1}}$ and $\boldsymbol{\lambda}^*_{\mathrm{R1}}$ in (13.8). A subsequent sustaining bifurcation produces

$$\boldsymbol{\lambda}_{\mathrm{F1}} = \frac{2}{n}(\underbrace{\hat{\boldsymbol{f}}_1, \ldots, \hat{\boldsymbol{f}}_1}_{n/2 \text{ times}}) = \boldsymbol{\lambda}_{\mathrm{H2}} \quad \text{with } \hat{\boldsymbol{f}}_1 = \hat{\boldsymbol{h}}_2 = (\boldsymbol{\lambda}_{\mathrm{R1}}, \boldsymbol{0}), \quad \boldsymbol{0} \in \mathbb{R}^n.$$

In the double bifurcation $\eta_{F0} \Rightarrow \eta_{F1}$ depicted in Fig. 13.5, both spatial periods $T_{xy}$ and $T_{dia}$ are doubled. Such doubling is called *full doubling of spatial period* herein.

**Secondary and Further Bifurcations**
Secondary and further bifurcations for the full doubling can be dealt with similarly to the half doubling. The secondary and further double bifurcations produce

$$\boldsymbol{\lambda}_{F\ell} = \boldsymbol{\lambda}_{H2\ell} \quad (\ell = 1, \ldots, k). \tag{13.14}$$

These double bifurcations engender a cascade of spatial period doubling between foursquare patterns associated with a chain of groups

$$\langle r, s, p_1, p_2 \rangle \Rightarrow \langle r, s, p_1^2, p_2^2 \rangle \Rightarrow \langle r, s, p_1^4, p_2^4 \rangle \Rightarrow \cdots \Rightarrow D_2,$$

where ($\Rightarrow$) indicates spatial period doubling at a double bifurcation point. There is another cascade, which occurs between oblique ones

$$\langle r, s, p_1 p_2, p_1^{-1} p_2 \rangle \Rightarrow \langle r, s, (p_1 p_2)^2, (p_1^{-1} p_2)^2 \rangle \Rightarrow \cdots \Rightarrow D_2.$$

Figure 13.5 depicts the mixed occurrences of half doubling and full doubling. Twice repeated occurrences of half doubling correspond to a single occurrence of full doubling, as implied by (13.14). Such mixed occurrences make the transition of spatial patterns on the square lattice more complicated than that of the racetrack economy.

## 13.5 Period Doubling for an Economic Geography Model

We investigate the period doubling on the racetrack economy and the square lattice in the numerical analysis of the replicator dynamics in the Forslid and Ottaviano model (2003) (Sect. 2.2). This investigation is supported by the bifurcation mechanism of spatial period doubling discussed in Sect. 13.4. In addition to regular polygonal invariant patterns studied above, we encounter other invariant patterns and non-invariant patterns (Fig. 13.6).

We set the lattice size as $n = 2^m$ $(m = 1, 2, 3, 4)$, the length of an inter-place road as $\tilde{L} = 1/n$, and parameter values as $(\sigma, \mu) = (10.0, 0.4)$, satisfying the no-black-hole condition $\mu < \sigma - 1$ in (8.9).

**Racetrack Economy**
The solution curves for the racetrack economy are plotted in Fig. 13.7. The horizontal lines A–E represent solution curves for invariant patterns that are stable for some range of $\tau$ (shown by the solid line).

Non-invariant patterns correspond to non-horizontal (nearly vertical) bifurcating curves and are sometimes stable. The horizontal lines are connected by these non-horizontal curves to form a stair-like zigzag curve.

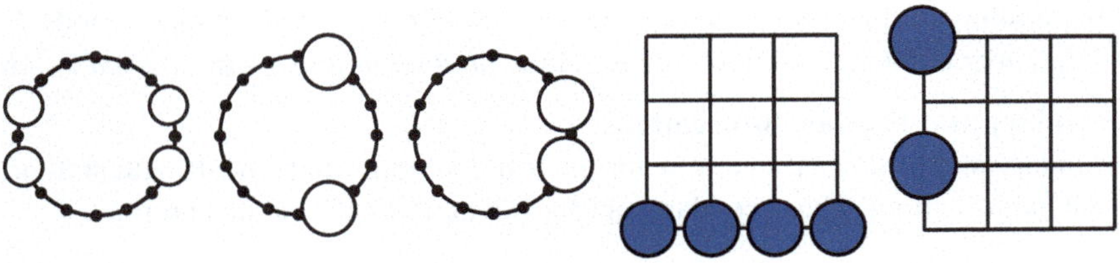

**Fig. 13.6** Examples of other invariant patterns on a racetrack economy and a square lattice (the area of a circle denotes a population size; (◯): agglomerated place; (•): place with no population for the racetrack economy)

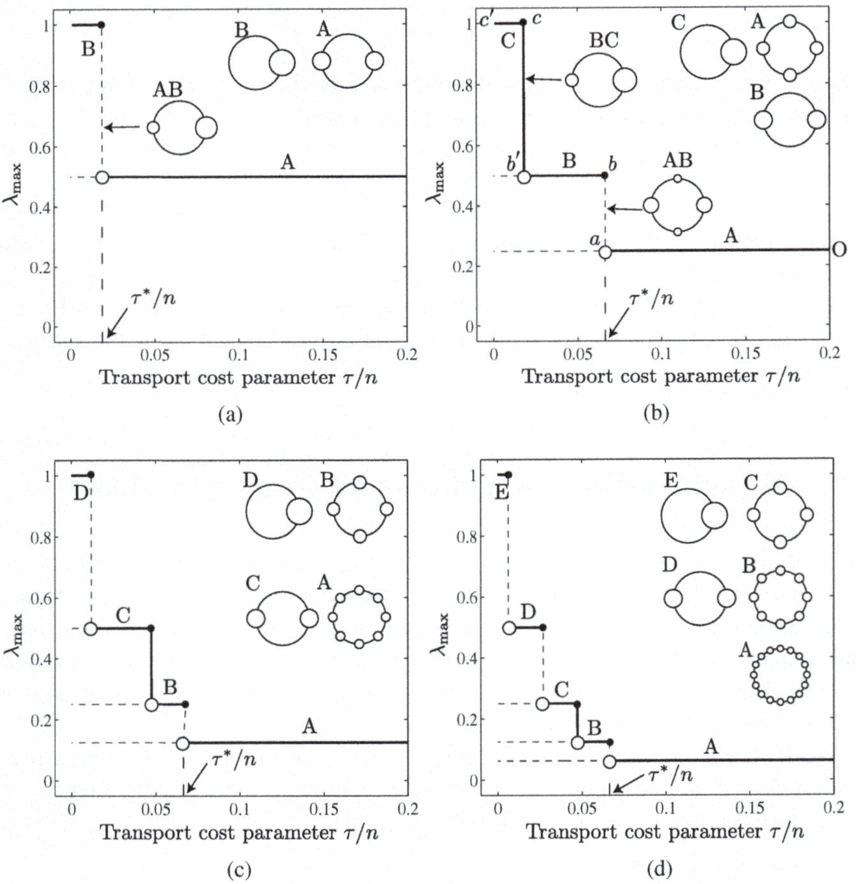


**Fig. 13.7** Solution curves for the racetrack economy with $n = 2$, 4, 8, and 16 (Ikeda et al., 2018) (solid line: stable; dashed curve: unstable; (∘): simple bifurcation point; (•): sustain point; $\lambda_{max} = \lambda_1$). **(a)** $n = 2$. **(b)** $n = 4$. **(c)** $n = 8$. **(d)** $n = 16$

A spatial period-doubling cascade among stable invariant patterns occurs in all cases ($n = 2, 4, 8, 16$). There is a stable, uniform state A for $\tau > \tau^*$. For $n = 4$ (Fig. 13.7b), an unstable transient state ab with two large places and two small places emerges at the pitchfork bifurcation point a at $\tau = \tau^*$. This state regains stability at the sustain point b in state B of two concentrated places and two extinguished places. After that, at the breaking point b′, a stable transient state b′c emerges en route to a stable full agglomeration (state C starting from the sustain point c). For larger $n$ ($n = 8$ or 16), the variety of stable invariant patterns increases with more occurrences of spatial period doubling in agreement with the theoretical prediction (Sect. 13.3).

**Square Lattice**

Figure 13.8 shows solution curves for the square lattice. The horizontal lines denote solution curves for spatial period-doubling invariant patterns, which are either

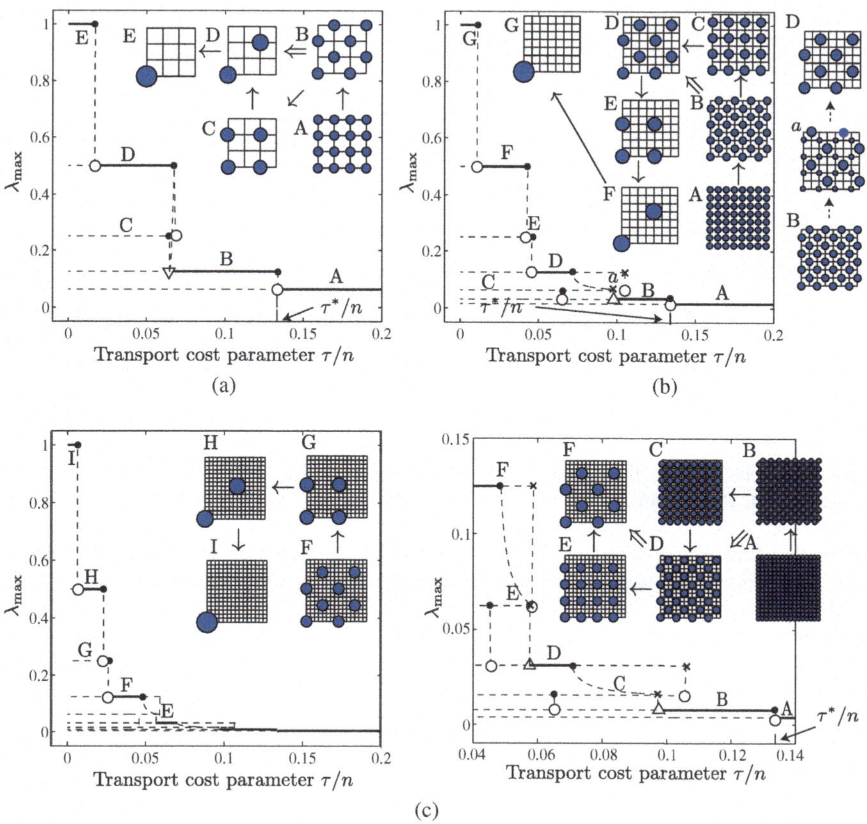


**Fig. 13.8** Solution curves for the square lattice with $n = 4$, 8, and 16 (Ikeda et al., 2018) (solid line: stable; dashed one: unstable; (∘): simple pitchfork bifurcation point; (•): sustain point; (△): double bifurcation point; (▽): triple bifurcation point; (×): sustaining bifurcation point; $\lambda_{max} = \lambda_1$). **(a)** $n = 4$. **(b)** $n = 8$. **(c)** $n = 16$

foursquare or oblique. These patterns, except for the full agglomeration, have either a simple, double, or triple bifurcation point[2] shown, respectively, by (∘), (△), (▽). All simple bifurcation points on stable curves are unstable pitchfork (tomahawk) bifurcation points. There are several sustaining bifurcation points shown by (•) and (×) ((•) denoting sustain point, in particular).

All oblique patterns (Fig. 13.4b) are well-posed, satisfying the condition in (10.5) that the break point is smaller than the sustain point, and have stable equilibria. However, the foursquare patterns (Fig. 13.4a) are either ill-posed without stable equilibria (C for $n = 4$, 8, and 16 and E for $n = 16$) or well-posed with very short durations of stable equilibria (E for $n = 8$ and G for $n = 16$). Thus, the oblique ones are superior in stability. In addition to these invariant patterns, there are non-invariant ones, which are shown by non-horizontal curves in Fig. 13.8. All these non-invariant ones are unstable.

As $\tau$ decreases, a *spatial period-doubling cascade* between the invariant patterns A–I takes place leading to fewer larger agglomerations. Half spatial period doubling occurs between two stable invariant patterns A and B in all cases ($n = 4$, 8, and 16).

## References


Forslid, R., & Ottaviano, G. I. P. (2003). An analytically solvable core–periphery model. *Journal of Economic Geography, 3*, 229–240.

Ikeda, K., Akamatsu, T., & Kono, T. (2012). Spatial period-doubling agglomeration of a core–periphery model with a system of cities. *Journal of Economic Dynamics and Control, 36*(5), 754–778.

Ikeda, K., & Murota, K. (2014). *Bifurcation theory for hexagonal agglomeration in economic geography*. Springer.

Ikeda, K., & Murota, K. (2019). *Imperfect bifurcation in structures and materials* (3rd ed.) Springer.

Ikeda, K., Onda, M., & Takayama, Y. (2018). Spatial period doubling, invariant pattern, and break point in economic agglomeration in two dimensions. *Journal of Economic Dynamics and Control, 92*, 129–152.

Kogure, Y., Ikeda, K., & Aizawa, H. (2021). Group-theoretic bifurcation mechanism of economic agglomerations on a square lattice. *International Journal of Bifurcation and Chaos, 31*(13), 2130040.

Vanderbauwhede, A. (1980). *Local Bifurcation and Symmetry*. Habilitation Thesis, Rijksuniversiteit Gent.


[2] This triple bifurcation point occurs as a coincidence of simple and double bifurcation points.

# Chapter 14
# Equidistant Economy

**Abstract** This chapter elucidates the bifurcation mechanism of an equidistant economy through the analysis of the symmetric group $S_N$. Preexisting results for the direct bifurcation are combined with the rules for secondary and further bifurcations. Invariant patterns in the replicator dynamics in the equidistant economy are obtained. The mechanism of breaking and sustaining bifurcations from these patterns is elucidated. Sustaining bifurcations are unique features of the replicator dynamics. The stability of bifurcating branches is analyzed, demonstrating that most of them are unstable immediately after bifurcation. Numerical analysis of economic geography models demonstrates that direct bifurcating curves connect the curves of invariant patterns, thereby creating a mesh-like network. This network grows rapidly as the size $N$ of $S_N$ increases. Chapter 9 is a prerequisite for sustaining bifurcations. This chapter is a foundation of the global system in Chap. 11.

**Keywords** Bifurcation · Economic geography model · Equidistant economy · Group-theoretic bifurcation analysis · Invariant pattern · Replicator dynamics · Stability · Sustaining bifurcation · Symmetric group

## 14.1 Introduction

In an equidistant economy, transportation costs between any pair of places are assumed to be identical. This economy is studied in a special case of three places[1] and under a general setting.[2] The mechanism of the direct bifurcation of this economy is readily available through bifurcation analysis of the symmetric group $S_N$ (Golubitsky and Stewart, 2002 and Elmhirst, 2004). Hopf bifurcations with $S_N$-

[1] See, e.g., Fujita et al. (1999), Castro et al. (2012), and Commendatore et al. (2015).

[2] See, e.g., Puga (1999), Tabuchi et al. (2005), Oyama (2009), Zeng and Uchikawa (2014), and Gaspar et al. (2020).

 
K. Ikeda, Y. Takayama, *Bifurcation Theory of Pattern Formation in Economic Geography*, Interdisciplinary Applied Mathematics 61,
https://doi.org/10.1007/978-3-031-71686-7_14

symmetry are studied (e.g., Golubitsky & Stewart, 1985 and Diaz & Rodrigues, 2007).

This chapter elucidates the bifurcation mechanism of this economy. The economy displays a complex, mesh-like network of solution curves. We investigate the mechanism of this network by (1) using a group-theoretic analysis of the symmetric group $S_N$ and (2) referring to invariant patterns in the replicator dynamics (Sect. 14.2.1). Then, we investigate the secondary (sustaining) bifurcation of these invariant patterns. Theoretical stability analysis demonstrates that some branches are asymptotically stable.

We conduct the numerical analysis of an equidistant economy with three, four, and eight places in two spatial economic models (Forslid & Ottaviano, 2003 and Pflüger, 2004). There are direct bifurcating curves connecting the curves of invariant solutions, thereby displaying mesh-like networks constituting the weft of invariant patterns. Almost all bifurcating solutions are found to be unstable.

This chapter is organized as follows: The bifurcation mechanism of the equidistant economy is put forth in Sect. 14.2. Section 14.3 studies the stability of bifurcating branches. Numerical bifurcation analyses for spatial economic models are conducted in Sect. 14.4.

## 14.2 Bifurcation Mechanism

The bifurcation mechanism of secondary and further bifurcations in an *equidistant economy* with $S_N$-symmetry is investigated, summarizing (Aizawa et al., 2020). The mechanism of the direct bifurcation of the uniform state (Golubitsky & Stewart, 2002 and Elmhirst, 2004) is also included in Sect. 14.2.2 for completeness.

### *14.2.1 Replicator Dynamics and Invariant Patterns*

We consider the governing equation

$$\boldsymbol{F}(\boldsymbol{\lambda}, \phi) = \boldsymbol{0} \tag{14.1}$$

in the replicator dynamics with a sufficiently smooth function $\boldsymbol{F}(\boldsymbol{\lambda}, \phi) = (F_i(\boldsymbol{\lambda}, \phi) \mid i = 1, \dots, N)$, defined as

$$F_i(\boldsymbol{\lambda}, \phi) = (v_i(\boldsymbol{\lambda}, \phi) - \bar{v}(\boldsymbol{\lambda}, \phi))\lambda_i \quad (i = 1, \dots, N). \tag{14.2}$$

Here, $\phi$ is the bifurcation parameter (trade freeness in (2.11)), $\boldsymbol{\lambda}$ is the independent variable vector, and $\bar{v} = \sum_{i=1}^{N} \lambda_i v_i$.

A corner solution, for which one or more places have zero population, can be expressed through an appropriate permutation of components of $\boldsymbol{\lambda}$ as

$$\boldsymbol{\lambda} = (\boldsymbol{\lambda}_m^+, \mathbf{0}_n) \qquad (1 \le m \le N-1,\ m+n=N)$$

with $\boldsymbol{\lambda}_m^+ = \{\lambda_i > 0,\ \sum_{1=1}^m \lambda_i = 1 \mid 1 \le i \le m\}$ and the $n$-dimensional zero vector $\mathbf{0}_n = (0, \ldots, 0)$.

There are several invariant patterns which are solutions to the governing equation for any $\phi$, that is, the uniform state

$$\boldsymbol{\lambda}^* = \frac{1}{N}\mathbf{1}_N$$

and a series of *core–periphery patterns*

$$\boldsymbol{\lambda}_m^{\mathrm{CP}} = \left(\frac{1}{m}\mathbf{1}_m, \mathbf{0}_n\right) \qquad (1 \le m \le N-1, \quad m+n=N) \tag{14.3}$$

with $\mathbf{1}_m = \underbrace{(1, \ldots, 1)}_{m \text{ times}}$. The core–periphery pattern encompasses the full agglomeration $\boldsymbol{\lambda}^{\mathrm{FA}} = (1, \mathbf{0}_{N-1})$ as a special case where $m = 1$.

**Proposition 14.1** *The uniform state and the core–periphery patterns in* (14.3) *are invariant patterns in an equidistant economy.*

***Proof*** The use of Lemma 9.1 with $G' = \mathrm{S}_N$ for the uniform state and with $G' = \mathrm{S}_m$ for the core–periphery patterns proves this proposition. □

## 14.2.2 Direct Bifurcation from the Uniform State

The mechanism of the direct bifurcation from the uniform state $\boldsymbol{\lambda}^* = \frac{1}{N}\mathbf{1}_N$ in an equidistant economy was elucidated through the bifurcation analysis of a *symmetric group* $\mathrm{S}_N$ expressing the symmetry of this economy (Golubitsky & Stewart, 2002 and Elmhirst, 2004). This analysis is presented consistently with our formulation.

The uniform state has a *Jacobian matrix* of the form:

$$J = A_N(a, b) = \begin{pmatrix} a & b & \cdots & b \\ b & \ddots & \ddots & \vdots \\ \vdots & \ddots & \ddots & b \\ b & \cdots & b & a \end{pmatrix} \tag{14.4}$$

with

$$a = \frac{\partial}{\partial \lambda_i}(v_i - \bar{v}) \quad (1 \le i \le N); \qquad b = \frac{\partial}{\partial \lambda_j}(v_i - \bar{v}) \quad (1 \le i, j \le N,\ i \ne j).$$

That is, all diagonal entries of the Jacobian matrix $J$ are $a$, and all off-diagonal ones are $b$. When $a = b$, this state encounters a direct bifurcation point with $(N-1)$-times repeated zero eigenvalues of the Jacobian matrix $(M = N - 1)$. At this point, several *two-level hierarchy* states of the form

$$\boldsymbol{\lambda}_{m,n} = (\underbrace{u_1, \ldots, u_1}_{m \text{ times}}, \underbrace{u_2, \ldots, u_2}_{n \text{ times}}) = (u_1 \mathbf{1}_m, u_2 \mathbf{1}_n)$$

$$(1 \le m \le N-1, \quad m+n = N, \quad u_1 m + u_2 n = 1; \; u_1, u_2 > 0) \tag{14.5}$$

branch simultaneously in the incremental directions of

$$\delta\boldsymbol{\lambda}_{m,n} = \varepsilon \left( \mathbf{1}_m, -\frac{m}{n} \mathbf{1}_n \right) \qquad (1 \le m \le N-1, \quad m+n = N, \; \varepsilon \in \mathbb{R}), \tag{14.6}$$

as explained in Proposition 14.2 below. Consequently, $N$ places are split into $m$ places with equal population size and $n$ places with the other size. The bifurcation path (branch) is called *symmetric* if $\delta\boldsymbol{\lambda}_{m,n}$ and $-\delta\boldsymbol{\lambda}_{m,n}$ are *conjugate*, denoting the same state up to a permutation of place numbers. It is designated as *asymmetric* if they are non-conjugate.

**Proposition 14.2** *At a bifurcation point of the uniform state, the two-level hierarchy states in* (14.5) *branch in the directions of* (14.6). *The branch for a two-level hierarchy state is symmetric if $n = m = N/2$ ($N$ even) and is asymmetric otherwise.*

***Proof*** See Appendix and the literature (Elmhirst, 2004) for the proof. □

### 14.2.3 Bifurcation from a Two-Level Hierarchy State

The two-level hierarchy state $\boldsymbol{\lambda}_{m,n} = (u_1 \mathbf{1}_m, u_2 \mathbf{1}_n)$ in (14.5) has *breaking* and *sustaining bifurcation points*. At a sustaining bifurcation point, where $u_2 \mathbf{1}_n$ vanishes, this state exits to a corner solution expressing the core–periphery pattern $\boldsymbol{\lambda}_m^{\mathrm{CP}} = \left( \frac{1}{m} \mathbf{1}_m, \mathbf{0}_n \right)$ in (14.3).

To investigate the breaking bifurcation of the two-level hierarchy state, we refer to its Jacobian matrix, which takes the form of

$$J = \begin{pmatrix} A_m(a,b) & eE_{mn} \\ fE_{nm} & A_n(c,d) \end{pmatrix}.$$

Therein, $A_m(a,b)$ and $A_n(c,d)$ are defined by (14.4); $E_{mn} = {\mathbf{1}_m}^{\top} \mathbf{1}_n$ is an $m \times n$ matrix with all entries being equal to 1, and $a, b, \ldots, f$ are constants. A breaking bifurcation takes place when $a = b$ or $c = d$ holds. We focus on the case of $c = d$ from now on, whereas we can treat the other case of $a = b$ similarly.

At a *breaking bifurcation point* where $c = d$, *three-level hierarchy* states

$$\boldsymbol{\lambda}_{m,p,n-p} = (u_1\mathbf{1}_m, u_2\mathbf{1}_p, u_3\mathbf{1}_{n-p})$$
$$(1 \le p \le n-1,\ m+n=N,\ u_1 m + u_2 p + u_3(n-p) = 1;\ u_1, u_2, u_3 > 0) \tag{14.7}$$

branch simultaneously in the directions of

$$\delta\boldsymbol{\lambda}_{m,p,n-p} = \varepsilon\left(\mathbf{0}_m, (n-p)\mathbf{1}_p, -p\mathbf{1}_{n-p}\right) \quad (1 \le p \le n-1,\ m+n=N;\ \varepsilon \in \mathbb{R}), \tag{14.8}$$

as explained in Proposition 14.3. Consequently, $n$ identical places split into $p$ places with identical population size and $n-p$ places with the other size ($1 \le p \le n-1$).

**Proposition 14.3** *At a breaking bifurcation point of the two-level hierarchy state in* (14.5), *the three-level hierarchy states in* (14.7) *branch in the directions of* (14.8). *The branch for a three-level hierarchy state is symmetric if $p = n/2$ ($n$ even) and asymmetric otherwise.*

***Proof*** See Appendix for the proof. □

The three-level hierarchy state in (14.7) can exit to the corner solution at a *sustaining bifurcation point*. Alternatively, it can undergo further breaking bifurcations to arrive at an aggregated interior state with an $s$-level hierarchy ($2 \le s \le N$):

$$\boldsymbol{\lambda}_{m_1,\ldots,m_s} = (u_1\mathbf{1}_{m_1}, \ldots, u_s\mathbf{1}_{m_s})$$

with $\sum_{i=1}^{s} m_i = N$ and $\sum_{i=1}^{s} u_i m_i = 1$. Breaking bifurcations can proceed until reaching a completely aggregated interior state of $\lambda_1 > \lambda_2 > \cdots > \lambda_N$.

### 14.2.4 Bifurcation from Core–Periphery Patterns

To investigate the bifurcation from the *core–periphery pattern* $\boldsymbol{\lambda}_m^{\mathrm{CP}} = \left(\frac{1}{m}\mathbf{1}_m, \mathbf{0}_n\right)$ in (14.3), we refer to its Jacobian matrix:

$$J = \begin{pmatrix} A_m(a,b) & eE_{mn} \\ O & cI_n \end{pmatrix} \qquad (1 \le m \le N-1). \tag{14.9}$$

Here, $I_n$ is an $n \times n$ identity matrix and $c = v_i - \bar{v}$ ($m+1 \le i \le N$) with $v_{m+1} = v_{m+2} = \cdots = v_N$. The critical point of this pattern is either a breaking bifurcation point for $a = b$ with singular $A_m(a,b)$ or a sustaining bifurcation point for $c = 0$ with singular $cI_n$.

Before the main discussion, we introduce the concept of *half-branch*. There are two half-branches for the breaking bifurcations presented above in both directions of $\delta\boldsymbol{\lambda}_{m,n}$ and $-\delta\boldsymbol{\lambda}_{m,n}$ in (cf. (14.6)). By contrast, a half-branch exists in only one direction for a sustaining bifurcation point because no negative population is allowed.

We start with the simplest core–periphery pattern: full agglomeration $\boldsymbol{\lambda}^{\mathrm{FA}} = (1, \mathbf{0}_{N-1})$. The full agglomeration has only sustaining bifurcation points, at which several bifurcating solutions of the form

$$(1 - pu, u\mathbf{1}_p, \mathbf{0}_{N-p-1}) \qquad (1 \le p \le N-1,\ 0 < u < 1/p) \tag{14.10}$$

branch simultaneously in the directions of

$$\varepsilon \left(-1, \frac{1}{p}\mathbf{1}_p, \mathbf{0}_{N-p-1}\right) \qquad (1 \le p \le N-1,\ \varepsilon > 0).$$

**Proposition 14.4** *Full agglomeration* $\boldsymbol{\lambda}^{\mathrm{FA}} = (1, \mathbf{0}_{N-1})$ *has sustaining bifurcation points with the half-branches in* (14.10) *and does not have either a limit point or a breaking bifurcation point.*

***Proof*** See Appendix for the proof. □

Other core–periphery patterns $\boldsymbol{\lambda}_m^{\mathrm{CP}} = \left(\frac{1}{m}\mathbf{1}_m, \mathbf{0}_n\right)$ $(m \ge 2)$ (cf. (14.3)) have both breaking and sustaining bifurcation points, which engender several bifurcating solutions, as expounded in the following propositions. Their proofs are similar to those of Propositions 14.3 and 14.4.

**Proposition 14.5** *At a breaking bifurcation point of the core–periphery pattern* $\boldsymbol{\lambda}_m^{\mathrm{CP}}$ $(m \ge 2)$*, three-level hierarchy states branch simultaneously as*

$$(u_1\mathbf{1}_p, u_2\mathbf{1}_{m-p}, \mathbf{0}_n) \quad (1 \le p \le m-1,\ u_1 p + u_2(m-p) = 1,\ u_1, u_2 > 0). \tag{14.11}$$

*The bifurcation path for a three-level hierarchy state is symmetric if* $p = m/2$ *(*$m$ *even) and asymmetric otherwise.*

**Proposition 14.6** *At a sustaining bifurcation point of the core–periphery pattern in* (14.3)*, there emerge half-branches:*

$$(u_1\mathbf{1}_m, u_2\mathbf{1}_p, \mathbf{0}_{n-p}) \quad (1 \le p \le n,\ u_1 m + u_2 p = 1,\ u_1, u_2 > 0). \tag{14.12}$$

The branches in (14.11) and (14.12) can encounter breaking and sustaining bifurcation points successively to arrive at an aggregated state with an $s$-level hierarchy:

$$(u_1\mathbf{1}_{m_1}, \ldots, u_{s-1}\mathbf{1}_{m_{s-1}}, \mathbf{0}_{m_s}) \quad \left(2 \le s \le N;\ \sum_{i=1}^{s} m_i = N,\ \sum_{i=1}^{s-1} u_i m_i = 1\right).$$

**Example 14.1** Figure 14.1a and b provides the hierarchies of spatial patterns of an equidistant economy for $N = 3$ and $N = 4$, respectively. We express a symmetric branch with a thick arrow and an asymmetric one with a thin one. In each case, the subhierarchy for interior solutions at the top is connected to that for corner solutions at the bottom. There is a recurrent property: The hierarchy of $N = 3$ in (a) is the subhierarchy of corner solutions for $N = 4$ in (b). For an arbitrary number $N$ of places, the subhierarchy of corner solutions is given by the hierarchy of $N - 1$ places (see Fig. 14.1c). As shown by this recurrent property, the hierarchy grows rapidly as $N$ increases. In turn, the bifurcation mechanism becomes progressively complicated. In an economic geography model, the bifurcation mechanism for three places (Fig. 14.1a) is applied to the study of the agglomeration mechanism (Gaspar et al., 2021). □

## 14.3 Asymptotic Stability of Branches

We investigate the stability of half-branches from a sustaining bifurcation point of the core–periphery pattern $\boldsymbol{\lambda}_m^{\mathrm{CP}} = \left(\frac{1}{m}\mathbf{1}_m, \mathbf{0}_n\right)$ in (14.3). We recall its Jacobian matrix in (14.9):

$$J = \begin{pmatrix} A_m(a,b) & eE_{mn} \\ O & cI_n \end{pmatrix} \qquad (1 \leq m \leq N-1,\ m+n = N)$$

and consider its sustaining bifurcation point at $\phi = \phi_{\mathrm{c}}$ with a singular $cI_n$ $(c = 0)$ and a non-singular $A_m(a,b)$ $(a \neq b)$. Incremental variables $(\boldsymbol{y}, \boldsymbol{x}, \psi)$ from this point are defined as

$$\boldsymbol{\lambda} = \left(\frac{1}{m}\mathbf{1}_m, \mathbf{0}_n\right) + (\boldsymbol{y}, \boldsymbol{x}), \qquad \phi = \phi_{\mathrm{c}} + \psi \tag{14.13}$$

with $\boldsymbol{y} = (y_1, \ldots, y_m)$ and $\boldsymbol{x} = (x_1, \ldots, x_n)$ $(x_i \geq 0 \mid i = 1, \ldots, n)$.

By the *Lyapunov–Schmidt reduction*,[3] we can derive the bifurcation equation as follows:

- Express the governing equation $\boldsymbol{F}(\boldsymbol{\lambda}, \phi) = \mathbf{0}$ in (14.1) in terms of these incremental variables $(\boldsymbol{y}, \boldsymbol{x}, \psi)$.
- Obtain $\boldsymbol{y} = \boldsymbol{y}(\boldsymbol{x}, \psi)$ by solving the first $m$ components of $\boldsymbol{F} = \mathbf{0}$, where $A_m(a,b)$ is non-singular.
- Eliminate $\boldsymbol{y} = \boldsymbol{y}(\boldsymbol{x}, \psi)$ from the last $n$ components of $\boldsymbol{F} = \mathbf{0}$.

[3] See, e.g., Golubitsky et al. (1988) and Ikeda and Murota (2019).

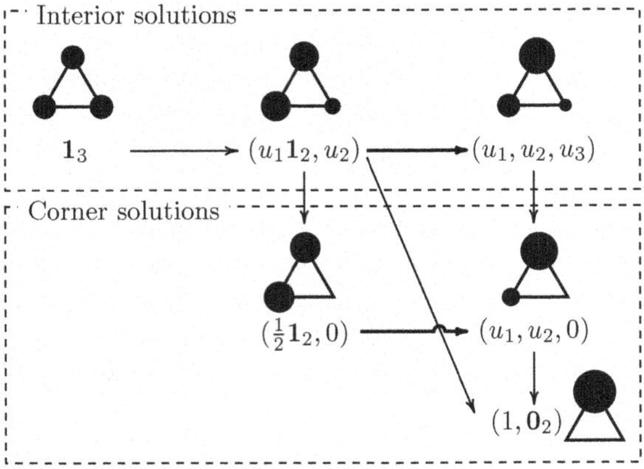


(a)

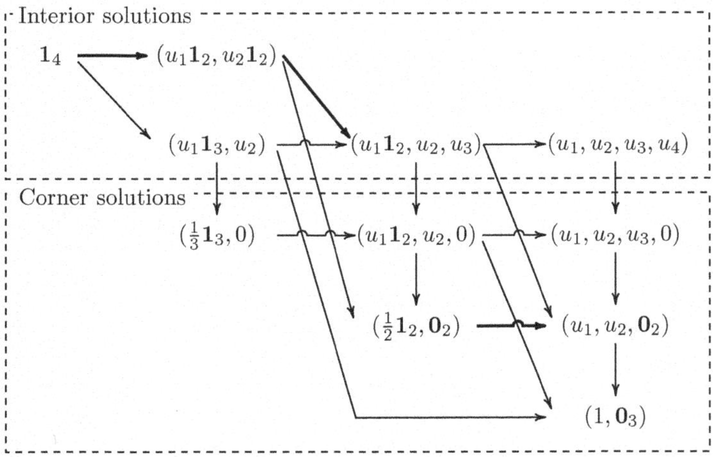


(b)

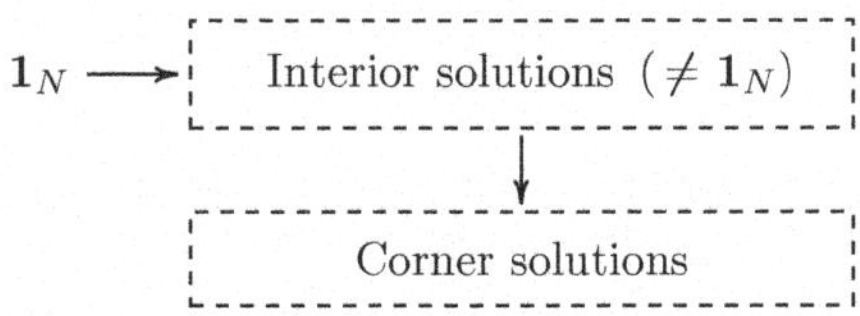


(c)

**Fig. 14.1** Bifurcation mechanism of an equidistant economy expressed by hierarchies of geometrical patterns (Aizawa et al., 2020) (the symmetric branch is expressed by a thick arrow and an asymmetric one by a thin one). (**a**) $N = 3$. (**b**) $N = 4$. (**c**) An arbitrary $N$

By a factored form (14.2) in the replicator dynamics, we obtain the *bifurcation equation* of a factored form:

$$G_i = x_i \cdot \hat{G}_i(\boldsymbol{x}, \psi) \quad (1 \leq i \leq n) \tag{14.14}$$

for some function $\hat{G}_i$. Then, solutions $(\boldsymbol{x}, \psi)$ of $\hat{G}_1 = \cdots = \hat{G}_n = 0$ give bifurcating solutions through $\boldsymbol{y} = \boldsymbol{y}(\boldsymbol{x}, \psi)$ and (14.13).

At a sustaining bifurcation point of the core–periphery pattern $\boldsymbol{\lambda}_m^{\mathrm{CP}}$, there emerge several half-branches in (14.12):

$$\boldsymbol{\lambda}_p = (u_1 \mathbf{1}_m, u_2 \mathbf{1}_p, \mathbf{0}_{n-p}) \quad (1 \leq p \leq n,\ u_1 m + u_2 p = 1,\ u_1, u_2 > 0). \tag{14.15}$$

We see that these branches are associated with

$$\boldsymbol{x} = \varepsilon\left(\mathbf{1}_p, \mathbf{0}_{n-p}\right) \qquad (\varepsilon > 0), \tag{14.16}$$

like the proof of Proposition 14.3.

We can expand $\hat{G}_i$ into a power series to arrive at $G_i \approx x_i \left(\alpha\psi + \sum_{j=1}^{n} \beta_j x_j\right)$ for some constants $\alpha$ and $\beta_i$. Owing to the symmetry (equivariance) of the system of equations $G_i$ $(i = 1, \ldots, n)$, a permutation $x_i \leftrightarrow x_j$ must lead to a permutation $G_i \leftrightarrow G_j$. This entails $\beta_j = \beta$ $(j = i)$ and $\beta_j = \gamma$ $(j \neq i)$ for some constants $\beta$ and $\gamma$. Then, we have

$$G_i \approx x_i \left(\alpha\psi + \beta x_i + \gamma \sum_{j \neq i}^{n} x_j\right) \qquad (i = 1, \ldots, n), \tag{14.17}$$

$$\frac{\partial G_i}{\partial x_j} \approx \begin{cases} \alpha\psi + 2\beta x_i + \gamma \sum_{j \neq i}^{n} x_j & (i = j), \\ \gamma x_i & (i \neq j). \end{cases} \tag{14.18}$$

The substitution of $\boldsymbol{x} = \varepsilon\left(\mathbf{1}_p, \mathbf{0}_{n-p}\right)$ in (14.16) into (14.17) leads to

$$G_1 = \cdots = G_p \approx \varepsilon\left[\alpha\psi + \{\beta + (p-1)\gamma\}\varepsilon\right], \qquad G_{p+1} = \cdots = G_n = 0.$$

Consequently, a set of equations $G_i = 0$ $(i = 1, \ldots, n)$ is satisfied by the asymptotic bifurcating $\psi$ versus $\varepsilon$ curves:

$$\psi \approx -\frac{\beta + (p-1)\gamma}{\alpha}\,\varepsilon \qquad (1 \leq p \leq N-1). \tag{14.19}$$

Substituting $\boldsymbol{x} = \varepsilon\left(\mathbf{1}_p, \mathbf{0}_{n-p}\right)$ in (14.16) into (14.18) and using (14.19), we obtain

$$\hat{J} = \left\{\frac{\partial G_i}{\partial x_j}\right\} = \varepsilon\begin{pmatrix} A_p(\beta,\gamma) & \gamma E_{pq} \\ O & (\gamma-\beta)I_{n-p} \end{pmatrix}.$$

The eigenvalues of the first diagonal block $\varepsilon A_p(\beta,\gamma)$ give $e_1$ (repeated once) and $e_2$ (repeated $p-1$ times), and the eigenvalues of the second diagonal block $\varepsilon(\gamma - \beta)I_{n-p}$ give $e_3$ (repeated $n-p$ times) in (14.20).

**Lemma 14.1** *The eigenvalues of $\partial \boldsymbol{G}/\partial \boldsymbol{x}$ for the branches of the core–periphery pattern $\boldsymbol{\lambda}_m^{\mathrm{CP}}$ are given asymptotically as*

$$\begin{cases} e_1 \approx \{\beta + (p-1)\gamma\}\varepsilon & \textit{(repeated once)}, \\ e_2 \approx -(\gamma-\beta)\varepsilon & \textit{(repeated } p-1 \textit{ times)}, \\ e_3 \approx (\gamma-\beta)\varepsilon & \textit{(repeated } n-p \textit{ times)}. \end{cases} \tag{14.20}$$

The associated half-branches are asymptotically stable if all eigenvalues in (14.20) are negative. Note that $p = 1$ and $p = n$ are exceptional cases where $e_2$ or $e_3$ is absent, respectively. Otherwise, there are $e_2$ and $e_3$ with opposite signs, and the branches are asymptotically unstable. As expounded below, asymptotically stable half-branches can exist in these exceptional cases. In the description of stability, we use the following assumption, which accords with the numerical results of sustaining bifurcation of the full agglomeration $\boldsymbol{\lambda}^{\mathrm{FA}} = (1, \mathbf{0}_{N-1})$ in spatial economic models (Sect. 14.4).

**Assumption 14.1** *The pre-bifurcation core–periphery pattern is stable for $\psi > 0$ ($\phi > \phi_{\mathrm{c}}$).*

**Proposition 14.7** *Under Assumption 14.1, the asymptotic stability of half-branches that emerge from $\boldsymbol{\lambda}_m^{\mathrm{CP}} = \left(\frac{1}{m}\mathbf{1}_m, \mathbf{0}_n\right)$ $(1 \le m \le N-1)$ is classified into three distinct cases:*

*(i) A three-level hierarchy state $(u_1\mathbf{1}_m, u_2, \mathbf{0}_{n-1})$ is the only stable half-branch for $\gamma < \beta < 0$. It resides in $\psi < 0$.*
*(ii) A two-level hierarchy state $(u_1\mathbf{1}_m, u_2\mathbf{1}_n)$ is the only stable half-branch for $\beta < \min(\gamma, -(n-1)\gamma)$. It resides in $\psi < 0$.*
*(iii) All half-branches are unstable for $\gamma \neq \beta$ and $\beta > 0$ or $-(n-1)\gamma < \beta < 0$.*

***Proof*** See Appendix for the proof. □

Among the plethora of half-branches in (14.15), at most one branch is stable. Figure 14.2 depicts the classification of the parameter space $(\beta, \gamma)$ into these three distinct cases. For the full agglomeration state $\boldsymbol{\lambda}^{\mathrm{FA}} = (1, \mathbf{0}_{N-1})$, the stability of half-branches is readily available by setting $m = 1$ and $n = N - 1$ in Proposition 14.7 and in Fig. 14.2.

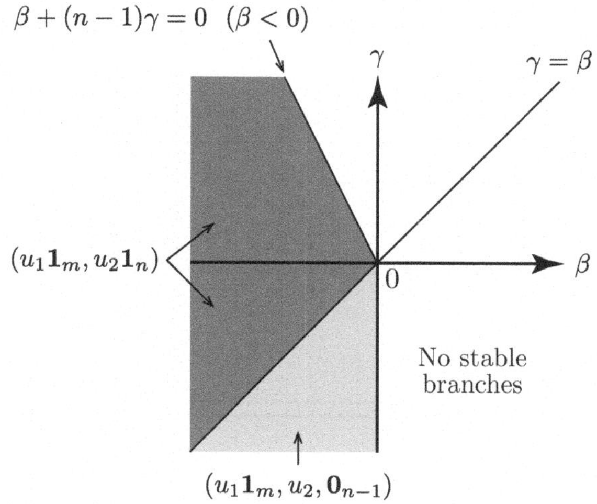


**Fig. 14.2** Classification of stability of half-branches for the core–periphery pattern $\boldsymbol{\lambda}_m^{\mathrm{CP}} = \left(\frac{1}{m}\mathbf{1}_m, \mathbf{0}_n\right)$ in the parameter space $(\beta, \gamma)$

## 14.4 Bifurcation Behavior of Economic Geography Models

We present the results of numerical bifurcation analysis of the FO model (Sect. 2.2) and the Pf model (Pflüger, 2004), referring to theoretical results presented in the previous sections. The values of the parameters of these models are set as $(\sigma, \mu) = (6.0, 0.4)$ for the FO model and as $(\sigma, \mu) = (4.0, 0.6)$ for the Pf model.

We employ the following bifurcation analysis procedure (Ikeda et al., 2019): (1) Obtain all invariant patterns and find all bifurcation points on the solution curves for these patterns. (2) Obtain all direct, secondary, and further bifurcating solution curves from these invariant patterns and find all bifurcation points on these curves. By using this procedure, one can obtain all possible bifurcating solutions connected to invariant solutions.

Figure 14.3 reports the bifurcation diagrams for the FO and Pf models with $N = 3$, 4, and 8 equidistant places. The horizontal axis is the trade freeness $\phi$ (cf. (2.11)), and the vertical axis is $\lambda_{\max}(\boldsymbol{\lambda}) = \max(\lambda_1, \ldots, \lambda_N)$. The horizontal lines express solution curves for invariant patterns (Proposition 14.1): uniform state $\boldsymbol{\lambda}^* = \frac{1}{N}\mathbf{1}_N$ and core–periphery patterns $\boldsymbol{\lambda}_m^{\mathrm{CP}} = \left(\frac{1}{m}\mathbf{1}_m, \mathbf{0}_n\right)$ $(1 \leq m \leq N-1)$. The solid (dashed) curves correspond to stable (unstable) solutions to the governing equation (14.1); ($\circ$) denotes breaking bifurcation points, ($\bullet$) denotes sustaining bifurcation points, and ($\circledcirc$) represents a limit point of $\phi$. There are no bifurcation points on these bifurcated curves; accordingly, there are no secondary bifurcations. Therefore, Fig. 14.3 encompasses all possible solution curves connected to the curves of invariant patterns.

The breaking bifurcation point A ($\circ$) is located at the right end of the solid horizontal line with $\lambda_{\max} = \frac{1}{N}$ of the stable uniform state $\boldsymbol{\lambda}^* = \frac{1}{N}\mathbf{1}_N$. Several *two-level hierarchy* states $\boldsymbol{\lambda}_{m,n} = (u_1\mathbf{1}_m, u_2\mathbf{1}_n)$ branch from this point (Proposition 14.2). These states connect the breaking bifurcation point A ($\circ$) of the uniform

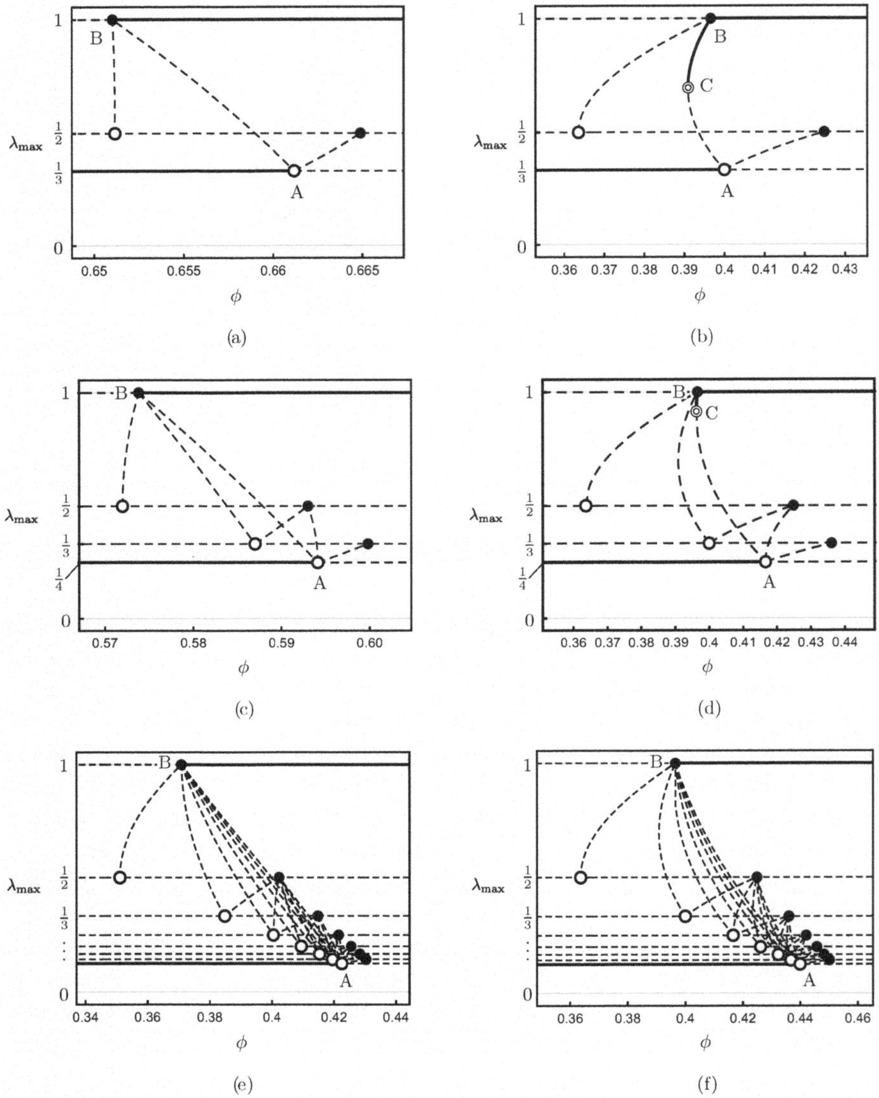


**Fig. 14.3** Bifurcation diagrams for $N = 3$, 4, and 8 (Aizawa et al., 2020) (solid curve: stable; dashed curve: unstable; (○): breaking bifurcation point; (•): sustaining bifurcation point; (◎): limit point of $\phi$). (**a**) Forslid and Ottaviano (2003), $N = 3$. (**b**) Pflüger (2004), $N = 3$. (**c**) Forslid and Ottaviano (2003), $N = 4$. (**d**) Pflüger (2004), $N = 4$. (**e**) Forslid and Ottaviano (2003), $N = 8$. (**f**) Pflüger (2004), $N = 8$

state $\boldsymbol{\lambda}^* = \frac{1}{N}\mathbf{1}_N$ with $N-1$ sustaining bifurcation points (•) of the core–periphery patterns $\boldsymbol{\lambda}_m^{\rm CP} = \left(\frac{1}{m}\mathbf{1}_m, \mathbf{0}_n\right)$ with $\lambda_{\max} = \frac{1}{m}$ $(1 \le m \le N-1)$.

From the breaking bifurcation points (∘) of the core–periphery patterns, there emerge several branches with *three-level hierarchy* states in (14.11):

$$(u_1\mathbf{1}_p, u_2\mathbf{1}_{m-p}, \mathbf{0}_n) \qquad (1 \le p \le m-1,\ u_1 p + u_2(m-p) = 1).$$

Each of these states connects a breaking bifurcation point (∘) with a sustaining bifurcation point (•). Such breaking and sustaining bifurcations occur recurrently until reaching the full agglomeration $\boldsymbol{\lambda}^{\rm FA} = (1, \mathbf{0}_{N-1})$ at the sustaining bifurcation point B, which resides at the left end of the solid horizontal line, representing stable full agglomeration. Direct bifurcating curves connecting invariant solutions thereby display mesh-like networks, which resemble threads constituting the weft of invariant patterns and warp of non-invariant ones.

Among the invariant patterns, only the uniform state and full agglomeration have stable equilibria. All the branches from the uniform state $\boldsymbol{\lambda}^* = \frac{1}{N}\mathbf{1}_N$ are unstable immediately after bifurcation (see Elmhirst, 2004 for a theoretical study of the stability). A branch BC in Fig. 14.3b from the full agglomeration is stable as predicted in Proposition 14.7 (ii); other non-invariant curves are unstable.

Figure 14.4 depicts the hierarchy of geometrical patterns for the current numerical analyses. The hierarchy for $N = 3$ is a "subset" of that for $N = 4$. These hierarchies correspond to the subsets of theoretical hierarchies summarized by Fig. 14.1. As demonstrated by Fig. 14.3e and f for $N = 8$, the hierarchy grows rapidly as $N$ increases.

# Appendix: Theoretical Details

## *Proof of Proposition 14.2*

We consider $\mathrm{S}_N$-invariant uniform state $\boldsymbol{\lambda}^* = \frac{1}{N}\mathbf{1}_N$ and a (possible bifurcated) state with the symmetry of $\Sigma = \mathrm{S}_m \times \mathrm{S}_n$ $(m+n=N)$. Denote by $\delta\boldsymbol{\lambda} = (\alpha\mathbf{1}_m, \beta\mathbf{1}_n)$ (for some variables $\alpha$ and $\beta$) an incremental variable vector in the latter state. By using the orthogonality between subspaces for $\mathrm{S}_N$ and $\mathrm{S}_m \times \mathrm{S}_n$ $(m+n=N)$, we obtain

$$\boldsymbol{\lambda}^*\,\delta\boldsymbol{\lambda}^\top = \frac{1}{N}\mathbf{1}_N\,(\alpha\mathbf{1}_m, \beta\mathbf{1}_n)^\top = \frac{1}{N}(\alpha m + \beta n) = 0.$$

Hence, $\beta = -\frac{m}{n}\alpha$ and

$$\delta\boldsymbol{\lambda} = \alpha\left(\mathbf{1}_m, -\frac{m}{n}\mathbf{1}_n\right) \tag{14.21}$$

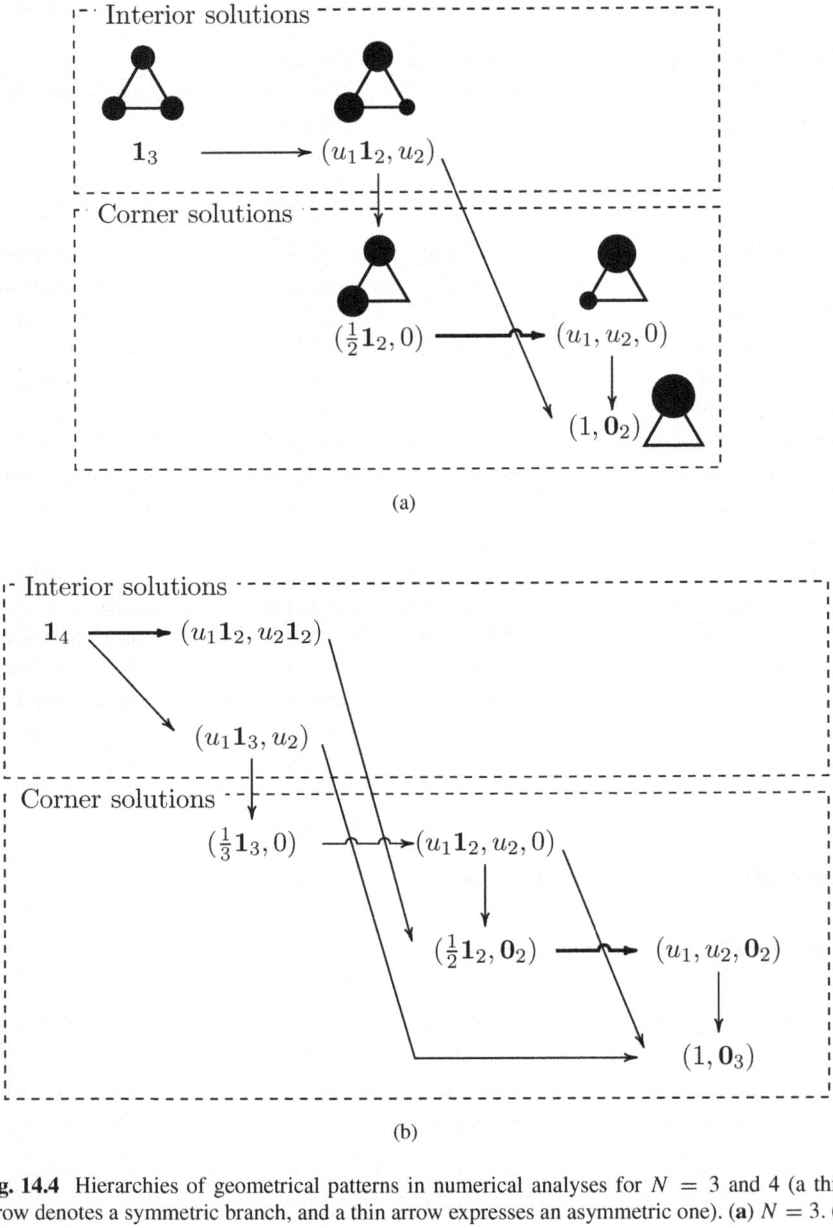


**Fig. 14.4** Hierarchies of geometrical patterns in numerical analyses for $N = 3$ and 4 (a thick arrow denotes a symmetric branch, and a thin arrow expresses an asymmetric one). (**a**) $N = 3$. (**b**) $N = 4$

spans a one-dimensional space. Then, by the equivariant branching lemma, a bifurcating solution with $\Sigma$-symmetry generically exists in the direction (14.21), i.e., (14.6). A bifurcating solution takes the form:

$$\boldsymbol{\lambda} = \frac{1}{N}\mathbf{1}_N + \delta\boldsymbol{\lambda} = \left(\left(\frac{1}{N} + \alpha\right)\mathbf{1}_m, \left(\frac{1}{N} - \alpha\frac{m}{n}\right)\mathbf{1}_n\right) = (u_1\mathbf{1}_m, u_2\mathbf{1}_n)$$

with $u_1 = \frac{1}{N} + \alpha$ and $u_2 = \frac{1}{N} - \alpha\frac{m}{n}$, thereby showing (14.5).

The branch for (14.21) is symmetric if $m = n$ because $\delta\boldsymbol{\lambda} = \alpha\,(\mathbf{1}_m, -\mathbf{1}_m)$ and $-\delta\boldsymbol{\lambda} = \alpha\,(-\mathbf{1}_m, \mathbf{1}_m)$ ($N = 2m$) are symmetric (conjugate and identical up to the permutation of components). It is asymmetric (non-conjugate) otherwise because the number of positive components is different for $\delta\boldsymbol{\lambda}$ and $-\delta\boldsymbol{\lambda}$.

## *Proof of Proposition 14.3*

Consider the uniform state $\boldsymbol{\lambda}^* = \frac{1}{N}\mathbf{1}_N$ with the symmetry of $\mathrm{S}_N$, a two-level hierarchy state $\boldsymbol{\lambda}_{m,n} = (u_1\mathbf{1}_m, u_2\mathbf{1}_n)$ with the symmetry of $\mathrm{S}_m \times \mathrm{S}_n$, and a three-level hierarchy state with the symmetry of $\Sigma = \mathrm{S}_m \times \mathrm{S}_{n_1} \times \mathrm{S}_{n_2}$ ($m+n = N$, $n_1+n_2 = n$). An incremental variable vector for this state with $\mathrm{S}_m \times \mathrm{S}_{n_1} \times \mathrm{S}_{n_2}$ symmetry can be denoted by

$$\delta\boldsymbol{\lambda} = (\alpha\mathbf{1}_m, \beta\mathbf{1}_{n_1}, \gamma\mathbf{1}_{n_2}) \tag{14.22}$$

for some variables $\alpha$, $\beta$, and $\gamma$. By using the orthogonality between subspaces for $\mathrm{S}_N$, $\mathrm{S}_m \times \mathrm{S}_n$, and $\mathrm{S}_m \times \mathrm{S}_{n_1} \times \mathrm{S}_{n_2}$, we obtain

$$\boldsymbol{\lambda}^*\,\delta\boldsymbol{\lambda}^\top = \frac{1}{N}\mathbf{1}_N\,(\alpha\mathbf{1}_m, \beta\mathbf{1}_{n_1}, \gamma\mathbf{1}_{n_2})^\top = \frac{1}{N}(\alpha m + \beta n_1 + \gamma n_2) = 0,$$

$$\boldsymbol{\lambda}_{m,n}\,\delta\boldsymbol{\lambda}^\top = (u_1\mathbf{1}_m, u_2\mathbf{1}_n)\,(\alpha\mathbf{1}_m, \beta\mathbf{1}_{n_1}, \gamma\mathbf{1}_{n_2})^\top = u_1\alpha m + u_2\,(\beta n_1 + \gamma n_2) = 0.$$

Thus, we get $\alpha = 0$ and $\gamma = -\frac{n_1}{n_2}\beta$. Then, (14.22) becomes $\delta\boldsymbol{\lambda} = \beta\left(\mathbf{0}_m, \mathbf{1}_{n_1}, -\frac{n_1}{n_2}\mathbf{1}_{n_2}\right)$. Because this is spanned by a one-dimensional space, by the equivariant branching lemma (Sect. 13.2), there generically exists a bifurcating solution with $\Sigma$-symmetry in this direction $\delta\boldsymbol{\lambda}$, which leads to (14.8) by setting $\beta = \varepsilon(n-p)$, $n_1 = p$, and $n_2 = n - p$. Equation (14.7) and symmetry/asymmetry of the branch can be proved similarly to the proof for Proposition 14.2.

## Proof of Proposition 14.4

The eigenvalue of the Jacobian matrix of the full agglomeration $\boldsymbol{\lambda}^{\mathrm{FA}} = (1, \mathbf{0}_{N-1})$ for a limit point or a breaking bifurcation point is given by $A_1 = a = \frac{\partial(v_1 - \bar{v})}{\partial \lambda_1} = -v_1$. Because $v_1$ generically non-zero (negative), a limit point or a breaking bifurcation point does not exist. The proof for the half-branches of the sustaining bifurcation point is similar to that presented for Proposition 14.3.

## Proof of Proposition 14.7

For $p = 2, \ldots, n-1$, eigenvalues $e_2 \approx -(\gamma - \beta)\varepsilon$ and $e_3 \approx (\gamma - \beta)\varepsilon$ in (14.20) exist and have opposite signs for $\gamma \neq \beta$. Accordingly, the associated branches are unstable for $\gamma \neq \beta$. From (14.18), the Jacobian matrix for the pre-bifurcation state has an $(n-1)$-times repeated eigenvalue $\alpha\psi$. Because the pre-bifurcation state is stable for $\psi > 0$, we have $\alpha < 0$.

For (i), setting $p = 1$ in (14.20), we get the stability conditions of $e_1 \approx \beta\varepsilon < 0$ and $e_3 \approx (\gamma - \beta)\varepsilon < 0$, i.e., $\gamma < \beta < 0$ since $\varepsilon > 0$. Then, from $\alpha < 0$ and (14.19), which reduces to $\psi \approx -\frac{\beta}{\alpha}\varepsilon$ in this case, it is apparent that $\psi < 0$.

For (ii), setting $p = n$ in (14.20), we get the stability conditions of $e_1 \approx \{\beta + (n-1)\gamma\}\varepsilon < 0$ and $e_2 \approx -(\gamma - \beta)\varepsilon < 0$, i.e., $\beta < \gamma$ and $\beta < -(n-1)\gamma$. Then, from (14.19), which becomes $\psi \approx -\frac{\beta + (n-1)\gamma}{\alpha}\varepsilon$ in this case, we see that $\psi < 0$.

In summary, a unique stable branch exists in cases (i) and (ii). In the remaining parameter space of $\beta > 0$ or $-(n-1)\gamma < \beta < 0$, designated as case (iii), there are no stable branches.

## References


Aizawa, H., Ikeda, K., Osawa, M., & Gaspar, J. M. (2020). Breaking and sustaining bifurcations in $\mathrm{S}_N$-invariant equidistant economy. *International Journal of Bifurcation and Chaos, 30*(16), 2050240.

Castro, S. B. S. D., Correia-da-Silva, J., & Mossay, P. (2012). The core-periphery model with three regions and more. *Papers in Regional Science, 91*(2), 401–418.

Commendatore, P., Kubin, I., & Sushko, I. (2015). Typical bifurcation scenario in a three region identical new economic geography model. *Mathematics and Computers in Simulation, 108*, 63–80.

Diaz, A. P. S., & Rodrigues, A. (2007). Hopf bifurcation with $\mathrm{S}_N$-symmetry. *Nonlinearity, 22*, 627–666.

Elmhirst, T. (2004). $\mathrm{S}_N$-equivariant symmetry-breaking bifurcations. *International Journal of Bifurcation and Chaos, 14*(3), 1017–1036.

Forslid, R., & Ottaviano, G. I. P. (2003). An analytically solvable core-periphery model. *Journal of Economic Geography, 3*, 229–340.

Fujita, M., Krugman, P., & Venables, A. J. (1999). *The spatial economy: Cities, regions, and international trade*. MIT Press.

Gaspar, J. M., Castro, S. B. S. D., & Correia-da-Silva, J. (2020). The footloose entrepreneur model with a finite number of equidistant regions. *International Journal of Economic Theory, 16*(4), 420–446.

Gaspar, J. M., Ikeda, K., & Onda, M. (2021). Global bifurcation mechanism and local stability of identical and equidistant regions: Application to three regions and more. *Regional Science and Urban Economics, 86*, 103597.

Golubitsky, M., & Stewart, I. (1985). Hopf bifurcation in the presence of symmetry. *Archive for Rational Mechanics and Analysis, 87*, 107–65.

Golubitsky, M., & Stewart, I. (2002). *The symmetry perspective*. Birkhäuser.

Golubitsky, M., Stewart, I., & Schaeffer, D. G. (1988). *Singularities and Groups in Bifurcation Theory* (Vol. II). Springer.

Ikeda, K., Kogure, Y., Aizawa, H., & Takayama, Y. (2019). Invariant patterns for replicator dynamics on a hexagonal lattice. *International Journal of Bifurcation and Chaos, 29*(6), 1930014.

Ikeda, K., & Murota, K. (2019). *Imperfect bifurcation in structures and materials* (3rd edn.). Springer.

Oyama, D. (2009). Agglomeration under forward-looking expectations: Potentials and global stability. *Regional Science and Urban Economics, 39*(6), 696–713.

Pflüger, M. (2004). A simple, analytically solvable, Chamberlinian agglomeration model. *Regional Science and Urban Economics, 34*, 565–573.

Puga, D. (1999). The rise and fall of regional inequalities. *European Economic Review, 43*(2), 303–334.

Tabuchi, T., Thisse, J.-F., & Zeng, D.-Z. (2005). On the number and size of cities. *Journal of Economic Geography, 5*(4), 423–448.

Zeng, D.-Z., & Uchikawa, T. (2014). Ubiquitous inequality: The home market effect in a multicountry space. *Journal of Mathematical Economics, 50*, 225–233.

# Index




K. Ikeda, Y. Takayama, *Bifurcation Theory of Pattern Formation in Economic Geography*, Interdisciplinary Applied Mathematics 61,
https://doi.org/10.1007/978-3-031-71686-7

The manufacturer's authorised representative in the EU is Springer Nature Customer Service Centre GmbH, Europaplatz 3, 69115 Heidelberg, Germany. If you have any concerns regarding our products, please contact ProductSafety@springernature.com

Printed and bound by CPI Group (UK) Ltd, Croydon, CR0 4YY
30/09/2026
02225569-0006